Cities in Globalization

Despite traditionally being a strong research topic in urban studies, inter-city relations had become grossly neglected until recently, when they were placed back on the research agenda with the advent of studies of world/global cities. More recently the 'external relations' of cities have taken their place alongside 'internal relations' within cities to constitute the full nature of cities.

This state-of-the-art collection of essays on how and why cities are connecting to each other in a globalizing world provides evidence for a new city-centred geography that is emerging in the twenty-first century. *Cities in Globalization* covers four key themes, beginning with the different ways of measuring a 'world city network', ranging from analyses of corporate structures to airline passenger flows. The second theme is the recent European advances in studying 'urban systems' which are compared to the Anglo-American city networks approach. These chapters add conceptual vigour to traditional themes and provide findings on European cities in globalization. Thirdly the political implications of these new geographies of flows are considered in a variety of contexts: the localism of city planning, specialist 'political world cities' and the 'war on terror'. Finally, there are a series of chapters that critically review the state of our knowledge on contemporary relations between cities in globalization.

Cities in Globalization provides an up-to-date assembly of leading American and European researchers reporting their ideas on the critical issue of how cities are faring in contemporary globalization and is highly illustrated throughout with over 40 figures and tables.

Peter J. Taylor is Professor of Geography at Loughborough University.

Ben Derudder is Research Assistant at the Geography Department of Ghent University.

Pieter Saey is Associate Professor of Social and Political Geography at the Geography Department of Ghent University.

Frank Witlox is Associate Professor of Economic Geography at the Geography Department of Ghent University.

Questioning Cities
Edited by Gary Bridge, *University of Bristol*, UK and
Sophie Watson, *The Open University*, UK

The 'Questioning Cities' series brings together an unusual mix of urban scholars. Rather than taking a broadly economic approach, planning approach or more socio-cultural approach, it aims to include titles from a multi-disciplinary field of those interested in critical urban analysis. The series thus includes authors who draw on contemporary social, urban and critical theory to explore different aspects of the city. It is not therefore a series made up of books which are largely case studies of different cities and predominantly descriptive. It seeks instead to extend current debates through, in most cases, excellent empirical work, and to develop sophisticated understandings of the city from a number of disciplines including geography, sociology, politics, planning, cultural studies, philosophy and literature. The series also aims to be thoroughly international where possible, to be innovative, to surprise, and to challenge received wisdom in urban studies. Overall it will encourage a multi-disciplinary and international dialogue, always bearing in mind that simple description or empirical observation which is not located within a broader theoretical framework would not – for this series at least – be enough.

Global Metropolitan
Globalizing cities in a capitalist world
John Rennie Short

Reason in the City of Difference
Pragmatism, communicative action
and contemporary urbanism
Gary Bridge

In the Nature of Cities
Urban political ecology and the
politics of urban metabolism
*Edited by Nik Heynen, Maria Kaika
and Erik Swyngedouw*

Ordinary Cities
Between modernity and development
Jennifer Robinson

Urban Space and Cityscapes
Perspectives from modern and
contemporary culture
Edited by Christoph Lindner

City Publics
The (dis)enchantments of urban
encounters
Sophie Watson

Small Cities
Urban experience beyond the
metropolis
Edited by David Bell and Mark Jayne

Cities and Race
America's new black ghettos
David Wilson

Cities in Globalization
Practices, policies and theories
*Edited by Peter J. Taylor, Ben
Derudder, Pieter Saey and Frank
Witlox*

Cities in Globalization

Practices, policies and theories

Edited by Peter J. Taylor, Ben Derudder, Pieter Saey and Frank Witlox

 Routledge
Taylor & Francis Group

LONDON AND NEW YORK

First published 2007
by Routledge
2 Park Square, Milton Park, Abingdon, Oxon OX14 4RN

Simultaneously published in the USA and Canada
by Taylor & Francis Inc
270 Madison Ave, New York, NY 10016

Routledge is an imprint of the Taylor & Francis Group

© 2007 Peter J. Taylor, Ben Derudder, Pieter Saey and Frank Witlox

Typeset in Times New Roman by
Prepress Projects Ltd, Perth, UK
Printed and bound in Great Britain by
TJI Digital, Padstow, Cornwall

British Library Cataloguing in Publication Data
A catalogue record for this book is available from the British Library

Library of Congress Cataloging in Publication Data

ISBN10: 0-415-40984-5
ISBN13: 978-0-415-40984-1

This book is dedicated to the most inspirational urbanist of the twentieth century, Jane Jacobs (1916–2006)

Contents

Figures

Tables

Contributors

Arthur S. Alderson is Professor of Sociology at Indiana University.

Jonathan V. Beaverstock is Professor of Economic Geography at Loughborough University and Co-Director of the Globalization and World Cities study group and network (GaWC).

Jason Beckfield is Assistant Professor of Sociology at the University of Chicago.

Roberto Camagni is Professor of Economics and of Urban Economics at Politecnico di Milano and President of the European Regional Science Association (ERSA).

Ben Derudder is Research Assistant at the Geography Department of Ghent University and a Research Fellow of the Globalization and World Cities (GaWC) group and network.

Stephen Graham is Professor of Human Geography at the University of Durham.

Phil Hubbard is Reader in Urban Social Geography at Loughborough University.

G.A. (Bert) van der Knaap is Professor of Economic Geography at the Erasmus University in Rotterdam.

Paul L. Knox is Dean of the College of Architecture and Urban Studies at Virginia Tech.

Stefan Krätke is Professor of Economic and Social Geography at the European University Viadrina in Frankfurt (Oder).

Denise Pumain is Professor of Geography at the University of Paris.

Céline Rozenblat is Professor of Geography at the University of Montpellier.

Pieter Saey is Professor of Social Geography, Political Geography and Spatial Dynamics at Ghent University.

Saskia Sassen is the Ralph Lewis Professor of Sociology at the University of Chicago and Centennial Visiting Professor at the London School of Economics.

Richard G. Smith is Lecturer in Human Geography at the University of Wales in Swansea.

Peter J. Taylor is Professor of Human Geography at Loughborough University and Co-Director of the Globalization and World Cities study group and network (GaWC).

Frank Witlox is Professor of Economic Geography at Ghent University.

Herman van der Wusten is Emeritus Professor at the University of Amsterdam.

Preface

This collection of essays emanates from a conference on contemporary inter-city relations held at Ghent University on 13–14 January 2005. The conference marked the end of Peter J. Taylor's tenure of a Francqui Visiting Professorship and we are grateful to the Francqui Foundation for their contribution to the funding of the conference. Contributors were all invited, which enabled us to both focus content and broaden horizons. The conference was consciously designed to follow a previous world cities conference in 1993 at Sterling, Virginia, sponsored by Virginia Tech, which produced the successful volume *World Cities in a World-Economy*, widely recognized as a benchmark publication in the field. We have similar aspirations here but after a decade or so of additional researches we are now in a position to focus on one critical aspect of world cities – the way they relate to one another. Our second aspiration was to broaden the scope of ideas beyond the Anglo-American core of world city studies that monopolized the previous volume. At the Ghent conference invitations balanced Anglo-American papers with key European contributions from Belgium, France, Germany, Italy and the Netherlands. In addition the conference was a multidisciplinary event including leading presenters from economics, geography, planning/architecture and sociology.

Of course, as well as being able to design content and scope, inviting speakers meant that we had quality control: we managed to bring together leading researchers in the field. We think this has directly translated into the book before you through its consistent high quality of chapters. The chapters are not transcripts of the conference presentations – this book is not a publication of proceedings – each author has revised and developed his or her paper to aid the editors in creating an integrated whole. We have found reviewing the papers a stimulating experience and we thank our authors for doing such a good job in bringing their papers to the high standard we demanded. Whether we have indeed been able to produce a second benchmark collection of essays only time will tell, but we are certain that we have put together a state of the art book on understanding relations between cities in the early twenty-first century.

Peter J. Taylor, Ben Derudder, Pieter Saey and Frank Witlox, February 2006

Prologue

A lineage for contemporary inter-city studies

Peter J. Taylor

INTRODUCTION: BYPASSING THE SOCIAL SCIENCE TRINITY

Social science has not been kind to cities. What I mean by this is that cities as social entities – how they feature in social change – have been relatively neglected. This may be because studying cities does not fit easily into the 'process-defined' division of social science into disciplines, the core trinity of economics, sociology and political science that purport to match the three basic spheres of human activities. Thus the study of cities has been largely relegated to marginal areas of social science, notably human ecology/urban sociology and urban geography/ city planning. In recent years these areas have been brought together as 'urban studies'; but notice that they remain a collection of 'studies' (implying theoretically challenged understanding) rather than a new focused 'discipline' (implying theoretical sophistication). But the neglect still remains curious not least because social science was invented to understand modern society and the concrete manifestation of this societal type has been large-scale urbanization. I suppose it could be argued that the urban has become ubiquitous and therefore cities are taken for granted in social science. But, again, this argument does not stand up, since states as social entities feature prominently in understanding social change while being typically taken for granted in social science. Clearly being taken for granted can be either a sign of immanent power or an indicator of irrelevance.

Although these proffered reasons for neglecting cities in social science are partial explanations at best, they do lead us to ask the right questions. Why does the defining of disciplines through separating social processes downgrade the importance of cities? What is this immanent power that states have in social science – and why don't cities have it? The answer to both these questions is that social practices that are modern have been naturalized. Thus the particular modern separation of politics from economy and civil society has resulted in the embedding of the core trinity into how social change is constituted as political, economic or societal processes. Cities may be the 'platforms' on which the processes unfold but the idea of 'urban process' or 'city as process' is off the radar. Similarly, owing to their origins in nineteenth-century reform movements, the social sciences have been premised upon bounded spheres of human action – society, economy and polity are treated as inherently 'national'. Thus the core trinity of the social sciences are the creations of, and subsequently the creatures of, the state. This embedding of social science in political space has produced ontological winners and losers that mimic the actual political winners and losers in the making of modernity: the centralization of states (winners) at the expense of, among others, the autonomy of cities (losers).

The nationalization of social knowledge has had curious effects: for instance, urban studies in social science might have been marginal but this did not mean that understanding cities avoided the embedded statism. This is most clearly seen in the study of inter-city relations. The main research thrust in this area has been the national urban systems school, which flourished from the late 1950s through to the early 1980s (Bourne and Simmons 1978). This research bounded cities; they were studied as 'cities in states'. The result was that cities were arranged into 'national urban hierarchies' as if they had no relations to other cities beyond their state's boundary. In this way, inter-city relations were deemed to reflect the practice of states as the organization of territory through hierarchies. Peter Hall's (1966) *The World Cities* was a major exception to these state-centric urban studies but it was not until the 1980s that the current world cities school of research began in earnest. In this school, cities are interpreted as transnational social entities, the new economic foci as 'cities in globalization'. Today, this school dominates how inter-city relations are understood.

Inevitably, there is much intellectual continuity between the national urban systems and world cities schools. The bounding has been removed but how the form of inter-city relations is conceived has been more resilient: 'urban' and 'hierarchy' continue to appear to be natural bedfellows, albeit at a grander scale. Quite simply, my view is that, to get beyond the pernicious influence of social science on urban studies, it is necessary to transgress modernity. I interpret this to mean bringing in a geohistorical perspective that treats cities as transhistorical social entities. I have identified just such a lineage of twentieth-century inter-city studies, the work of five authors briefly outlined below. This work has never constituted a rival 'school' of ideas but it seems to me that it provides a more appropriate antecedent for contemporary studies of cities in globalization than the national urban systems school. Never part of mainstream urban studies, my selected authors, all social science 'outsiders' in their different ways, nonetheless provide much intellectual raw material for contemporary global urban studies.

The five authors, Henri Pirenne, Fernand Braudel, John Hicks, Jane Jacobs and Giovanni Arrighi, share one critical interest, a historical curiosity about how cities have been party to critical social change. This is expressed as two related projects: first, as an analytical historical description of the rise of modern Europe through its cities (Pirenne, Braudel, Arrighi), and second, as the development of a materialist theory of cities (Hicks, Jacobs). However, Pirenne, as the earliest of the authors, has the privileged position of providing the foundations to both projects.[1]

THE CITY TRAJECTORY TO MODERN EUROPE

History emerged as a modern discipline at about the same time as social science and has been equally state-centric through its telling of national stories. The resulting meta-history of Europe is essentially territorialist. It begins with the demise of the great Roman Empire (in the west) due to invasions by 'tribes', several of which are deemed to have later become 'modern nations'. The initial turmoil (the 'Dark Ages') gives way to a restitution of territorial order through reformulation of the 'Roman Empire' – symbolically Charlemagne was crowned Emperor by the Pope in 800 – and creation of new kingdoms that, in various ways, lead eventually to the formation of modern nation-states. This is the 'Europe of nations', a spatial mosaic, that is currently undermining political integration of the European Union. Obviously this nationalist meta-history cannot be under-estimated for its political importance but it is not the only meta-history of Europe. There is an alternative network meta-history that may be more relevant to understanding contemporary globalization; it is a history of formation of city networks rather than nation-states.

The starting point is that it is incorrect to view the Roman Empire as a single territorial monolith; it was formed by a city-state and remained 'legally, and in practice, a federation of city-states' (Grimal 1983: 109). In the east these were old 'allied cities' tied to Rome through treaties, and in the west they were largely new colonial cities planted following conquest (ibid.: 329). The result was a world-economy centred on the Mediterranean operating through inter-city networks of production and trade (Pirenne 1925: chapter 1; Braudel 1984: 25, 33). It was this world that was invaded by Germanic tribes, leading to the deposing of the last Roman Emperor in the west in 476. But for Pirenne (1925: 10) this political event was not a decisive episode because the Roman economy of cities continued (ibid.: 12). This 'Mediterranean commonwealth', to use Pirenne's phrase (ibid.: 3, 10), only collapsed with the rise of Islam and its conquest of the eastern, southern and western shores of the Mediterranean. This resulted in pushing the centre of post-Roman Europe northwards away from the old world-economy: the Carolingian empire of Charlemagne was 'a closed state, a state without foreign markets, living in a condition of almost complete isolation' (ibid.: 29). The conclusion in terms of social change is inescapable: Charlemagne represents regression, not revival (ibid.: 40). It can be shown that this was a delinked territorial backwater by comparison with eastern Europe where southern Russia centred on Kiev was expanding economically through links to Constantinople and Baghdad – the Russians had a trading quarter in the former, a concept inconceivable for Charlemagne's Germans. Pirenne contrasts Kiev with Charlemagne's 'capital', Aix-la-Chapelle (ibid.: 54). According to Barraclough (1979: 109), the latter was 'in no sense a city' having 'a population of no more than two or three thousand' before being destroyed by Vikings, which is in stark contrast to the Viking creation of Kiev as a great trading city (Price 2000) reaching a population estimated at 45,000 by 1000 (Chandler 1987: 15). In 1000 there was one city in western Europe comparable to

Kiev – Venice. Also in Constantinople's orbit, Venice was western Europe's largest city at this time, also with an estimated population of 45,000 (ibid.). This is the source of the upturn in Europe's trajectory in the network meta-history, in contrast to the privileging of Charlemagne in the conventional territorialist meta-history.

While the western Roman Emperor was resident in little Aix, the eastern Emperor was living in a Constantinople with a population estimated at 250,000 in 800, rising to 300,000 in 1000 (Chandler 1987: 14, 15). Although in Italy, Venice was legally part of the latter empire, which provided a formal link to a massive urban market. It was the taking economic advantage of this political link that precipitated the rise of Venice as a great trading city (Pirenne 1925: 82–93). From Venice, commercial life spread to Lombardy in the tenth century (ibid.: 87) to create what Braudel (1984: 96) calls the 'southern pole' – northern Italy – of the 'first European world-economy'. The other, weaker, 'northern pole' – Flanders – was created by tenth century commercial links (the Viking Baltic Sea–Black Sea route) to Constantinople via Kiev. With merchant colonies gradually established in cities (Pirenne 1925: 121) across both commercial poles, this led to the Europe-wide commercial revival of the tenth and eleventh centuries that culminate in the thirteenth-century 'commercial revolution' (Spufford 2002). In short, medieval Europe became 'a region of cities' (Pirenne 1925: 103).

The two 'poles' of this economic region were initially connected through a series of fairs at approximately the midpoint between them: the famous Champagne Fairs of the thirteenth century. These were an annual programme of six two-month fairs starting in Lagny-sur-Marne, moving on to Bar-sur-Aube, on to Provins, on to Troyes, back to Provins and finally back to Troyes (Braudel 1984: 111). Each fair was divided into two periods: the first month was given over to commodity trading dominated by textiles moving from north to south, and the second month was for financial transactions dominated by Italian bankers. Note that these temporary economic nodes were not major cities; they were just 'fair towns'. But in their heyday they were 'a rendezvous for the whole of Europe, for the offerings of both North and South' (ibid.: 111). However, this arrangement collapsed in the fourteenth-century recession to be replaced by direct links between the two poles: by ship from the Mediterranean direct to Bruges, and through opening up the eastern Alpine passes (St Gotthard and Brenner) to the Rhinelands and central Europe. Braudel has called the Champagne fairs 'an interlude' (ibid.: 111). I prefer the term 'transition': from a 'region of cities' as two loosely connected urban zones to a European city network.

This new city network was an interlocking network, with 'families', 'houses', 'nations' and firms of traders and financiers present across the European cities to 'interlock' them into a cohesive network. For instance, in the fifteenth century the famous Medici Bank of Florence had branches in Venice, Genoa, Milan and Rome, plus representatives in London, Bruges, Geneva and Lyons, plus a correspondent link in Lubeck (the chief city of the Hanseatic League) (Kindleberger 1993: 45–6). A century later, the equally famous Fuggers of Augsburg had 18 'factories' across Europe from Denmark to Naples (ibid.: 47). This was the inter-city network upon which modern Europe was subsequently built.

In debates over the 'transition from feudalism to capitalism' this commercial reality is often underplayed (Hilton 1976). I have tried to illustrate this directly in Table 0.1 in which European cities are evaluated in world terms. This statistical description of the Pirenne thesis shows that by 1300 Europe was a city-rich region and this was consolidated in subsequent centuries (Table 0.1). Using Chandler's (1987: 513, 514) city population estimates, I can make the following statements. In 1400 (i.e. before the beginning of Wallerstein's (1979) modern world-system), there were 18 western European cities with populations of 40,000 or more, all ranked in the top 80 cities worldwide. By 1500, this number had risen to 22 such large cities, also all within the world top 80. In other words, the urban legacy of medieval Europe to the European modern world-system was approximately one quarter of the leading cities in the world. And the derivation is Venetian, not Roman, in either its classical or medieval imperial form.

It is Arrighi (1994), drawing heavily on Braudel (1984) and also Pirenne (1953), who brings Pirenne's story forward into the modern world-system and through to contemporary globalization. He develops the concept of systematic cycles of accumulation to show how cities as international financial centres have steered the capitalist world-economy (aka modern world-system). His starting point is the basic formula for individual investment of capital MCM′, where M is initial money capital, C is commodity capital (the investment) and M′ is the expanded money capital. Arrighi's basic insight is to convert this formula into a collective description of capitalist development. Thus MCM′ defines a cycle of accumulation involving two phases. MC is a general conversion of flexible, fluid capital into commodity production to generate an epoch of material expansion. When the market for the commodities becomes problematic there is a general reversion to fluid flexible capital, CM′, creating an epoch of financial expansion. This can be illustrated by the 'antecedent' cycle of capital accumulation prior to the modern world-system in the fifteenth century in which the cities referred to in the previous paragraph operated. The MC phase lasted to mid-century, and was an era of inter-city cooperation as cities developed niches in expanding divisions of labour and shared knowledge to reduce risks. The CM′ phase of trade contraction resulted initially in inter-city competition (notably the Venice–Genoa war) for pieces of the smaller commercial cake. But incessant war provides new outlets for capital (Venice and Milan make war tools) and Medici Florence flourishes as an international financial centre: 'high finance' is invented. With these new outlets for capital, 'a new kind of cooperation' (Arrighi 1994: 93) arises. Thus the fifteenth century fits the MCM′ model as a cycle of capital accumulation.

Arrighi (1994) identifies four overlapping systematic accumulation cycles in the capitalist world-economy: the Genoese cycle from the late fifteenth century to the early sixteenth century, the Dutch cycle from the late sixteenth century through the eighteenth century, the British cycle from the second half of the eighteenth century to the early twentieth century, and the American cycle from the late nineteenth century to the present. The Genoese cycle is least known because of the separation of political and economic leadership – the dichotomous agency consisted of Iberian protection/power alongside Genoese trade/profit. The

Table 0.1 The rise of city-rich medieval Europe, 1000–1500

City	Population (000s)	World rank	Interpretation of cities as new entrants
1000			
Venice	45	51	The starting point for a new "Europe of cities"
1100			
Venice	55	38	
Milan	42	58	Expansion to start Braudel's 'southern pole'
1200			
Paris	110	13	First major 'political city'
Venice	70	28	
Milan	60	35	
Cologne	50	42	Start of 'northern pole'
Leon	40	68	First Spanish political centre
London	40	68	Anglo-Norman political centre
Rouen	40	68	Anglo-Norman political centre
1300			
Paris	228	4	
Venice	110	10	
Seville	90	18	Former Islamic Spain
Genoa	85	23	Rise of Venice's great rival
Florence	60	33	Expansion of 'southern pole'
Milan	60	33	
Cologne	54	39	
Bruges	50	40	Expansion of 'northern pole'
Rouen	50	40	
London	45	57	
Ghent	42	58	Expansion of 'northern pole'
Bologna	40	59	Expansion of 'southern pole'
Caffa	40	59	Expansion of 'southern pole'
Córdoba	40	59	Former Islamic Spain
Naples	40	59	Political centre of South Italy
Prague	40	59	Imperial political centre
1400			
Paris	280	4	
Venice	110	14	
Prague	95	16	
Milan	80	24	
Rouen	80	24	
Caffa	75	27	Expansion of 'southern pole'
Ghent	70	33	
Genoa	66	38	
Florence	61	40	
Bruges	60	41	
Seville	60	41	
Lisbon	55	51	

City	Population (000s)	World rank	Interpretation of cities as new entrants
1400 (Continued)			
London	50	53	
Troki	50	53	
Valencia	45	73	Former Islamic Spain
Bologna	43	77	Expansion to Baltic region
Cologne	40	78	
Naples	40	78	
1500			
Paris	185	8	
Venice	115	18	
Naples	114	19	
Milan	89	21	
Ghent	80	24	
Florence	70	29	
Granada	70	29	Former Islamic Spain
Prague	70	29	
Genoa	62	37	
Bruges	60	39	
Bologna	55	52	
Lisbon	55	52	
London	50	55	
Rouen	50	55	
Smolensk	50	55	Expansion to Baltic region
Seville	46	72	
Brescia	40	79	Expansion of 'southern pole'
Cremona	40	79	Expansion of 'southern pole'
Lyons	40	79	Rise of French trading centre
Valencia	40	79	
Valladolid	40	79	Old Castile political centre
Vienna	40	79	Rival imperial centre

Source: populations and rankings are derived from Chandler (1987) and include all European cities with populations estimated at 40,000 and above. The choice of this threshold is forced upon me by the data but nevertheless seems to be a reasonable size to define major cities at these times. European cities are from 'western Christendom', i.e omitting Islamic Iberia and orthodox eastern Europe.

inter-city relations (Genoa, Venice, Milan, Florence, Lyons, Augsburg, Seville, Antwerp) were 'basically cooperative' in the first half of the sixteenth century but this changed with the financial crisis of 1557 after which Genoa managed Spain's permanent financial crisis as part of its 'discrete rule over Europe' (Braudel 1984: 164). The international financial centres that followed Genoa all conform to a geographical arrangement that modernity has conditioned us to expect – the congruence of economic and political functions within the same sovereign territory: Genoa/Iberia is followed by Amsterdam/Dutch Republic, London/UK and New York/USA respectively. These last three states are the hegemonic states whose cycles define the modern world-system (Wallerstein 1984: chapter 4; Arrighi 1990, 1994: chapter 1) culminating in three episodes of financial prowess. Thus they are closely related to the accumulation cycles. Of particular relevance here, however, is Braudel's (1984) interpretation of this cyclical sequence as a movement from city-centred economies to state-centred economies (with the Amsterdam/Dutch arrangement a 'half-way house' (ibid.: 175)): I will have more to say about this reading of social change in the next section.

For Arrighi, the contemporary relevance of his model is that current descriptions of recent restructuring of the world-economy can be seen through this historical perspective as not at all unique. For instance, Harvey's (1989) historical transition from Fordism–Keynesianism to flexible accumulation, which overcomes rigidity leading to financialization of capital, is the latest CM' phase, another epoch of financial expansion, the latest 'rebirth' of capital (Arrighi 1994: 4). From my perspective, there is a second important implication of the model. As economic globalization has provided new outlets for capital, the competitive phase of inter-city relations that dominated the 1970s and 1980s (generally, bad times for cities) has given way to a new cooperation (since 1990, there have been good times for cities). The outcome is a contemporary world city network (Taylor 2004).

TOWARDS A MATERIALIST THEORY OF INTER-CITY RELATIONS

Pirenne, Braudel and Arrighi are primarily concerned for understanding the concrete historical ordering of social change; Hicks (1969) and Jacobs (1970, 1984) are, perhaps, more abstract in their use of city activities to elucidate social change. This is to simplify greatly because these sources interrelate: Jacobs explicitly uses Pirenne's thesis in her work and Arrighi draws on Hicks's work. But this is good because it means that any semblance to a materialist theory of inter-city relations I derive from Hicks and Jacobs will be easily related to cities in history.

The useful section of Hick's (1969) work for my purpose here is his description of the working of a mercantile economy. He identifies the key agents as traders; when there are enough merchants to form a community they constitute a mercantile economy centred on cities. Why cities? This is explained in what he calls his 'pure theory of commercial development' (ibid.: 62). He starts with two traders and two regions, each with a surplus of a different good. The inter-regional trade exchanging the goods is a voluntary trade leading to 'All-round Advantage' (ibid.: 44): consumers in each region get a new good, traders develop two new markets and can make high profits. How can such a simple commercial arrangement be expanded? Obviously the traders can invest in more of the goods to sell but the catch is that their profits are automatically squeezed: they increase demand in each region-as-supplier thus raising the price they buy at, while increasing supply in each region-as-market thus reducing the price they sell at. If simple trade leads to diminishing returns then investment must be made elsewhere, to a different trade. But this requires new contacts and new knowledge: enter cities to provide a level of trade diversification to get around the diminishing returns. In addition, cities make for efficient trade expansion; there are also increasing returns for trade with size. The key point is that Hicks identifies two types of Marshall's external economies: 'There are some which arise from the expansion of the individual city . . . or trading centre; and there are some that arise from the multiplication of trading centres' (ibid.: 47). Both benefits are in reducing costs and risks by pooling knowledge. The first is an 'intra-city type' of external economy defining economic clusters in cities, the second is an 'inter-city type' defining economic networks of cities. Thus, both cities and inter-city relations are derived as necessary for the expansion of profit in trade.

For Hicks (1969) this conclusion is about the operation of the mercantile economy, which is just part of a model of 'economic states of society' (ibid.: 6). His theory of economic history moves from customary and command (non-market) economies through to mercantile economies (in the era of city-states) before markets penetrating different spheres of activity culminating in the modern industrial economy (in nation-states). Thus, here again there is the assumed economic transition from city-state to territorial state that was noted in Braudel and Arrighi above. But does the formation of large states, taking over cities, really result in the 'economy' automatically 'moving up' to the state level? An answer can be found in an early study of fifteenth-century state centralization (at the expense of cities) by Braudel (1972: 309):

> The victorious states could not take control of and responsibility for everything. They were cumbersome machines inadequate to handle their new superhuman tasks. The so-called territorial economy of textbook classification could not stifle the so-called urban economy. The cities remained the driving forces. States that included these cities had to come to terms with them and tolerate them. The relationship was accepted the more naturally since even the most independent cities needed the use of the space belonging to territorial states.

Although the contemporary state is very different from the early modern state, they are still quite 'cumbersome machines' when it comes to expanding economic life, which remains a 'superhuman task'. This is Jane Jacobs' (1984) view, with which I completely concur. She argues that states 'are political and military entities' and there is therefore no reason to assume that they are also 'the basic salient entities of economic life' (ibid.: 31). This idea that states define economies she calls the 'mercantilist tautology' since it derives from early modern economic theories but it has been subsequently accepted as the unexamined premise of all schools of economic thought. In contrast, Jacobs argues that cities and city-regions are the basic entities of economic life and therefore it follows that so-called 'national economies' are 'grab bags' of different city economies brought together by political sovereignty (ibid.: 32). Interestingly, Jacobs comes to this position through using Pirenne's thesis on the Venetian origins of an 'urban Europe'. But this idea that the national economy is a myth does cause a rethink for Braudel's and Arrighi's transition from city to national economy. The three 'hegemonic states' that were the United Provinces, Kingdom and States were the product of vibrant cities that do not add up to the whole so-called 'national economy': hegemony was produced largely in the cities of central and southern Holland (not, for instance, in the outer provinces of Gelderland, Overijssel, Friesland and Groningen), largely in the industrial cities of northern England (not, for instance, in the outer reaches of western Ireland, northern Scotland and central Wales) and largely in the 'manufacturing belt' of the USA (not, for instance, in the American 'South'). Concentrations of dynamic cities prospering through both intra-city and inter-city externalities create economic leadership; it is not created by states although the latter will be in a position to exploit the situation politically.

Jacobs begins where Hicks finishes: 'A city seems always to have implied a group of cities, in trade with one another' (Jacobs 1970: 35). Cities are distinguished from other settlements not simply by size, their essence is their 'more complicated and diverse' economies (ibid.: 50). This collection of economic complexities enables economic expansion to occur through the 'immense, even awesome, economic force' that is import replacement (ibid.: 150). Import replacement combines the 'little movements at the hubs' (Hicks's intra-city externalities) with the 'great wheels of economic life' (Hicks's inter-city externalities). For Jacobs, the decline of the Roman world (in the west) was precisely because this combination ceased (as Pirenne showed) (ibid.: 176). And, again following Pirenne, it was only when Venice was able to replace imports from Constantinople (ibid.: 174), and then Lombard cities to replace imports from Venice, that the 'great wheels of commerce' began to operate once more in western Europe. Jacobs describes in detail (ibid.: chapter 5) how this process operates and I will not explain it here except to note that it is premised on economic growth and inter-city cooperation and thus is consistent with Arrighi's argument on changing forms of inter-city relations. Her ideas have been used as the theoretical base for understanding the contemporary world city network (Taylor 2004: chapter 2).

The theoretical contributions of Hicks and Jacobs are similar but distinctive; the former shows the advantages of cities for commercial activities, the latter discovers the city mechanism that expands economic life. They are similar because neither treats cities as agents; the processes are created by economic agents – merchants, firms – who create intra-city clusters and inter-city relations whilst doing their routine work. This must be the correct starting point for a materialist theory of inter-city relations: as shown historically above, city networks are interlocking networks and this is how I have specified the contemporary world city network (Taylor 2001), produced and reproduced by global service firms operating as 'interlockers' (Taylor 2004). The dissimilarities between Hicks and Jacobs are in the way they move from agent to urban outcome. For Hicks, the city outcome is abstractly derived but made historically contingent – city economies are transcended by national industrial economies. Jacobs, however, produces a more fundamental argument so that cities transcend states transhistorically. Both arguments can be accommodated to contemporary economic globalization: using Hicks there is a 'return of the city' (not city-state) because global business requires the transnational externalities that can only be found in world/global cities; using Jacobs, cities never 'went away' as critical economic entities but now the national mask is lowered to reveal import replacement through cities generating a global space of flows that is economic globalization. Both positions are far superior to an 'upscaling from the national' interpretation of globalization but, choosing between them, I subscribe to the Jacobs position.

CONCLUSION: PROCESS BEYOND THE SOCIAL SCIENCE TRINITY

In this brief discussion my intention has been to initiate an alternative basis for studying inter-city relations that both is superior to the national urban systems legacy and appears plausible under conditions of contemporary globalization. By plausible I mean that the argument is more than 'urban study'; it is process-based and therefore credible in the traditional social science disciplinary sense. But the ideas I have introduced are outside this disciplinary mind-set: the process I deal with is not any of the economic, political and societal processes that define the social science trinity. Going beyond the trinity is made possible because 'process' is no longer divorced from 'physical' time or space (Hubbard *et al.* 2004). The lineage I have defined is a social time; it results from social activities but also moulds those activities. Similarly, social processes create social spaces, and vice versa. They cannot be separated. Castells (1996: 386) is most explicit on this when he argues that a city is 'not a place but a process'. In other words, economic, political and societal change does not happen in cities; it happens through cities: it is shaped by cities and in turn it shapes cities. Hence this does not imply yet another call for 'inter-disciplinary' research; rather it is, in the spirit of Wallerstein (1979: ix), a uni-disciplinary argument for a single geohistorical social science through which to build a credible understanding of both cities and globalization.

NOTE

1 I appreciate that Pirenne's work on cities has been the source of much controversy over the years both with historians (Hivighurst 1958/1976; Van Werveke 1963; Verhulst 1989) and, in association with Jacobs, with social scientists (Bairoch 1988; Hansen 2000). This prologue is not the place to rehearse this debate: I take the position of Spruyt (1994: 61–2) that Pirenne's 'thesis regarding the connection between urbanization and trade has stood the test of time'.

REFERENCES

Arrighi, G. (1990) 'The three hegemonies of historical capitalism', *Review*, 13: 365–408.

Arrighi, G. (1994) *The Long Twentieth Century: Money, Power and the Origin of our Times*, London: Verso.

Bairoch, P. (1988) *Cities and Economic Development*, Chicago: University of Chicago Press.

Barraclough, G. (1979) *The Times Atlas of World History*, London: Times Books.

Bourne, L.S. and Simmons, J.W. (eds) (1978) *Systems of Cities*, New York: Oxford University Press.

Braudel, F. (1972) *The Mediterranean and the Mediterranean World in the Age of Phillip II*, vol. 1, London: Collins.

Braudel, F. (1984) *The Perspective of the World*, London: Collins.

Castells, M. (1996) *The Rise of Network Society*, Oxford: Blackwell.

Chandler, T. (1987) *Four Thousand Years of Urban Growth: An Historical Census*, Lewiston: Edwin Mellen.

Grimal, P. (1983) *Roman Cities*, Madison: University of Wisconsin Press.

Hall, P. (1966) *The World Cities*, London: Heinemann.

Hansen, M.G. (2000) 'Introduction: the concepts of city-state and city-state culture', in M.G. Hansen (ed.) *A Comparative Study of Thirty City-State Cultures*, Copenhagen: Historisk-filosofiske Skrifter.

Harvey, D. (1989) *The Condition of Postmodernity*, Oxford: Blackwell.

Hivighurst, A.E. (ed.) (1958/1976) *The Pirenne Thesis: Analysis, Criticism, and Revision*, Boston: Heath.

Hicks, J. (1969) *A Theory of Economic History*, Oxford: Clarendon.

Hilton, R. (ed.) (1976) *The Transition from Feudalism to Capitalism*, London: Verso.

Hubbard, P., Kitchen, R. and Valentine, G. (2004) 'Editor's introduction', in P. Hubbard, R. Kitchen and G. Valentine (eds) *Key Thinkers on Space and Place*, Beverly Hills, CA: Sage.

Jacobs, J. (1970) *The Economy of Cities*, New York: Vintage.

Jacobs, J. (1984) *Cities and the Wealth of Nations*, New York: Vintage.

Kindleberger, C.P. (1993) *A Financial History of Western Europe*, New York: Oxford University Press.

Pirenne, H. (1925) *Medieval Cities: Their Origins and the Revival of Trade*, Princeton: Princeton University Press.

Pirenne, H. (1953) 'Stages in the social history of capitalism', in R. Bendix and S. Lipset (eds) *Class, Status and Power: A Reader in Social Stratification*, Glencoe, IL: Free Press.

Price, N. (2000) 'Novgorod, Kiev and their satellites: the city-state model and the Viking age politics of European Russia', in M.H. Hansen (ed.) *A Comparative Study of Thirty City-State Cultures*, Copenhagen: Historisk-filosofiske Skrifter.

Spufford, P. (2002) *The Merchant in Medieval Europe*, London: Thames and Hudson.

Spruyt, H. (1994) *The Sovereign State and its Competitors*, Princeton, NJ: Princeton University Press.

Taylor, P.J. (2001) 'Specification of the world city network', *Geographical Analysis*, 33: 181–94.

Taylor, P.J. (2004) *World City Network: a Global Urban Analysis*, London: Routledge.

Van Werveke, H. (1963) 'The rise of the towns', in M. Poston, E. Rich and E. Miller (eds) *The Cambridge Economic History of Europe*, 3: 3–41.

Verhulst, A. (1989) 'The origin of towns in the Low Countries and the Pirenne thesis', *Past and Present*, 122: 3–35.

Wallerstein, I. (1979) *The Capitalist World-Economy*, Cambridge: Cambridge University Press.

Wallerstein, I. (1984) *Politics of the World-Economy*, Cambridge: Cambridge University Press.

1 Introduction

Cities in globalization

Peter J. Taylor, Ben Derudder, Pieter Saey and
Frank Witlox

Cities are inherently complex and diverse. Their multifaceted nature means that
there are very many routes to investigating their importance for understanding
social change. All this is as true for their contemporary manifestation as 'world'
or 'global' cities as at any time in the past. In fact, it can be argued that, given the
growth in the scale of 'city operations', cities today are moving towards an apo-
gee of complexity and diversity. Be that as it may, contemporary cities have certainly thrown up many challenges to urban studies, both old (e.g. gross material
inequalities) and new (e.g. sustainability of 'third world mega-cities'). The facet
of cities that we focus upon here is a mixture of old and new: inter-city relations.
Cities have always existed in relation to one another and contemporary cities are
by no means an exception. Quite the opposite: since the 1970s, transport and
communication/computing technologies have been fostering an intensification,
expansion and extension of inter-city relations. In this way cities have become
central to how many people understand contemporary globalization. This is our
subject matter here.

Our use of the term 'cities in globalization' to portray this tendency is not
conceptually neutral. The most common terms used to describe current cities with
worldwide relations have been 'world cities' and 'global cities'. These terms are
commonly used in the chapters below but we do need to make it clear that we do
not necessarily wish to carry on board the baggage that comes with both terms.
'World city' is associated with John Friedmann (1986) and his 'world city hierarchy'; this particular arrangement of cities we leave for empirical investigation,
not presumption. 'Global city' is associated with Saskia Sassen (1991) and her
contention that there are a limited number of major cities that are strategically
global in their functions; this particular arrangement of cities presumes a division
of cities rather than a continuity of cities in terms of global functions, which,
again, is a question we prefer to leave to empirical investigation. The key point
with both concepts is that they may be used to suggest 'lesser opposites' such as
'sub-global' and 'non-world cities'. Since we understand globalization processes
to be pervasive in the contemporary world, we consider this train of thought to
be potentially misleading. We challenge anybody to find a contemporary city or
town that shows no evidence of globalization processes in the activities that oc-

cur within it. Given that we expect nobody to be able to meet this challenge, we conclude that all cities today can be characterized to some degree as both 'world' and 'global' in nature. Hence, they are all 'cities in globalization'.

The crux of this argument focuses on the 'degree' of globalization experienced in cities. Following Castells (1996), we treat cities as processes, locales through which the social (social space) is constituted as spaces of flows. Globalization is a bundle of processes that constitute a global social space through cities. But there are other bundles of processes constituting spaces of flows at other geographical scales. World-regional scale outcomes are a ubiquitous finding in world city network studies using a variety of techniques and measures of flows (Shin & Timberlake 2000; Smith and Timberlake 2002; Derudder *et al.* 2003; Taylor 2004a,b; Derudder and Witlox 2005). And, of course, national-scale processes remain important to varying degrees, most clearly seen in the case of the USA (Taylor and Lang 2005). Finally, any trip to a shopping centre or mall illustrates the continuing importance of local spaces of flows to the vitality of cities and city-regions. In other words, focusing on the 'global' or worldwide functions can only be a partial view of cities, one of varying importance among cities. Further, these 'global processes' can never be separate from the processes concentrated at other scales ('cities in world-regions', 'cities in states' and 'cities in hinterlands'). Research focus on one scale should be seen, therefore, always as a pedagogic decision, a way of circumscribing reality to create a manageable, and still coherent, study agenda. The proof of the pudding is in the eating and the world cities literature has remained vibrant over two decades; proof enough. The 'cities in globalization' argument is a simple caution not to over-interpret world and/or global cities.

Previously we have referred to inter-city relations on several occasions; but, specifically, what are these relations between cities in globalization? Here again we follow Castells's (1996) concept of space of flows that he describes as occurring at three levels: infrastructure, social relations and elite relations. Like most social science, our studies are mainly about social relations, but we consider also infrastructural relations and elite relations, both economic and political. One key effect of using this process approach is that we understand inter-city relations not to be primarily the product of the cities themselves. There are 'official' inter-city links created by city administrations (from city twinning to city collectives like Eurocities) but these pale into insignificance when compared with the plethora of 'non-official' inter-city links: it is private agents that constitute, by and large, the spaces of flows that are our subject matter. Not wholly economic in nature; nevertheless, it is economic agents (firms) who shape most of the inter-city relations in reality, and this is mirrored in this volume. A key consensus amongst all our authors is that they identify agents within cities and not simply cities as agents.

We have divided the chapters into four sections: World City Networks, Inter-City Relations as Networks and Systems, Politics in Inter-City Relations, and Rethinking Cities in Globalization. Like all such divisions, the allocations are to some extent arbitrary: our authors tackle complex subjects that cannot easily be pigeon-holed into simple categories. For instance, as we would expect from leading thinkers in the field, all authors do some rethinking in developing their arguments. Our approach has been to group together chapters on the basis of what we think to be the main thrust of the contribution to the literature. One of the long-term criticisms of the world cities literature is that we actually know less about relations between cities in globalization than has often been assumed. In the first part, key researchers who have begun to seriously tackle this problem present some of their recent researches. This part is intended to give readers a 'feel' for contemporary inter-city relations and its study through a variety of different research routes, substantive and methodological. In the second part, contributions focus upon how networks and systems concepts have been combined to provide distinctive depictions of inter-city relations. These chapters introduce continental European contributions to understanding inter-city relations. The third part includes chapters that are explicitly concerned with political questions relating to inter-city relations. Politics is an under-researched but necessary component of inter-city relations that is introduced here in quite different ways. Finally, the book concludes with Part IV, which focuses on rethinking how we might study inter-city relations. These chapters aspire to suggest future research agendas through mixes of developing new ideas and repackaging some old ideas.

The chapters in Part I all have a substantial empirical core that is worldwide in scope. The first three chapters each illustrate a level of Castells's (1996) space of flows, beginning with social relations and moving on to a key infrastructural relation before focusing on elite relations. We begin with Alderson and Beckfield's longitudinal study of corporate presence in cities from 1981 to 2000. This follows their cross-sectional study of 2000 (Alderson and Beckwith 2004), which is the largest-scale published study of world cities. They combine social network analysis with the world cities literature to test ideas on whether the growth of world cities is accentuating core–periphery patterns or eroding them. Their findings clearly support increased worldwide divergence. Previously the only world cities literature to have attempted longitudinal studies have used airline passenger data. In Chapter 2 Witlox and Derudder review past limitations of using such data, and show fresh possibilities through introducing a new database. This overcomes many of the important problems previous researchers have encountered to enable unique passenger origin-destination network analysis to be undertaken. The elites that Beaverstock studies in Chapter 3 are investment bankers. He shows how these highly professional, and knowledge-rich, staff of global investment banks require face-to-face contacts generating inter-city expert labour mobility. His focus on the firm leads to a call for intensive research on 'micro-network systems'. In the final chapter of Part I Knox uses the interlocking network model of inter-city relations (Taylor 2001, 2004a) to look at the recent globalization of design services. He finds a dominance of large firms in a limited number of world-regions where capital is being invested heavily in real estate, notably in Pacific Asia and the Middle East Gulf states.

The pedigree of the chapters in Part I can be traced back to *World Cities in a World-Economy* and its Anglo-American research origins. Running parallel with such studies there has been a very strong research tradition on inter-city relations among researchers in continental Europe. These have borne fruit in several 'Europe of Cities' policy and research programmes through the European Union (Bagnasco and Le Galès 2000; Ipenburg and Lambrechts 2001; Faludi 2002; ES-PON 2005) and in research of a rather speculative nature in the new members of the Union (e.g. Centre for European Studies 2003). Here we present four chapters on City Networks and Systems by influential authors from the Netherlands, Italy, France and Germany, countries that have been particularly prominent in EU urban thinking. Van der Knaap shows how inter-city research has changed as the world has changed resulting in central place theory giving way to 'network systems'. He emphasizes the role of government in network formation and the variety of networks, reflected at different scales as polynuclear city, network city and urban networks. Camagni describes the network paradigm that is a key product of the southern European tradition of spatial analysis. This provides a new logic of synergy and complementarity in city networks that can act as policy tools: it has been adopted in the EU spatial strategy document, the ESDP – European Spatial Development Perspective. Rozenblat and Pumain present their approach to urban systems using ownership networks which are now multinational networks. They derive principles of spatial ordering to describe an emerging European system of cities. Finally, Krätke develops the thesis that Europe's economic territory is a process of metropolitanization of economic development potentials and innovation capacities. The focus is upon development paths of city regions and their transnational inter-linkages.

As reflected in the first two parts, the world cities literature has focused largely on economic globalization: political processes have been relatively neglected. To be sure, there are scattered references to a range of political questions: on particular 'political world cities' (e.g. Elmhorn 2001), on city planning policies (e.g. Newman and Thornley 2005), on citizenship (e.g. Isin 2000), and on global civil society/global governance (e.g. Sassen 2002; Taylor 2005). However, as this short list shows, there is no coherent pattern of research on politics and world cities. This may be because political science/international relations research inevitably focuses on states, resulting in most political researchers ignoring cities. Thus in Part III we have three chapters on quite different topics. Hubbard discusses the limitations of sedentary thinking in urban politics under conditions of contemporary globalization. Cities exist in spaces of flows and therefore city politics has to include a politics of flows: politics needs to be upscaled and network externalities harnessed from flows through the city. Van der Wusten is concerned for the role of cities in multilateral political organizations, which he terms political central places. He focuses on the politics of four European cities in four small countries – Geneva, The Hague, Brussels and Vienna – that are disproportionately important as loci for international relations. Finally Graham relates contemporary cities to the 'war on terror'. He describes a new imagination that is an urban geography of terror, a discourse that opposes 'terror cities' to 'homeland cities'. The result is the creation of strongly anti-cosmopolitan politics, an anathema to the future vibrancy of cities.

From the heydays of central place theory and national urban systems research in the 1960s and 1970s until fairly recently the study of inter-city relations was relatively neglected, largely because of the paucity of relational data. This book symbolizes a turnaround in this situation in recent years – so much so that the last part is dedicated to searching out new ways forward from a reasonably sound foundation. Certainly this rethinking is necessary since we remain at the very beginning of our understanding of inter-city relations in contemporary globalization. Sassen concentrates on telematics and how these affect cities in what she terms a 'global digital age'. This introduces a new complexity into her global city analyses: there are myriad different circuits that are transnational but also fragmented and often quite transient. Smith argues that we should move away from the conventional political economy basis of world city research to more post-structural approaches. He favours bringing Deleuze to centre stage to solve a 'theoretical lacuna' in the world cities literature by using philosophies of connection, actor-network theory and non-representation theory. Derudder's rethinking takes a conceptual turn: whereas it has become commonplace to bemoan the paucity of data, he argues that we also have to get our concepts in order. He compares, contrasts and critiques the three main approaches to worldwide inter-city research: world city, global city and global city-region. Taylor's chapter locates his interlocking network model of the world cities network in a broad geohistorical argument about inter-city relations. Transhistorical processes of town-ness and city-ness are defined in relation to the hinter-work and net-work that operates through cities. These are related to world-systems core–periphery concepts. Finally, Saey promotes some methodological and ethical rethinking. He takes five writings from the world cities literature (by Alderson and Beckfield, Camagni, Friedmann, Smith, and Taylor) that represent different methodological standpoints and uses them as 'stepping stones' to contrast and critique the various research approaches.

We do not claim that the chapters that follow provide a comprehensive collection of world cities studies in the early twenty-first century – that would be impossible in one volume – but they do represent an unusually rich range of empirics, concepts, theories and ideas. We provide a state-of-the-art collection of key writers on inter-city relations in contemporary globalization.

REFERENCES

Alderson, A.S. and Beckfield, J. (2004) 'Power and Position in the World City System', *American Journal of Sociology*, 109: 811–51.

Bagnasco, A. and Le Galès, P. (2000) *Cities in Contemporary Europe*, Cambridge: Cambridge University Press.

Faludi, A. (ed.) (2002) *European Spatial Planning*, Cambridge, MA: Lincoln Institute of Land Policy.

Castells, M. (1996) *The Rise of Network Society*, Oxford: Blackwell.

Centre for European Studies (2003) 'European Space in the face of enlargement', in *The West to East European Trajectory Project*, Europe XXI, vol. 8, Warsaw: Stanislaw Leszczycki Institute of Geography and Spatial Organization.

Elmhorn, C. (2001) *Brussels: A Reflexive World City*, Stockholm: Almqvist and Wiksell.

Derudder B., Taylor, P.J., Witlox, F. and Catalano, G. (2003) 'Hierarchical tendencies and regional patterns in the world city network: a global urban analysis of 234 cities', *Regional Studies*, 37: 875–96.

Derudder, B. and Witlox, F. (2005) 'An appraisal of the use of airline data in assessing the world city network: a research note on data', *Urban Studies*, 42: 2371–88.

ESPON (European Spatial Planning Observation Network) (2005) 'The role, specific situation and potentials of urban areas as nodes in a polycentric development (2002–04)', *Project 1.1.1.*, http://www.espon.lu/online/documentation/projects/thematic.

Friedman, J, (1986) 'The world city hypothesis', *Development and Change*, 17: 69–83.

Ipenburg, D. and Lambrechts, B. (2001) 'Polynuclear urban regions in North West Europe, a survey of key actor views', *EURBANET Report 1, Housing and Urban Policy Studies*, 18.

Isin, E.F. (ed.) (2000) *Democracy, Citizenship and the Global City*, London: Routledge.

Newman, P. and Thornley, A. (2005) *Planning World Cities: Globalization and Urban Politics*, London: Palgrave Macmillan.

Sassen, S. (1991) *The Global City*, Princeton, NJ: Princeton University Press.

Sassen, S. (2002) 'Global cities and diasporic networks: microsites in Global Civil Society', in M. Glasius, M. Kaldor and H. Anheier (eds) *Global Civil Society 2002*, Oxford: Oxford University Press.

Shin, K.H. and Timberlake, M. (2000) 'World cities in Asia: cliques, centrality and connectedness', *Urban Studies*, 37: 2257–85.

Smith, D.A. and Timberlake, M. (2002) 'Hierarchies of dominance among world cities: a network approach', in S. Sassen (ed.) *Global Networks, Linked Cities*, London: Routledge, pp. 117–41.

Taylor, P.J. (2001) 'Specification of the world city network', *Geographical Analysis*, 33: 181–94.

Taylor, P.J. (2004a) *World City Network: a Global Urban Analysis*, London: Routledge.

Taylor, P.J. (2004b) 'Regionality in the world city network', *International Social Science Review*, 181: 361–72.

Taylor, P.J. (2005) 'New political geographies: global civil society and global governance through world city networks', *Political Geography*, 24: 703–30.

Taylor, P.J. and Lang, R.E. (2005) *US Cities in the World City Network*, Washington, DC: The Brookings Institution (Metropolitan Policy Program, Survey Series).

Part I World city networks

2 Globalization and the world city system

Preliminary results from a longitudinal data set

Arthur S. Alderson and Jason Beckfield

INTRODUCTION

The phenomenon of globalization has renewed interest in thinking about the place and role of cities in the international system. Recent literature proposes that the fate of cities (and their residents) has become increasingly tied to their position in international flows of investment and trade. Where traditional thinking on the city (e.g. Hawley 1950; Duncan *et al.* 1960) treats processes of urbanization as regional or national phenomena, a new urban sociology argues for situating the city in the larger context of the development of the world economy (Friedmann 1986; Knox and Taylor 1995; Sassen 2001). Globalization is argued to be generating a new geography of centrality and marginality that cuts across the old divides in the world between the rich developed countries and the poor underdeveloped countries, and between East and West (Sassen 1994: 4). In particular, developments of the past few decades are seen as producing a new global hierarchy of cities, at the apex of which are located what have variously been referred to as 'world cities' (Friedmann 1986) or 'global cities' (Sassen 2001). Such cities are argued to be the key nodes or command points that exercise power over other cities in a system of cities and, thus, the world economy.

We address these arguments empirically by building on the strong affinity between the literature on world cities and social network analysis. As conceptualized in the literature, the power of world cities is inherently relational: cities do not have power in and of themselves, they have power to the extent that they function as command points and centers of planning and thus establish the framework in which other cities operate in the world economy. Similarly, social network analysts suggest that power is best viewed as a consequence of patterns of social relations that generate opportunities and constraints: some actors are favored because they occupy positions that are more favorable than others (Granovetter 1973; Padgett and Ansell 1993; Guiffre 1999). Moreover, network analysts have developed a set of tools that enable those interested in pursuing the world city hypothesis to assess the power and prestige of individual cities.[1] With such methods, one can address some of the fundamental hypotheses that animate research on the world city system.[2]

In earlier research (Alderson and Beckfield 2004), we used data on the head-quarters and branch locations of the world's 500 largest multinational firms in 2000 to generate a relational data set in which ties are those between headquarter and subsidiary cities. More recently, we have introduced a longitudinal compo-nent by gathering the same data for 1981 (the earliest available). As assembled, this produces two matrices, each linking more than 3,000 cities around the globe. In this chapter, we present the first results of this work. Using the techniques detailed below, we address a central problematic of world city research in this chapter: How has the global restructuring of the past two decades altered the world city system? In the course of addressing this question, we also address the more general question of the link between the global urban hierarchy and the world-system.

WORLD CITY HYPOTHESES AND GLOBALIZATION

As suggested above, we are guided in this research by three classic statements of the world city hypothesis, those of Stephen Hymer (1972), John Friedmann (1986), and Saskia Sassen (1991). As we review these arguments in detail else-where (see Alderson and Beckfield 2004), we briefly note here the broad similari-ties and differences in these statements as they speak to the evolution of the world city system in the context of globalization. All three see globalization as funda-mentally altering the world city system and, thus, the geography of inequality. However, they differ strongly on exactly how this is changing. Hymer implicitly invokes the 'Matthew Effect,' arguing that globalization will largely reproduce existing patterns of inequality and dependency. While alteration of the global ur-ban hierarchy is likely (e.g. the emergence of new centers of production in the South and the simultaneous decline of old centers of production in the North), Hymer predicted that the map of the world city system would match on increas-ingly tightly to the map of the world-system – the 'core' will grow increasingly core, and the 'periphery' will grow increasingly peripheral. Friedman and Sassen offer a rather different vision of the impact of globalization on the world city system. They present the possibility of a new geography of centrality and mar-ginality, one that cuts across long-standing North/South and East/West divides in the world system. Friedmann (1995) characterizes the world city system as a dynamic hierarchy with ranks and entrance criteria that are, in principle, open: To the extent that cities can attract investment and capture more of the command and control functions of the world economy, their status in the urban hierarchy will rise. While there is every expectation that cities such as New York, London, and Tokyo will emerge as cities of the first rank in any empirical analysis, 'cities may rise into the ranks of world cities, they may drop from the order, and they may rise or fall in rank' (ibid.: 26). Sassen (1994) paints a similar picture: Areas in the developed world once conceptualized as 'core' are being peripheralized, while areas once 'peripheral' are joining the core of the world city system (e.g. the de-cline of Detroit and the rise of São Paulo). For both, then, globalization opens up the possibility of substantial slippage between the map of the contemporary world city system and established maps of the world-system. The new urban hierarchy generated by globalization cuts across the old North/South and East/West divides in the world system.

DATA AND METHODS

To assess change in the structure of the world city system and to determine the degree to which a new geography of centrality and marginality may have emerged in recent years, we have assembled longitudinal data on a key relation linking cities into a world-system of cities: that between multinational enterprises and their subsidiaries. Using sources such as *Fortune* magazine's Global 500, we identify the world's 500 largest multinationals (in terms of revenue) in 1981 and 2000. We then use the Directory of Corporate Affiliations (National Register 1981, 2000) to collect information on the headquarter and branch locations of each firm in both periods.[3]

Power and prestige in the world city system

We assess the power of world cities in light of three measures of point centrality (namely outdegree, closeness, and betweenness). To understand the differences between these measures, it may be useful to consider the two networks illustrated in Figure 2.1. Assume, for instance, that the relation illustrated in Figure 2.1 involves the exchange of resources between cities. Examining the star network, one would conclude that city A occupies a favorable structural position, while cities B–G occupy the same, equally unfavorable position. In the circle network, by contrast, all cities appear equally advantaged or disadvantaged. Why is city A

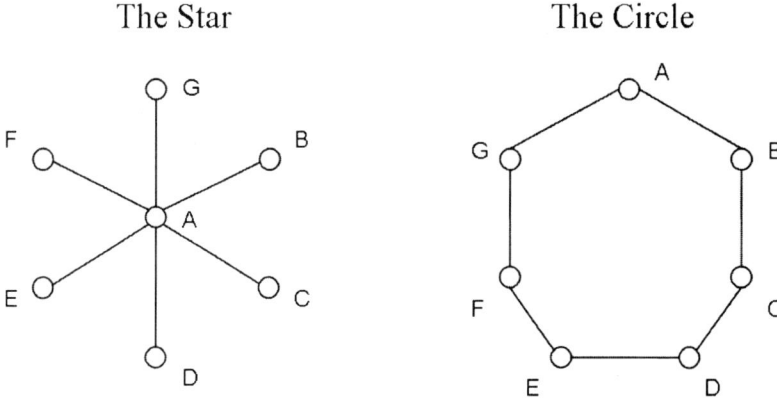

Figure 2.1 Graphs to illustrate centrality measures

advantaged in the star network? Freeman's (1979) now classic treatment of centrality in social networks suggests three distinct reasons.

Outdegree centrality

City A in the star network is advantaged because it is more active than cities B–G. As such, city A has more alternatives. If city B refuses an exchange with A, A can rely on resources from C–G. B–G, by contrast, are less active. They are isolated from direct involvement with others in the network and have no alternatives to exchange with A. In this sense, city A is more powerful than cities B–G. In the circle network, by contrast, all cities are equally active and thus equally advantaged or disadvantaged. With directional data, it is often important to distinguish between outdegree (ties sent) and indegree (ties received). With a relation of the sort explored in this chapter, the outdegree of each city is a fairly straightforward measure of power or influence: cities that send more ties are cities that have captured more of the control functions of the world economy (i.e. display more 'world-cityness' than others).

Closeness centrality

City A in the star network is advantaged because it is closer to more cities than cities B–G. City A is adjacent to all other cities while B–G are two steps from all other cities (except A). Consequently, city A has greater power in the sense that it is more independent than the others (or, alternatively, in the sense that it can avoid being controlled by others). For resources to pass from city B to city E, they must pass through A. In contrast, city A can directly communicate with B–G. This gives city A in the star network a distinct structural advantage. In the circle network, all actors are equally close and are thus, again, equally advantaged or disadvantaged.

Betweenness centrality

Finally, city A in the star network is advantaged because it stands between all of the other pairs of actors. It thus has greater power in the sense that it brokers all exchanges. If city B wishes to exchange resources with city E, it must do so through city A. City A thus has the power to coordinate action and to withhold or distort information to its advantage. This, mirrored by the fact that city A needs no broker for exchanges with B–G in the star network, gives it a distinct structural advantage. In the circle network, each city lies between each other pair of actors and A–G are, again, equally advantaged or disadvantaged.

Indegree centrality

While the world city hypothesis emphasizes the power of world cities, it also suggests that, in addition to being influential, world cities are prominent or *prestigious*

– they are sought out by other cities, have ties directed to them, and are chosen over others. Put differently, 'world-cityness' involves choices received in addition to choices made. A simple and straightforward measure of a city's prestige is its indegree – the number of ties it receives. Cities that have high indegree are prestigious in precisely the above sense.

Actor degree, closeness, and betweenness were calculated using the network analysis program UCINET (Borgatti *et al.* 2002).

Changes in the structure of the world city system

Having established a ranking of cities in terms of their power and prestige in 1981 and 2000, we turn to address three proximate questions regarding the impact of global restructuring and the possible decoupling of the world city system from the world-system:

- How extensively has the structure of the world city system been altered across the era of globalization?

Hymer (1972), Friedmann (1986), and Sassen (2001) all share the expectation that globalization is producing a novel global hierarchy of cities. To address this, we calculate the *Spearman rank-order correlation* between each of our measures of power and prestige in 1981 and 2000. This allows us, in a global fashion, to assess the degree to which the deck has been 'reshuffled' over the last two decades (i.e. the degree to which cities have exchanged ranks within the world city system).

- Have power and prestige in the world city system grown more concentrated across the era of globalization?

Hymer (1972) and Sassen (2001) share the expectation that power will become increasingly concentrated in the hands of multinationals located in an increasingly small number of 'world' or 'global' cities as the world-economy develops. To address this, we calculate and track the change from 1981 to 2000 in a standard measure of inequality – the *coefficient of variation* – for each of our measures of power and prestige.

- Have cities decoupled from 'traditional' political geography in the course of globalization?

Friedmann (1986) and Sassen (2001) argue that globalization is generating a new geography of centrality and marginality that cuts across the old core/periphery and East/West divides in the world-system. To address this claim, we match each city to its country and assign it (1) to *core, peripheral,* or *semiperipheral* status on the basis of Bollen's (1983) revision and update (Bollen and Appold 1993) of the scheme originally presented by Snyder and Kick (1979); and (2) to

Africa, Latin America, or *Asia* (with North American and Europe as the reference group). We then estimate a series of regressions based on the *conditional change model*:

$$Y_t = \alpha + \beta_1 X_t + \beta_2 Y_{t-1} + \varepsilon_t,$$ (2.1)

where the dependent variable Y_t is predicted from its lagged value Y_{t-1} and from a vector of independent variables X measured at the same time point. A key feature of this model is that it can be expressed in terms of ΔY by subtracting Y_{t-1} from (2.1) to yield (Finkel 1995: 7):

$$\Delta Y = \alpha + \beta_1 X_t + (\beta_2 - 1) Y_{t-1} + \varepsilon_t$$ (2.2)

From this expression, it is clear that β_1 can be interpreted as *the effect of X on change in Y, controlling for the initial values of the dependent variable* (thus the name, *conditional* change model). The effect of the lagged dependent variable is simply the stability effect of Y_{t-1} on Y in (2.1) minus 1 and has a variety of possible substantive interpretations (Finkel 1995: 7–11). With this model, we are able to assess the extent to which the restructuring that the world city system has undergone over the 1981–2000 period cuts across region and the position of nation-states in the interstate system.

RESULTS

How has the global restructuring of past two decades altered the world city system?

Of the senses of power discussed in the literature on world cities, degree centrality is arguably the most prominent. World cities are variously defined as 'headquarter cities' (Hymer 1972), as 'basing points in the spatial organization and articulation of production and markets' (Friedmann 1986), and as 'concentrated command points in the organization of the world economy' (Sassen 2001). In this sense, outdegree is an unambiguous indicator of 'world cityness': Cities that send more ties are cities that have captured more of the control functions of the world economy. As one can note from Table 2.1, the ranking of cities on outdegree in 1981 is only moderately correlated with that ranking in 2000 ($r = 0.672$), indicating a measurable 'reshuffling' of the global urban hierarchy over the last two decades.

Outdegree centrality identifies as powerful or influential those cities that are most active and visible (i.e. send the largest number of ties). Closeness centrality, by contrast, identifies as powerful those cities with the shortest paths to others in the network (quantified as the inverse average distance between a given city and all others). By this criterion, the restructuring of the world city system over the 1981–2000 period appears even more substantial, the correlation between the rankings of cities on closeness in 1981 and in 2000 being just 0.570.

Table 2.1 Spearman correlations (*r*) and coefficients of variation (*COV*) for measures of power and prestige in the world city system, 1981 vs. 2000

r Outdegree 1981/Outdegree 2000	0.672
COV Outdegree 1981	10.981
COV Outdegree 2000	13.018
r Closeness 1981/Closeness 2000	0.570
COV Closeness 1981	0.105
COV Closeness 2000	0.095
r Betweenness 1981/Betweenness 2000	0.646
COV Betweenness 1981	9.555
COV Betweenness 2000	12.672
r Indegree 1981/Indegree 2000	0.617
COV Indegree 1981	5.866
COV Indegree 2000	5.957

Betweenness centrality identifies as powerful those cities that lie on the paths connecting other cities. Actors with high betweenness have greater power in the sense that they serve as brokers and can control the flow of information through the network. The correlation between the rankings of cities on betweenness in 1981 and in 2000 is considerably less than perfect ($r = 0.646$), again indicating substantial alteration in the global urban hierarchy.

A city's indegree can be interpreted as an indicator of its *prestige* in the sense that cities with high indegree have been chosen over others. Just as various versions of the world city hypothesis predict substantial change in the distribution of power within the system in the context of globalization, they also suggest the redistribution of prestige as firms find it increasingly important to establish ties to those cities in which global control is produced and reproduced. The results presented in Table 2.1 again lend support to the world city hypothesis. Far from being stable, the ranking of cities in terms of prestige by 2000 was only moderately correlated with that ranking in 1981 ($r = 0.617$).

Viewed together, these results lend support to Hymer (1972), Friedmann (1986) and Sassen's (2001) view that globalization is producing a novel global urban hierarchy. Given knowledge of the position of cities in the system's distribution of power and prestige in 1981, one's ability to predict their location in 2000 is far from perfect. There is thus measurable 'slippage' between the 1981 and 2000 maps of the world city system.

The world city hypothesis predicts more than just the emergence of a new urban hierarchy. It also (especially Hymer and Sassen's versions) describes that hierarchy as one in which power is (increasingly) concentrated in the hands of multinationals located in an (increasingly) small number of 'world' or 'global' cities. In other words, cities are not only expected to exchange positions within the hierarchy in the context of global restructuring, but the global distribution of power and prestige is predicted to *grow increasingly unequal*. We assess this dimension of the world city hypothesis by tracking changes in the coefficients of variation in each of our measures of power and prestige. The coefficient of variation (COV) – calculated as the standard deviation/mean – is a standard measure of inequality. Smaller values indicate less relative dispersion. The COV is often used, for instance, by economists as a measure of 'σ-convergence,' or of whether the distribution of national income across economies is becoming more or less equitable over time (e.g. Barro and Sala-i-Martin 1995).

The results presented in Table 2.1 are consistent with the world city hypothesis. The distribution of outdegree and betweenness (i.e. of two of our three measures of power) grew more unequal over the 1981–2000 period. The COV of outdegree grew from 10.981 to 13.018 and the COV of betweenness grew from 9.555 to 12.672. Substantively, this means that the control functions of the world-economy did indeed become more centralized (outdegree) while fewer cities 'mediate' between others over time (betweenness). The COV of closeness, on the other hand, marginally declines from 0.105 to 0.095. Substantively, this means that the average city is no more distant from others in the system in 2000 than it was in 1981. This should be interpreted in light of the findings for betweenness: With fewer cities mediating between others any pair of cities should, *ceteris paribus*, be closer. The results for our measure of prestige are also interesting. The COV of indegree grows from 5.866 to 5.957. The distribution of prestige is thus growing more unequal over time – a subset of cities received a larger proportion of all ties sent in 2000 than was the case in 1981.

Has the world city system decoupled from the world-system?

Consistent with the world city hypothesis, the above analysis suggests that recent decades have witnessed the emergence of a new urban hierarchy and that power and prestige within this hierarchy have grown increasingly concentrated. What are we to make of these facts? Smith and Timberlake (1995: 81) have suggested that in studying the structure of the world city system 'we stand to learn a great deal more about the nature of the world-system itself.' Relations between cities, they propose, undergird the structure of the world system and help to reproduce a global economic order that exhibits a core/periphery structure at the level of the interstate system. As such, we might expect change in the structure of the world city system to lead change in the structure of the world system: Any revision of the global hierarchy is likely to be manifest *first* in the alteration of relations between cities. Thus, for instance, if the world-system is indeed in the midst of an important shift in power from West to East (e.g. Arrighi 1994; Frank 1998), this should be observable in change in the structure of the global urban hierarchy well before it is revealed in change in the structure of trade between nations (Smith and White 1992).

The literature on world cities is rich with claims regarding the effects of the most recent round of globalization on the world-system. Friedmann (1986) and Sassen (2001) argue that the restructuring that the world economy has undergone in the past two to three decades has generated an urban hierarchy that cuts across the traditional divides in the world-system. To the extent that this is true, we would expect to observe substantial slippage between the map of the contemporary world city system drawn out above and the map of the world system. Hymer (1972), in contrast, thought that globalization would largely reproduce pre-existing cross-national patterns of inequality and dependency. In his view, we should expect the standing of cities in the world city system to match rather closely the position of their nation-states in the world-system.

To explore the degree to which a new geography of centrality and marginality may have emerged, we relate *change* in the power and prestige of individual cities over the 1981–2000 period to the *world-system position* of the countries in which they are located and to the *world-region* in which they are located. In all models below, we also control for the population of each city. Interestingly, we find that cities in the network that are located in non-core countries are, on average, *ten times* larger than those located in core countries. As we show below, there is a strong relationship between the population or raw demographic prominence of a city and its power and prestige in the city system. It is therefore critical that we control for population in testing for effects of world system position; otherwise any effect of world system position would likely be confounded by the large differences in average city size across world system positions.

The results for the measures of power and prestige are presented in Tables 2.2 and 2.3. For each measure, we present three models. In the first, we introduce indicator variables for location in a semiperipheral or peripheral country (with location in a core country as the reference category) and city population (natural logarithm (ln)). In the second, we introduce indicator variables coding for location in Africa, Latin America, or Asia (with location in North America or Europe as the reference category). All of these variables are combined in a third model.

In all of the models in which it appears, population (ln) has a significant positive effect on change in power or prestige over the 1981–2000 period. Outdegree (ln), closeness (ln), betweenness (ln), and indegree (ln) grew significantly more in larger cities than it did in smaller cities. The shifting position of cities in the global urban hierarchy is thus correlated with their raw demographic prominence (Chase-Dunn and Manning 1999; see also Derudder 2003; Short 2004).

World-system position is explicitly ordered (Wallerstein 1974; Chase-Dunn and Grimes 1995): the core region of the world economy is more powerful than the semiperipheral region, and the semiperiphery is more powerful than the periphery. At the extreme, then, if Friedmann and Sassen are correct in their argument that the contemporary urban hierarchy cuts across the traditional divides in the world-system, the semiperiphery and periphery indicators should not significantly predict change in power and prestige. Alternatively, if Hymer is correct in his prediction that globalization would simply re-peripheralize the underdeveloped world, change in power and prestige across the 1981–2000 period will be significantly ordered by world system position.

The results for outdegree, closeness, betweenness, and indegree are most consistent with the latter set of expectations. When change in power and prestige is

Table 2.2 Regression models of measures of network centrality: OLS estimates for 3,023 cities

Variable	Outdegree 2000 (ln)			Closeness 2000 (ln)		
	(1)	(2)	(3)	(4)	(5)	(6)
Outdegree 1981 (ln)	0.929 ***	0.986 ***	0.927 ***			
	(0.016)	(0.014)	(0.016)			
Closeness 1981 (ln)				0.011 ***	0.019 ***	0.011 ***
				(0.001)	(0.001)	(0.001)
Population (ln)	0.076 ***		0.078 ***	0.023 ***		0.024 ***
	(0.007)		(0.007)	(0.001)		(0.001)
Semiperiphery	-0.090 *		-0.082	-0.015 **		-0.014 *
	(0.042)		(0.044)	(0.006)		(0.006)
Periphery	-0.231 ***		-0.215 ***	-0.037 ***		-0.039 ***
	(0.037)		(0.050)	(0.005)		(0.007)
Africa		-0.051	-0.096		0.033 ***	0.013
		(0.056)	(0.080)		(0.009)	(0.011)
Latin America		0.021	0.034		0.026 ***	0.010
		(0.045)	(0.069)		(0.007)	(0.009)
Asia		0.026	-0.037		0.024 ***	-0.013 *
		(0.026)	(0.042)		(0.004)	(0.006)
Constant	-0.694 ***	0.051 ***	-0.714 ***	3.309 ***	3.533 ***	3.302 ***
	(0.071)	(0.010)	(0.073)	(0.010)	(0.002)	(0.010)
Adjusted R^2	0.582	0.562	0.582	0.272	0.128	0.275

Note
Standard errors in parentheses.
*$p<0.05$, **$p<0.01$, ***$p<0.001$ (two-tailed tests)

Table 2.3 Regression models of measures of network centrality: OLS estimates for 3,023 cities

Variable	Betweenness 2000 (ln)			Indegree 2000 (ln)		
	(7)	(8)	(9)	(10)	(11)	(12)
Betweenness 1981 (ln)	0.727 ***	0.754 ***	0.726 ***			
	(0.012)	(0.010)	(0.012)			
Indegree 1981 (ln)				0.694 ***	0.839 ***	0.701 ***
				(0.012)	(0.011)	(0.013)
Population (ln)	0.010 ***		0.010 ***	0.149 ***		0.148 ***
	(0.001)		(0.001)	(0.007)		(0.007)
Semiperiphery	−0.015		−0.014	−0.009		0.011
	(0.008)		(0.009)	(0.037)		(0.039)
Periphery	−0.030 ***		−0.027 **	−0.125 ***		−0.053
	(0.007)		(0.010)	(0.033)		(0.045)
Africa		−0.007	−0.014		−0.029	−0.328 ***
		(0.011)	(0.016)		(0.054)	(0.070)
Latin America		−0.002	−0.003		0.170 ***	−0.047
		(0.009)	(0.014)		(0.044)	(0.061)
Asia		0.007	−0.001		0.275 ***	0.003
		(0.005)	(0.008)		(0.025)	(0.037)
Constant	−0.095 ***	0.006 **	−0.096 ***	−0.658 ***	0.785 ***	−0.647 ***
	(0.014)	(0.002)	(0.014)	(0.066)	(0.011)	(0.069)
Adjusted R^2	0.597	0.589	0.597	0.691	0.633	0.693

Note
Standard errors in parentheses.
*$p<0.05$, **$p<0.01$, ***$p<0.001$ (two-tailed tests)

significantly related to semiperipheral and peripheral position, the relationship is *negative* and the effects of world system position are *ordered*. The results of Model 1 indicate that the average outdegree of cities located in semiperipheral countries is roughly 9 percent lower than that of cities located in core countries (i.e. $\exp(-0.090) - 1 = -0.086$), whereas the outdegree of cities located in peripheral countries is 21 percent lower (i.e. $\exp(-0.231) - 1 = -0.206$). Remembering that the conditional change model can also be interpreted in terms of ΔY (see equation (2) above), the results of Model 1 mean, in addition, that the 'reshuffling' of cities and the rise in inequality in the distribution of outdegree documented above was biased in favor of cities located in core countries – controlling for initial levels of outdegree, semiperipheral cities had 9 percent less outdegree relative to core cities in 2000 than they did in 1981, while peripheral cities had 21 percent less. Effects of similar magnitude are found when indicator variables for region are introduced into the regression (Model 3).

The results for closeness, betweenness, and indegree tell a similar story. Controlling for initial levels of the dependent variable and relative to the average city located in a core country, the average city located in a semiperipheral country was less close (Model 4), less between (Model 7), and less prestigious (Model 10) in 2000 than it was in 1981 (significant at greater than the 10 percent level for closeness and betweenness). The average city located in a peripheral country grew roughly 4 percent less close, 3 percent less between, and 12 percent less prestigious. Taken together, the results presented in Tables 2.2 and 2.3 are more consistent with Hymer's view of globalization as reproducing existing cross-national patterns of inequality and dependency than with Friedmann's and Sassen's visions of a world city system in the grips of novel forms of global restructuring. Rather than cutting across the hierarchy of states in the interstate system, the urban hierarchy appears to map onto it increasingly well.

What of the effects of region? Is there any evidence of significant global restructuring in these terms? If so, which statement of the world city hypothesis might such evidence support? Interestingly, we find that the world region in which cities are located is not significantly related to their outdegree (Models 2 and 3). African, Latin American, and Asian cities did not experience a different rate of change in outdegree than North American or European cities across the 1981–2000 period. We draw the same conclusions for betweenness from Models 8 and 9. So while the 'reshuffling' of cities and the rise in inequality in the distribution of outdegree and betweenness was biased in favor of cities located in core countries, this occurred independent of any regional effects. The results for closeness indicate that, controlling for the initial level of the dependent variable and relative to North American and European cities, African, Latin American, and Asian countries grew significantly more close over the 1981–2000 period (Model 5). However, when population (ln) and world-system position are controlled (Model 6), the positive effects of Africa and Latin America are substantially attenuated and no longer significant and Asian cities are found to have grown significantly *less close* across the era of globalization. Finally, the results for indegree in Model 11 show that Latin American and Asian cities grew relatively more prestigious over the time period under consideration. When the other controls are introduced, however, these coefficients are no longer significant and African cities are found to have grown significantly less prestigious (Model 12). This effect is substantively large, with African cities, on average, having 28 percent less indegree in 2000 than they had in 1981, and this regional effect appears to account for the negative effect of peripheral status on prestige in Model 10.

CONCLUSIONS

In this chapter, we present the first results of a longitudinal analysis of data on what we view as a key relation linking cities into a world system of cities: that between multinationals and their subsidiaries. Our analysis of these data was motivated by two general concerns. First, although the literature on world cities is rich with claims regarding the effects of globalization on the structure of the world-system, few have attempted to trace out exactly how the structure of the world city system has changed over the course of the past two decades.[4] Second, and related, although scholars such as Friedmann and Sassen suggest that the world city system has decoupled from the world-system and that it has evolved in such a way as to cut across the traditional divides in the world-system, no one has systematically put such ideas to a test.

To these ends, we addressed three proximate questions. First, how extensively has the structure of the world city system been altered across the era of globalization? Our analysis of the Spearman correlations between each of our measures of power and prestige in 1981 and 2000 suggests that the 'deck' has been substantially 'reshuffled' over the last two decades. Some cities have risen in rank, others have fallen, and one's ability to predict the position of cities in the system's distribution of power and privilege in 2000 based on their standing in 1981 is modest. Second, have power and prestige in the world city system grown more concentrated across the era of globalization? To address this, we tracked the change from 1981 to 2000 in a standard measure of inequality – the *coefficient of variation* – for each of our measures of power and prestige. The results show that the distributions of two of our three measures of power (outdegree and betweenness) grew more unequal over time, while inequality in closeness declined marginally. The distribution of indegree, our measure of prestige, grew marginally more unequal over time. Consistent with the world city hypothesis, then, we find evidence that the distribution of power and prestige within the global urban hierarchy has grown increasingly unequal. Finally, have cities been decoupled from 'traditional' political geography in the course of globalization? Results of a series of conditional change regressions estimated for each of our measures of power and prestige support the position taken more than three decades ago by Stephen Hymer (1972). While there is certainly less than a one-to-one matching of cities onto nation-states in the world system, we find that the evolution of a city's power is shaped in a significant fashion by the world-system position of its country. Moreover, these world-system effects are ordered in a fashion consistent with the idea that cities located in core countries will, on average, grow to be more powerful and prestigious than cities located in semiperipheral countries, which, in turn, will be more powerful and prestigious than cities located in peripheral countries. We find no evidence for the new geography of centrality and marginality discussed by scholars such as Friedmann (1986) and Sassen (2001). Rather, we find just the opposite – the reproduction of the 'old' geography in an even more pronounced form. Our analysis of regional effects likewise reveals no evidence for an urban hierarchy that cross-cuts regional divides in novel ways. We find, for instance, no evidence of the rising Eastern hegemony described by Arrighi (1994) or Frank (1998). When cities located in Asia are different from those in North America or Europe in properly specified models, we find that it is only that they have grown relatively less close across the 1981–2000 period. African cities grew notably less prestigious – a finding that is consistent with many descriptions of the 'abandonment' or 'neglect' of Africa in the contemporary world economy.

In this chapter we have taken another step toward mapping the contemporary world city system. While we believe that we have made some progress in this regard, there are at least two promising directions for future research that would firm up our preliminary results and serve to advance the longitudinal analysis of the world city system. First, the results presented in this chapter pertain to the entire world city system. Although we see no *general* pattern whereby the new urban hierarchy cuts across North and South and East and West, this does not speak to the possibility of a substantial reordering in the past few decades at the very pinnacle of the urban hierarchy. In research under way, we are focusing our analysis on the most powerful and prestigious cities, asking whether the above findings hold for the class of headquarter/world/global cities that make up such a small fraction of the cities analyzed in this chapter. Second, our data set could be analyzed using the interlocking network model developed by Peter Taylor and the GaWC group (Taylor 2004). This would specify a three-layered network – a net level (the world economy), a nodal level (cities), and a subnodal level of inter-lockers (the multinational enterprises (MNEs)) – that would unbundle the equation of 'city' with 'firm' that the present analysis dictates (Taylor 2004). Methods designed for the interlocking network model (Taylor 2005), could then be used to validate our results for the entire system and for the subset of headquarter/world/global cities.

NOTES

1 We use the generic 'world city hypothesis' to refer to work following in the vein of John Friedmann and Saskia Sassen (briefly reviewed below) who argue that developments in the world economy are generating a global urban hierarchy with a distinctive, novel morphology. Although scholars in this tradition may vary appreciably on the particulars, they share the view that it is imperative to situate processes of urbanization in the context of the development of the world economy, rather than viewing them solely or even primarily as regional or national phenomena.

2 In the last decade there has been a flurry of work that, in different ways, grounds

the network metaphor that has defined thinking on the world city system (see, for instance, Smith and Timberlake 1995; Taylor *et al.* 2002; Taylor 2004; Derudder and Taylor 2005; Derudder and Witlox 2005).

3 In the interests of space, we have kept the discussion of data and methods to a minimum (see Alderson and Beckfield 2004 for details). Briefly, for each firm listed, we coded the location of the firm's headquarters and subsidiaries. We then constructed directional, valued matrices for 1981 and 2000 from this list. In 2000, there are 3,692 unique cities in the list, 1,745 of which do not appear in the 1981 list. These cities thus entered the system between 1981 and 2000 and necessarily had point centrality (see below) of zero in 1981. Owing to missing data, the total number of cities in the regression analyses is reduced to 3,023.

4 See Smith and Timberlake (2002) and Shin and Timberlake (2000) for important exceptions.

REFERENCES

Alderson, A.S. and Beckfield, J. (2004) 'Power and position in the world city system', *American Journal of Sociology*, 109: 811–51.

Arrighi, G. (1994) *The Long Twentieth Century: Money, Power, and the Origins of Our Times*, London: Verso.

Barro, R.J. and Sala-i-Martin, X. (1995) *Economic Growth*, Cambridge, MA: MIT Press.

Bollen, K.A. (1983) 'World system position, dependency, and democracy: the cross-national evidence', *American Sociological Review*, 48: 468–79.

Bollen, K.A. and Appold, S.J. (1993) 'National industrial structure and the global system', *American Sociological Review*, 58: 283–301.

Borgatti, S.P., Everett, M.G. and Freeman, L.C. (2002) *UCINET 6 for Windows: Software for Social Network Analysis*, Natick: Analytic Technologies.

Chase-Dunn, C. and Grimes, P. (1995) 'World systems analysis', *Annual Review of Sociology*, 21: 387–417.

Chase-Dunn, C. and Manning, S. (1999) 'City systems and world-systems: four millennia of city growth and decline', unpublished manuscript, Johns Hopkins University.

Derudder, B. (2003) 'Beyond the state: mapping the semi-periphery through urban networks', *Capitalism, Nature, Socialism*, 14: 91–120.

Derudder, B. and Taylor, P. J. (2005) 'The cliquishness of world cities', *Global Networks*, 5: 71–91.

Derudder, B. and Witlox, F. (2005) 'An appraisal of the use of airline data in assessing the world city network: a research note on data', *Urban Studies*, 42: 2371–88.

Duncan, O.D., Scott, W.R., Lieberson, S., Duncan, B. and Winsborough, H. (1960) *Metropolis and Region*, Baltimore, MD: Johns Hopkins University Press.

Finkel, S.E. (1995) *Causal Analysis with Panel Data*, Thousand Oaks, CA: Sage.

Frank, A.G. (1998) *ReOrient: Global Economy in the Asian Age*, Los Angeles, CA: University of California Press.

Freeman, L. (1979) 'Centrality in social networks I: conceptual clarification', *Social Networks*, 1: 215–39.

Friedmann, J. (1986) 'The world city hypothesis', *Development and Change*, 17: 69–84.

Friedmann, J. (1995) 'Where we stand: a decade of world city research', in P.L. Knox and P.J. Taylor (eds) *World Cities in a World-System*, New York: Cambridge University Press, pp. 21–47.

Granovetter, M. (1973) 'The strength of weak ties', *American Journal of Sociology*, 78: 1360–80.

Guiffre, K. (1999) 'Sandpiles of opportunity: success in the art world', *Social Forces*, 77: 815–32.

Hymer, S. (1972) 'The multinational corporation and the law of uneven development', in J.N. Bhagwati (ed.) *Economics and World Order*, New York: Macmillan, pp. 113–40.

Hawley, A. (1950) *Human Ecology*, New York: Ronald Press.

Knox, P.L. and Taylor, P.J. (1995) *World Cities in a World-System*, New York: Cambridge University Press.

National Register (1981 and 2000) *Directory of Corporate Affiliations*, Skokie, IL: National Register.

Padgett, J.F. and Ansell, C.K. (1993) 'Robust Action and the Rise of the Medici', *American Journal of Sociology*, 98: 1259–1319.

Sassen, S. (1991) *The Global City: New York, London, Tokyo*, Princeton, NJ: Princeton University Press.

Sassen, S. (1994) *Cities in a World Economy*, Thousand Oaks, CA: Pine Forge Press.

Sassen, S. (2001) *The Global City: New York, London, Tokyo*, 2nd edn, Princeton, NJ: Princeton University Press.

Shin, K.-H. and Timberlake, M. (2000) 'World cities in Asia: cliques, centrality and connectedness', *Urban Studies*, 37: 2257–85.

Short, J.R. (2004) 'Black holes and loose connections in a global urban network', *The Professional Geographer*, 56: 295–302.

Smith, D.A. and White, D.R. (1992) 'Structure and dynamics of the global economy: network analysis of international trade 1965–1980', *Social Forces*, 70: 857–93.

Smith, D.A. and Timberlake, M. (1995) 'Cities in global matrices: toward mapping the world-system's city-system', in P.L. Knox and P.J. Taylor (eds) *World Cities in a World-System*, New York: Cambridge University Press, pp. 79–97.

Smith, D.A. and Timberlake, M. (2002) 'Hierarchies of dominance among world cities: a network approach', in S. Sassen (ed.) *Global Networks, Linked Cities*, New York: Routledge, pp. 117–41.

Snyder, D. and Kick, E.L. (1979) 'Structural position in the world system and economic growth 1955–1970: a multiple-network analysis of transnational interactions', *American Journal of Sociology*, 84: 1096–1126.

Taylor, P.J., Walker, D.R.F. and Beaverstock, J.V. (2002) 'Firms and their global service networks', in S. Sassen (ed.) *Global Networks, Linked Cities*, New York: Routledge, pp. 93–116.

Taylor, P.J. (2004) *World City Network: A Global Urban Analysis*, London: Routledge.

Taylor, P.J. (2005) *Parallel Paths to Understanding Global Inter-city Relations*. GaWC research bulletin 143.

Wallerstein, I. (1974) *The Modern World-System*, New York: Academic Press.

3 Airline passenger flows through cities

Some new evidence

Frank Witlox and Ben Derudder

INTRODUCTION

In parallel with Gottmann's (1989: 62) observation that '[d]ependence on a network has become the general rule for the majority of substantial cities anywhere', it has become commonplace to describe the global urban system as a hallmark example of a global *network* (see, for instance, Castells 1996: 415). However, although most seminal theoretical contributions thus consider worldwide inter-city *relations* to be of crucial importance, it can be noted that this notion of a 'world city network' (WCN) has rarely been accompanied by genuine empirical network analyses (notable exceptions include Smith and Timberlake 2001, 2002; Alderson and Beckfield 2004; Derudder and Taylor 2005). The main reason for this dearth of empirical network analyses is simply the paucity of data detailing with intercity flows at the global level (Smith and Timberlake 1995a,b; Taylor 1997, 1999). There have recently been some attempts to rectify this situation, but the major exception to this empirical conundrum has been the compilation of information on communication networks in general and air transport in particular (see, for instance, Cattan 1995; Keeling 1995; Kunzmann 1998; Rimmer 1998; Smith and Timberlake 2001, 2002, 2005; Matsumoto 2004).

The main reason for this abundance of air transport-based analyses is that the potential usefulness of such data in this context appears to be quite clear: data are comparatively easy to obtain, air transport is traditionally organized through cities rather than through states, and transportation is, of course, mainly about connections and flows. However, despite the obvious appeal of airline data, it can be noted that previous airline analyses have been hampered by inadequate data. Earlier contributions by Taylor (1999) and Beaverstock *et al.* (2000) provided brief assessments of this inadequacy, but in this chapter we shall attempt a more systematic overview of the alleged data problems in previous studies. At the same time, we shall introduce a dataset that is partially able to circumvent a number of these problems.

This chapter is organized as follows. First, since most empirical studies relying upon airline data have a rather limited discussion of this data source's potential usefulness, we begin by addressing the overarching rationale in somewhat more detail. Second, we point to some obstacles in previous datasets for measuring worldwide inter-city relations, and argue that a clear-cut translation of these databases into spatial analyses of the WCN is therefore not a straightforward issue. Third, then, we outline the construction of an inter-city matrix based on the so-called Marketing Information Data Transfer (MIDT) database, which contains detailed flight information of bookings made through Global Distribution Systems (GDS). The discussion primarily focuses on how the information contained in this database might circumvent some of the difficulties described in the previous section.

RATIONALE FOR USING AIRLINE DATA TO ANALYSE THE WORLD CITY NETWORK

There is a whole range of specific WCN conceptualizations that each require a specific empirical grasp (see Chapter 15 in this volume), but in this chapter the notion of a WCN is employed as a shorthand concept for emphasizing that key cities in the global economy derive their status from a position in a great variety of networks. As Taylor (2004: 42) puts it, cities 'operate in a contemporary space of flows that enables them to have a global reach when circumstances require such connections.' For the sake of this chapter, therefore, the main issue is that, regardless of the broad diversity of WCN conceptualizations, authors have increasingly stressed the importance of a relational stance: it is generally acknowledged that major cities are increasingly (re)produced by what flows through them (people, information, knowledge, money, and cultural practices) rather than what is fixed within them, and it is therefore the presence of a vast bundle of global relations that lies at the root of world city formation (Allen 1999; Taylor 1999; Derudder *et al.* 2003). From an empirical point of view, the consequences of this clear-cut relational standpoint are self-evident. This is to say that empirical analyses of the WCN should reflect the relational perspective that lies at the root of its conceptualization. If we wish to take forward the view of world cities as a process (re)produced by global networking and connectivity, it is vital that relational data are sought after. This is most clearly explained by Smith and Timberlake (1995a,b), who stress the importance of constructing and analysing a variety of databases that take the form of 'cities in global matrices'.

Existing data sources in this field of research are generally poor (Short *et al.* 1996). In most cases, data on flows between cities are conspicuous by their absence, and with few exceptions empirical world city research has thus neglected the relations and linkages between world cities. The reasons for this are spelt out in detail by Taylor (1997, 1999), who traces the lack of suitable data back to the fact that most data collection agencies focus upon attribute data because they are generally easier to collect, while most demands seem to favour information in this format. Furthermore, where relational data are available (e.g. trade data), they are largely inadequate for world city research needs because they primarily cover states – the prime generators of data – and not cities. Thus, there are data on flows between countries ('state-istics'), but little on flows between cities located in different countries; as a consequence, empirical analyses of the world city network have long suffered from 'the iron grip of the nation-state on the social imagination' (Taylor 1996: x). There have been some exceptions to this empirical conundrum at the heart of WCN research. Generally speaking, these databases have been premised upon two different foundations, which may be labelled the corporate organization and the infrastructure solutions.

The corporate organization approach to the construction of global inter-city matrices is premised on the observation that inter-city relations are primarily created by firms that pursue global strategies and are thus prime world city agents. The most elaborate examples of this corporate organization approach are the research pursued by the Globalization and World Cities group and network (GaWC, http://www.lboro.ac.uk/gawc), and work by Alderson and Beckfield (2004). GaWC has developed a methodology for studying the formation of the WCN, which has largely been based on Sassen's (1991, 2000, 2001, 2002) work on place and production in the global economy. The GaWC's approach starts from the observation that advanced producer-service firms 'interlock' world cities through their intra-firm communications of information, knowledge, plans, directions, advice etc. to create a network of global service centres. Building on this specification of the WCN, information was gathered from global service firms about the size of their presence in a city and about any 'extraterritorial' functions of their offices. This information was converted into data that summarize the location strategies of 100 firms across 315 cities. Applying a formal social network methodology, this information was then converted into a 315 × 315 matrix, which enabled a formal quantitative social network analysis of the WCN. (For more details, see Taylor 2001, 2004; Taylor *et al.* 2002a,b; Derudder and Taylor 2005). By contrast, Alderson and Beckfield (2004) draw on Hymer (1972), Friedmann (1986), and Godfrey and Zhou (1999) to interpret world cities as command centres that host the headquarters of major MNEs. On this starting point, they analyse links between 3,692 cities based on the organizational locations of 446 of the largest MNEs and their subsidiaries in 2000.

Although there are considerable differences between the two analyses, it can be noted that both base their city-centred spatial analysis on an assessment of the location strategies of firms with transnational fields of activity (Taylor 2005). It is suggested that world city relations can be derived from intra-firm communications between different parts of their holdings: Alderson and Beckfield (2004) consider this to be a 'key relation' in 'an MNE-generated city system', while Taylor (2004) argues that it is 'firms through their office networks that have created the overall structure of the [world city] network.' Thus collecting data on the organizational geography of global corporate activity enables estimation of inter-city relations.

The research pursued from the direction of the infrastructure, in contrast, starts from the observation that advanced telecommunication and transportation infrastructures are unquestionably tied to key cities in the world economy. That is, the most important cities in the world economy also harbour the most important airports, while extensive fibre backbone telecommunications networks that support the Internet have been rolled out across the globe. The latter networks have predominantly been deployed within and between major cities, creating a planetary infrastructure web on which the global economy has come to depend almost as much as physical transport networks (Moss and Townsend 2000; Rutherford *et al.* 2004). These enabling telecommunication and transportation networks are the foundations on which the connectivity within the world city network is built, and it is therefore not surprising that the geographical structure of these networks has been used to invoke a spatial imagery of the world city system. Examples include the global urban hierarchy based on air traffic flows between cities (e.g. Cattan 1995; Keeling 1995; Kunzmann 1998; Rimmer 1998; Smith and Timberlake 2001, 2002; O'Connor 2003; Matsumoto 2004; Witlox *et al.* 2004; Derudder and Witlox 2005a), and those based on postal flows, telephone calls and Internet linkages (Marek 1992; Warf 1995; Graham and Marvin 1996; Hanley 2004).

The use of air travel data as common data of choice for analysing worldwide inter-city flows has been largely attributed to the fact that airline data are comparatively easy to get hold of and that air transport is traditionally organized through cities rather than through states. More detailed appraisals of the usefulness of airline data are elaborated in Smith and Timberlake (2001, 2002) and Keeling (1995: 118), who present five interrelated arguments for why airline linkages are a suitable empirical source for assessing the WCN:

(i)   global airline flows are one of the few indices available of transnational flows of inter-urban connectivity;
(ii) air networks and their associated infrastructure are the most visible manifestation of world city interaction;
(iii) great demand still exists for face-to-face relationships, despite the global telecommunications revolution;
(iv) air transport is the preferred mode of inter-city movement for the transnational capitalist class, migrants, tourists and high-value goods;
(v)   airline links are an important component of a city's aspirations to world city status.

OBSTACLES IN THE USE OF AIRLINE DATA

Although airline data have some obvious advantages over other data (e.g. the limited number of assumptions needed in acquiring network data), some of the previously employed data sources and frameworks for analysis have not fulfilled their potential. Here we engage in a more elaborate overview of biases that result from the use of standard airline data sources. It is not our intention to embark on an exhaustive overview of all airline analyses to date, but rather to sketch out the main obstacles in this context.

Lack of origin/destination data

A first major obstruction in the translation of airline data into urban analyses is induced by the lack of origin/destination information in the databases. Most airline data record the individual legs of a trip rather than the trip as a whole, so that in case of a stopover a significant number of real inter-city links are replaced by two or more links that reflect corporate strategy rather than relationships between key cities. Furthermore, the lack of origin/destination information makes geographically detailed assessments of the global urban system difficult because direct connections become less likely once one goes down the urban hierarchy. Global air traffic data should preferably also reflect real passenger movements instead of derived passenger movements through such proxy variables as scheduled flights or services (e.g. Official Airline Guides data). Moreover, air traffic data should also avoid 'false' origin movements due to certain customs and immigration regulations. With respect to the latter, Boberg and Collison (1989: 25), noted that in the United States passenger counts are taken as points where the passengers clear customs and immigration. In other words, a passenger who originates a trip in Chicago using a flight to Tokyo via Seattle will have Seattle listed as the origin if it is the clearance point for customs and immigration.

Another, though important, problem is that airline data cannot circumvent undervaluing a second-tier city that is close to a major world city because of the availability of other transportation modes. This is to say that highly structured high-speed rail networks may be considered a valid alternative to short-haul flights, which results in an underestimation of the importance of some short-distance inter-city links. In other words, Antwerp and Rotterdam, for instance, will be undervalued because both cities are well within the aerial catchment area of Brussels and Amsterdam.

To see how the lack of origin/destination data distorts a WCN analysis, we can refer to Keeling's (1995) world city map, which is based on an analysis of the dominant linkages in the global airline network. This map was created from a matrix of scheduled air services involving 266 cities with populations exceeding 1 million. Only non-stop and direct flights between two cities, however, were taken into account, so that the measures used are not necessarily a reflection of actual inter-city relations. It is likely that in such an analysis the importance of cities that function as airline hubs, such as Amsterdam (KLM) and Frankfurt (Lufthansa), is overestimated at the expense of the likes of Brussels and Berlin. Furthermore, direct links between, say, Brussels and Rio de Janeiro cannot be measured because passengers are likely to go through São Paulo to make the trip. Relations between second-tier cities are hence difficult to measure with a data source that only contains individual and direct trips. As a consequence, the urban system analyses by Keeling (1995), Smith and Timberlake (2001, 2002) and Matsumoto (2004) are biased towards first-tier cities and important hub cities.

A subtle form of state-centrism

A second obstacle to a clear-cut translation of air transport databases into global inter-city analyses can be traced to the fact that some of these data sources incorporate a subtle form of state-centrism. Despite their global aspirations, most analyses are based on databases that contain information on international flows.

This delicate bias towards inter-state rather than trans-state flows tends to undervalue relations between key cities that are situated in large and/or significant nation-states.

Rimmer (1998), for instance, bases his analysis on data on international passengers, an approach that downgrades US world cities in particular since connections such as Los Angeles–New York and Chicago–New York are not included. Hence, Chicago only appears on one of Rimmer's maps as a 'fourth-level' link to Toronto, while Dublin appears on all maps because of its 'first-level' link with London. Nobody would argue that Dublin is more important than Chicago as a world city; it only appears to be when one relies on international rather than global data. Another example of this problem is found in Smith and Timberlake (2002), who faced the absence of information on the volume of air passenger traffic between Hong Kong and London. This important global link did not feature in pre-1997 ICAO databases because a London to Hong Kong connection was considered to be 'national'. Admittedly, Smith and Timberlake circumvent the London–Hong Kong problem by making an estimation, while the relegation of US cities was tackled through the introduction of another data source that contained information on the major routes in the US (i.e. data provided by the Air Transport Association in Washington DC). Although this sidestepping of the most pressing holes in their initial database results in one of the most refined databases to date, in general this problem stays in place for major cities in a number of countries such as Canada and Brazil. A database detailing global rather than international air passenger flows would overcome the problems associated with the introduction of this subtle form of state-centrism.

Intersection with non-world city processes

A third obstacle to the straightforward application of air transport statistics arises from the observation that such data measure *general* flow patterns. Since airline statistics are unable to differentiate between specific flows within the various linkages, it is doubtful that the specific flows that define the WCN can straightforwardly be deduced from such data. The latter is, of course, a more general problem that pertains to the variety of theoretical WCN conceptualizations, but in this paragraph reference is made to specific, overarching distortions such as tourism.

In a mapping of the European urban hierarchy based on air passenger flows, Kunzmann (1998: 49) lists 14 airports that are secondary to the big three (London, Paris and Frankfurt), including Munich, Milan, Madrid and Palma de Mallorca. However, the high ranking of the last merely reflects its role as one of the most popular holiday destinations in Europe; nobody would argue that it is a major world city. Although it is likely that most researchers would agree that destinations such as Palma de Mallorca should be omitted from the analysis, such data manipulation becomes increasingly difficult when non-world city processes *intersect* with world city-formation. The rising importance of Miami, for instance, can be traced back to its control functions vis-à-vis the Caribbean (Grosfoguel 1995; Nijman 1996; Brown *et al.* 2002) *and* its function as a retirement centre and major holiday destination. Whether the latter intersection distorts the analysis depends, of course, on the exact – and therefore possibly debatable – specification of the characteristics of a world city.

In his initial formulation of 'The World City Hypothesis', John Friedmann (1986: 74) maintained that the major driving forces behind world city growth were found in a limited number of rapidly expanding sectors. Although Friedmann identified world cities as major tourist destinations, it seems that tourism is merely an ancillary function, since

> [m]ajor importance attaches to corporate headquarters, international finance, global transport and communications; and high level business services, such as advertising, accounting, insurance, and legal services. An ancillary function of world cities is ideological penetration and control. New York and Los Angeles, London and Paris, and to a lesser degree Tokyo, are centres for the production and dissemination of information, news, entertainment and other cultural artefacts.

Hence, although it is clear that cities such as New York, London, Los Angeles and Tokyo have become major tourist attractions in their own right, this is a secondary function at best, so that questions can be raised on the alleged importance of tourist flows in the various databases. We agree that trying to single out the tourist functions may be perceived by some as a questionable move, but it seems nonetheless clear that existing clear-cut tourist destinations should be omitted from the data. Irrespective of the potential controversy over this point, we maintain that airline linkages reflect myriad processes of which world city formation is only one element, so that educing the WCN from airline databases is not a straightforward matter. Admittedly, this problem may be the hardest to overcome, since we have no clear procedures for estimating the amount of world city traffic in overall air travel, while such a procedure would at the same time be open for debate and depend on the employed conceptualization of the WCN.

Airline data are not always provided within an appropriate framework

A final impediment to the use of airline data in WCN analyses is that the statistics that can be derived from them are not necessarily available and/or analysed within an appropriate relational framework.

Airline data are not necessarily provided in the relational form presupposed by WCN research. Cattan (1995), for instance, has determined the global importance of 90 major European cities in terms of their international exchanges. She presents an assessment of the European urban hierarchy based on the computation of various *attribute* measures, such as the number of international flights, the rate of international travel per head of population, the percentage of international traffic in overall traffic and the number of direct international links. Although these measures provide good proxies for ranking the connectivity of cities, they do not provide information on how overall connectivity can be disentangled into spatial patterns. As a consequence, whereas such an analysis may convey the hierarchical trends between the major European cities in the context of the WCN, the broader geography of this part of the network remains obscure. Only data in the form of inter-city matrices can unravel the spatiality behind overall connectivity.

AN INTERCITY MATRIX BUILT ON THE MIDT DATABASE

Ideally, air transport-based global urban analyses should not suffer from the above-mentioned drawbacks. In other words, the data used must be global. That is, data should (i) detail global rather than international flows in order to overcome the state-centrism problem; (ii) contain information on origin/destination travel (i.e. the full trip), thereby avoiding an overestimation of the importance of locations acting as hubs; (iii) comprise all passengers travelling on a specific origin–destination pair, which means that in principle as many airlines servicing a specific connection will have to be included as possible; and (iv) pertain to real passenger movements. In this section, we describe the construction of a geographically detailed inter-city matrix that is able to circumvent and/or overcome some of the previously identified problems. After a brief introduction on the content and the manipulation of the initial dataset, we explain in what respect this new dataset is able to overcome the limitations highlighted in the previous section. We do not engage in a thorough analysis of this database (see Derudder and Witlox 2005a,b), but mainly introduce it as an alternative to other, more commonly used air traffic databases.

The MIDT database

The MIDT database contains information on bookings made through so-called Global Distribution Systems (GDS) such as Galileo, Sabre, Worldspan, Amadeus, Topas, Infini and Abaccus. GDS are electronic platforms used by travel agencies to manage airline bookings (i.e. the selling of seats on flights offered by different airlines), hotel reservations and car rentals. Using a GDS-based database therefore implies that bookings made directly with an airline are excluded from the system. Airlines choose this direct booking option to avoid commissions charged by travel agencies. Direct bookings via the Internet are estimated to cost an airline $1, whereas bookings at travel agents cost an estimated $10 (Goetzl 2000). Southwest, EasyJet, Virgin and Ryanair, which are particularly low-cost carriers, have many direct sales, and consequently their flights do not feature prominently in GDS-based databases. However, in 1999, 80 per cent of all reservations continued to be made through GDS (Miller 1999). This suggests that our data source might provide a biased picture of airline transport. However, there is no reason to assume that the spatiality of the reservations made by direct bookings differs fundamentally from that for reservations made through a GDS.

With the cooperation of an airline, we were able to obtain a partial MIDT database that covers the period from January to August 2001 and contains information on a total of 3.7 million trips. Each MIDT record is made up of an entire airline trip, and comprises information on the IATA airport codes of origin/destination, the air carrier, the connecting airports (if any) and the number of passengers.

Airlines purchase the MIDT database for a variety of reasons, the most important of which is its ability to forecast demand. It is also a helpful tool for assessing the market share and the competitive position of an airline in a specific geographical area. In the context of our research, however, the database is used to construct a global inter-city matrix. The first step in the creation of this matrix (for more details see Witlox *et al.* 2004) is to transform the information because we are mainly interested in the total volume of passenger flows between cities. To achieve this, we relabelled the airport codes as city codes. These city codes are needed to compute meaningful intercity measures because a number of cities have more than one major airport. The particular airport used by a passenger is not important in this context because, for recording the London–New York relation, it is irrelevant whether a flight goes from Heathrow to JFK or from Gatwick to Newark. Having summed the directional information into a single measurement detailing the total volume of passengers, we created a global inter-city matrix that focuses on the most important cities in the world economy. That is, we omitted key holiday destinations and less important cities. For this, we used the tentative world city list compiled by GaWC, which contains 315 cities and includes the capital cities of all but the smallest states and numerous other cities of economic importance (Taylor *et al.* 2002a). Nine of these 315 cities were excluded either because they had no airport (e.g. Bonn and Kawasaki) or because the airport was not serviced in the period under consideration because of political instability (e.g. Kabul). This reconfiguration produced a 306×306 matrix that quantifies the relations between the most important cities in the world economy. Tables 3.1 and 3.2 and Figure 3.1 present an overview of the chief connections. Table 3.1 presents a 10×10 sub-matrix from the 306×306 matrix and Table 3.2 presents the 20 most important inter-city relations in the dataset. Figure 3.1 depicts the connections between the 30 most important cities in terms of total passenger flows. The size of the nodes varies with the total number of incoming or outgoing passengers; the size of the edges varies with the number of passengers flying between two cities. For reasons of clarity, only the most important links are shown.

Appraisal of the MIDT database

First, as the MIDT database contains origin/destination information, the overrating of the connectivity of airline hubs and first-tier world cities is minimized, which allows assessing the relational patterns in the lower rungs of the WCN in more detail (e.g. the downsizing of the importance of hub cities such as Amsterdam and Frankfurt). Second, the MIDT-based database does not distinguish between national and international flows, and can therefore be thought of as a truly global intercity matrix. The New York–Chicago link is appropriately treated in the same way as the New York–Toronto link, which further reduces the underestimation

Table 3.1 Excerpt of inter-city matrix of the number of passengers based on the MIDT database

City	Abidjan	Abu Dhabi	Accra	Addis Ababa	Adelaide	Ahmedabad
Abidjan						
Abu Dhabi	45					
Accra	15517	5				
Addis Ababa	2573	173	4227			
Adelaide	1	14	1	3		
Ahmedabad	0	35	1	2	2	
Algiers	860	103	16	18	0	0
Almaty	7	1	2	5	17	0
Amman	31	39562	58	126	28	2
Amsterdam	4784	9096	26289	2324	1952	414
Ankara	0	30	43	12	18	0
Antwerp	79	233	921	18	0	2
. . .	. . .	. . .	. . .	. . .	. . .	. . .

of second-tier cities in large and/or significant nation-states. Third, reconfiguring the database by using GaWC's detailed world city list excludes obvious holiday destinations, which results in a (modest) redirection of plain air flow-centred information to city-centred information. Fourth, this database contains relational information in a single global framework. This allows overall connectivities to be disentangled into spatial patterns, while analyses of area subsets can be carried out in the context of an overarching WCN, as suggested by Taylor and Derudder (2004).

Although this MIDT-based intercity matrix is able to overcome and/or circumvent some of the problems that have been associated with the use of airline data in

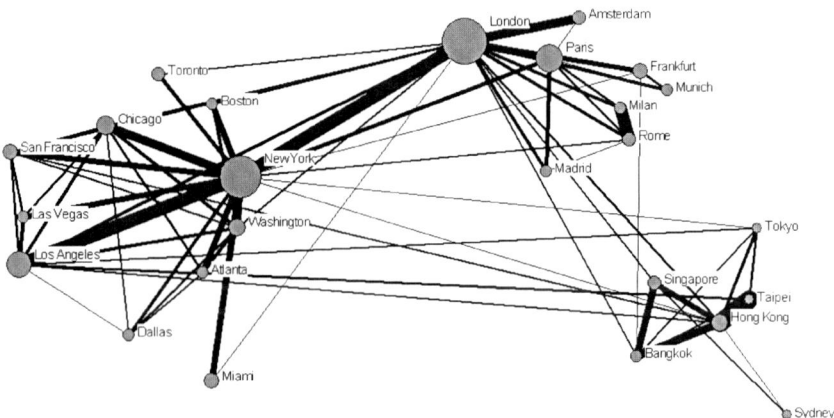

Figure 3.1 Most important cities and their links in the world city network (source: Derudder and Witlox 2005b)

Algiers	Almaty	Amman	Amsterdam	Ankara	Antwerp	...
						...
						...
						...
						...
						...
						...
						...
8						...
1103	96					...
664	6520	18059				...
283	1647	4152	8562			...
0	0	57	8307	2		...
...	...	...	...	...	...	

Table 3.2 Most important inter-city links in the MIDT database

Rank	Connection		Number of passengers
1.	Hong Kong	Taipei	2,138,484
2.	Los Angeles	New York	1,697,593
3.	London	New York	1,609,337
4.	Melbourne	Sydney	1,563,106
5.	Milan	Rome	1,518,767
6.	Cape Town	Johannesburg	1,406,897
7.	Los Angeles	San Francisco	1,375,660
8.	Amsterdam	London	1,242,550
9.	Chicago	New York	1,182,326
10.	Bangkok	Hong Kong	1,141,062
11.	London	Paris	1,060,999
12.	Dublin	London	1,050,940
13.	Marseilles	Paris	1,044,128
14.	Bangkok	Singapore	1,024,818
15.	Rio De Janeiro	Sao Paulo	992,775
16.	Boston	New York	988,976
17.	Miami	New York	955,838
18.	Atlanta	New York	935,265
19.	Las Vegas	Los Angeles	924,732
20.	New York	San Francisco	909,514

WCN analyses (see Derudder and Witlox 2005a,b), a number of problems remain for future research. The main problem is that it remains largely impossible to discern flows generated within the context of the WCN from 'other' flows, while the precise meaning of WCN flows is itself subject to a variety of meanings. Either way, it is clear that the importance of the New York–Miami route and particularly the New York–Las Vegas route, for instance, suggests more than a business link. The linkages related to obvious holiday destinations such as Palma de Mallorca have been deleted from the database, but this manipulation only works for airports that are obviously not world city airports. Furthermore, although problems with the availability of global origin/destination data addresses the undervaluation of second-tier world cities, airline data cannot avoid undervaluing a second-tier city that is close to a major world city. For example, a passenger travelling from Rotterdam to New York is likely to depart from Amsterdam because of (i) the short distance between Rotterdam and Amsterdam (less than 50 miles) and (ii) the importance of Amsterdam's Schiphol airport. This bias is exacerbated by the overall tendency to underestimate the connectivity between nearby cities because of the availability of other modes of transportation. Elaborate high-speed rail networks, for instance, may be an alternative to short-haul flights, which results in an underestimation of the importance of some short-distance intercity links. An overall solution to this underestimation problem may be to omit some cities from the database. Although this implies a further reduction in the geographical detail of the database, ensuing analyses would be more meaningful and robust. A final drawback of the dataset described in this paper is that, because it only covers a single period, it cannot be used to analyse the evolution of the WCN. Thus, unlike Smith and Timberlake (2002), we cannot track changes over time.

CONCLUSIONS

The world city system has frequently been described as an example of a global network. Although a theoretical discourse along the lines of a 'world city network' has already been widely invoked, it can be noted that this has only seldom been accompanied by genuine empirical network analyses. We have argued that the main reason for this dearth of full-fledged network analyses of the world city system is to be found in the paucity of data detailing inter-city flows at the global level. Although airline analyses have stood out in this context, previous data sources have not always been up to the task of mapping the WCN. In these data sources, cities like Chicago and New York tend to be undervalued since domestic transport is excluded, while cities like Miami and Honolulu tend to be overvalued because their importance reflects non-world city processes. The overall point, however, is that, despite many claims in this direction, one cannot straightforwardly deduce a 'global urban system' from airline databases. More generally, there are a series of complex and multifaceted problems associated with the use of airline data, and we have merely tried to single out some data-specific problems: the lack of origin/destination data, the use of international rather than global data, the intersection with non-world city processes such as tourism, and the fact that airline data are not always provided within an appropriate, relational framework. We showed how a new inter-city matrix complied on the basis of the Marketing Information Data Transfer (MIDT) database is able to overcome a number of these problems and leave only a few data-specific problems unresolved. We are confident that conducting exploratory analyses using the MIDT data contributes to an overall better understanding of the WCN. Not only is there an important progress in using enhanced relational air travel data, but there is the possibility to conduct a very detailed geographical analysis of multi-layered inter-city connectivities.

REFERENCES

Alderson, A.S. and Beckfield, J. (2004) 'Power and position in the world city system', *American Journal of Sociology*, 109: 811–51.

Allen, J. (1999) 'Cities of power and influence: settled formations', in J. Allen, D. Massey and M. Pryke (eds) *Unsettling Cities*, London: Routledge, pp. 181–228.

Beaverstock, J.V., Smith, R.G. and Taylor, P.J. (2000) 'World city network: a new metageography?', *Annals of the Association of American Geographers*, 90: 123–34.

Boberg, K.B. and Collison, F. M. (1989) 'International air transportation trends in the Pacific Basin', *Transportation Journal*, 3: 24–35.

Brown, E., Catalano, G. and Taylor, P.J. (2002) 'Beyond world cities: Central America in a space of flows', *Area*, 34: 139–48.

Castells, M. (1996) *The Rise of the Network Society*, Oxford: Blackwell.

Cattan, N. (1995) 'Attractivity and internationalisation of major European cities: the example of air traffic', *Urban Studies*, 32: 303–12.

Derudder, B. and Taylor, P.J. (2005) 'The cliquishness of world cities', *Global Networks*, 5: 71–91.

Derudder, B. and Witlox, F. (2005a) 'On the use of inadequate airline data in mappings of a global urban system', *Journal of Air Transport Management*, 11: 231–37.

Derudder, B. and Witlox, F. (2005b) 'An appraisal of the use of airline data in assessing the world city network: a research note on data', *Urban Studies*, 42: 2371–88.

Derudder, B., Taylor, P.J., Witlox, F. and Catalano, G. (2003) 'Hierarchical tendencies and regional patterns in the world city network: a global urban analysis of 234 cities', *Regional Studies*, 37: 875–86.

Friedmann, J. (1986) 'The world city hypothesis', *Development and Change*, 17: 69–83.

Godfrey, B.J. and Zhou, Y. (1999) 'Ranking world cities: multinational corporations and the global urban hierarchy', *Urban Geography*, 20: 268–81.

Goetzl, D. (2000) 'Southwest Airlines takes flight fare fight to the web', *Advertising Age*, 71 (40): 20–2.

Gottmann, J. (1989) 'What are cities becoming the centres of? Sorting out the possibilities', in R.V. Knight and G. Gappert (eds) *Cities in a Global Society*, London: Sage, pp. 58–67.

Graham, S. and Marvin, S. (1996) *Telecommunication and the City*. London: Routledge.

Grosfoguel, R. (1995) 'Global logics in the Caribbean city system: the case of Miami', in P.L. Knox and P.J. Taylor (eds) *World Cities in a World-System*, Cambridge: Cambridge University Press, pp. 156–70.

Hanley, R.E. (ed.) (2004) *Moving People, Goods, and Information in the 21st century. The Cutting-Edge Infrastructures of Networked Cities*, London: Routledge.

Hymer, S. (1972) 'The multinational corporation and the law of uneven development', in J. Bhagwati (ed.) *Economics and World Order from the 1970s to the 1990s*, London: Collier-MacMillan, pp. 113–40.

Keeling, D.J. (1995) 'Transport and the world city paradigm', in P.L. Knox and P.J. Taylor (eds) *World Cities in a World-System*, Cambridge: Cambridge University Press, pp. 115–31.

Kunzmann, K.R. (1998) 'World city regions in Europe: structural change and future challenges', in F.-C. Lo and Y.-M. Yeung (eds) *Globalization and the World of Large Cities*, Tokyo: United Nations University Press, pp. 37–75.

Marek, S. (1992) 'Private networks directory', *Satellite Communications*, September issue.

Matsumoto, H. (2004) 'International urban systems and air passenger and cargo flows: some calculations', *Journal of Air Transport Management*, 10: 241–9.

Miller, W.H. (1999) 'Airlines take to the internet', *Industry Week*, 248 (15): 130–4.

Moss, M.L. and Townsend, A.M. (2000) 'The internet backbone and the American metropolis', *The Information Society*, 16: 35–47.

Nijman, J. (1996) 'Breaking the rules: Miami in the urban hierarchy', *Urban Geography*, 17: 5–22.

O'Connor, K. (2003) 'Global air travel: toward concentration or dispersal?', *Journal of Transport Geography*, 11: 83–92.

Rimmer, P. J. (1998) 'Transport and telecommunications among world cities', in F.-C. Lo and Y.-M. Yeung (eds) *Globalization and the World of Large Cities*, Tokyo: United Nations University Press, pp. 433–70.

Rutherford, J., Gillespie, A. and Richardson, R. (2004) 'The territoriality of pan-European telecommunications backbone networks', *Journal of Urban Technology*, 11 (3): 1–34.

Sassen, S. (1991) *The Global City: New York, London, Tokyo*, Princeton, NJ: Princeton University Press.

Sassen, S. (2000) *Cities in a World Economy*, Thousand Oaks, CA: Pine Forge.

Sassen, S. (2001) *The Global City: New York, London, Tokyo*, 2nd edn, Princeton, NJ: Princeton University Press.

Sassen, S. (ed.) (2002) *Global Networks, Linked Cities*, London: Routledge

Shin, K.H. and Timberlake, M. (2000) 'World cities in Asia: cliques, centrality and connectedness', *Urban Studies*, 37: 2257–85.

Short, J.R., Kim, Y., Kuss, M. and Wells, H. (1996) 'The dirty little secret of world city research', *International Journal of Regional and Urban Research*, 20: 697–717.

Smith, D.A. and Timberlake, M. (1995a) 'Conceptualising and mapping the structure of the world system's city system', *Urban Studies*, 32: 287–302.

Smith, D.A. and Timberlake, M. (1995b) 'Cities in global matrices: toward mapping the world-system's city system', in P.L. Knox and P.J. Taylor (eds) *World Cities in a World-System*, Cambridge: Cambridge University Press, pp. 79–97.

Smith, D.A. and Timberlake, M. (2001) 'World city networks and hierarchies 1979–1999: an empirical analysis of global air travel links', *American Behavioral Scientist*, 44: 1656–77.

Smith, D.A. and Timberlake, M. (2002) 'Hierarchies of dominance among world cities: a network approach', in S. Sassen (ed.) *Global Networks, Linked Cities*, London: Routledge, pp. 117–41.

Smith, D.A. and Timberlake, M. (2005) 'Inter-urban links and flows: the contemporary global city network', *Proceedings of the XXVth Sunbelt Social Network Conference*, Redondo Beach (CA): International Network For Social Network Analysis, p. 76.

Taylor, P.J. (1996) 'Embedded statism and the social sciences: opening up to new spaces', *Environment and Planning A*, 28: 1917–28.

Taylor, P.J. (1997) 'Hierarchical tendencies amongst world cities: a global research proposal', *Cities*, 14: 323–32.

Taylor, P.J. (1999) 'So-called 'World Cities': the evidential structure within a literature', *Environment and Planning A*, 31: 1901–4.

Taylor, P.J. (2001) 'Specification of the world city network', *Geographical Analysis*, 33: 181–94.

Taylor, P.J. (2004) *World City Network: A Global Urban Analysis*, London: Routledge.

Taylor, P.J. (2005) Parallel Paths to Understanding Global Inter-city Relations. *GaWC Research Bulletin* 143 http://www.lboro.ac.uk/gawc/rb/rb143.html

Taylor, P.J. and Derudder, B. (2004) 'Porous Europe: European cities in global urban arenas', *Tijdschrift voor Economische en Sociale Geografie*, 95: 527–38.

Taylor, P.J., Catalano, G. and Walker, D.R.F. (2002a) 'Measurement of the world city network', *Urban Studies*, 39: 2367–76.

Taylor, P.J., Catalano, G. and Walker, D.R.F. (2002b) 'Exploratory analysis of the world city network', *Urban Studies*, 39: 2377–94.

Warf, B. (1995) 'Telecommunications and the changing geographies of knowledge transmission in the late 20th century', *Urban Studies*, 32: 361–78.

Witlox, F., Vereecken, L. and Derudder, B. (2004) 'Mapping the global network economy on the basis of air transport passenger flows', *GaWC Research Bulletin*, 157, http://www.lboro.ac.uk/gawc/rb/rb157.html.

4 World city networks 'from below'

International mobility and inter-city relations in the global investment banking industry

Jonathan V. Beaverstock

INTRODUCTION

As Castells's (2000) *The Network Society* and Sassen's (2001) *The Global City* are so influential, path-breaking and omnipresent in current globalization and world city discourses, little value would be taken from (yet another) critical analysis of the space of flows and the 'global city thesis' (see for example Beaverstock *et al.* 2000), especially when Taylor's (2004) *World City Network* has firmly squared all the circles in his reading of the Castellian and Sassen spatial logic. From Castells (2000: 443) we already know that the third layer of the space of flows is composed of 'the dominant managerial elites', who occupy the command and control functions of global cities. In a similar vein, Sassen (2000) has discussed how the concentration of advanced producer services in 'the new production complex' (2000: 71) constitutes a significant reproductive synergy for the intensification and concentration of command and control in the global city. Such an argument gives credence not only to the productive nature of 'face-to-face' contact, where 'time replaces weight . . . as a force of agglomeration' (ibid.: 72), but also to inter-city relations, both virtual and physical, as it is now a requirement for a professional workforce to deliver accurate information, advice, solutions and specialism across borders in the knowledge service economy (Beaverstock 2002). But, after briefly airing the main propositions of the new relational spatial logic of understanding world cities in a connected society, in the rest of this chapter it is important to bring value to the debate which unpicks one significant agent of inter-city relations: the firm. My intervention in this edited collection, therefore, is to discuss inter-city relations through the organizational role of the advanced producer service firm, which will add process-led 'flesh' to the skeletal frame of the world city network (Beaverstock *et al.* 2000; Taylor 2004). In my reading of inter-city relations, it is the firm and their client–supplier relations that are the dynamic 'shakers and movers' of the world city network, who have agency because they cannot ply their global business without expert labour fixed and fluid at the point of demand, interfacing with a global clientele where they sell expertise, bespoke solutions and reputation (see Beaverstock *et al.* 2002).

I wish to illustrate the role of the firm in making inter-city relations and world city networks through the example of international labour mobility within global investment banking. Global investment banks, like Goldman Sachs and J.P. Morgan, penetrate international markets through world city networks because they require a physical presence in the market to serve clients and co-locate with suppliers (e.g. lawyers) (see Gardener and Molyneux 1996). As global investment banks are akin to professional service firms (Lowendahl 1997), key organizational strategies are labour and mobility, where experts work in close proximity to colleagues (via intra-firm relations) and clients through transfer, secondment and business travel arrangements (Beaverstock forthcoming). Investment banks accumulate financial capital through the embodied knowledge of their expert staff in world city client networks. Expert staff have to physically travel to service clients in co-location in order to deliver their services between cities because such knowledge is not easily tradable in virtual form. Accordingly, labour mobility is an efficient mechanism to make the knowledge structures of world city networks. The rest of this chapter is divided into four sections. In the next section, it is important to discuss the organizational strategy of firms, who provide the structural processes for making world city networks; namely, transnational knowledge management. After all, firms are the agents that create, direct and manage flow and inter-city relations (Beaverstock *et al.* 2002). Following on from this initial conceptual platform, the discussion will then advocate the agency of global investment banks in making inter-city relations. In the third section, research findings will be reported that illustrate the key knowledge production process of labour mobility in several global investment banks, drawing upon interviews with firms and bankers themselves. Finally, several conclusions are reached on the role of international mobility in producing inter-city relations and world city networks 'from below'.

MAKING INTER-CITY RELATIONS: KNOWLEDGE MANAGEMENT AND LABOUR MOBILITY IN PROFESSIONAL SERVICE FIRMS

Transnational service corporations are integrated networks, characterized by multi-faceted linkages and relationships both intra- and inter-firm with clients and suppliers alike (Aharoni and Nachum 2000). At an organizational level, an important strategy for the firm is to circulate knowledge and best practice between subsidiaries whether it is from top (global headquarters) to bottom (subsidiary) and vice versa, or laterally between subsidiaries of the firm (including the HQ). Nohria and Ghoshal (1997: 4) refer to the firm as having 'a differentiated network', with 'local linkages' within national subsidiaries and linkages between headquarters and subsidiaries. But perhaps the most useful analysis of firm knowledge management can be derived from Bartlett and Ghoshal's (1998) *Managing across Borders*, which evaluates the development and diffusion of knowledge in four typologies of the firm, pertaining to all economic sectors; multinational, global, international and transnational. The key issue is that knowledge is developed, managed and diffused differently depending upon the typology of the firm (Table 4.1). For example, in the 'international' firm – akin to global investment banks – 'knowledge is 'developed at the centre and transferred to overseas units', because

Table 4.1 Organizational characteristics of the transnational professional service firms

Organizational characteristics	Multinational	Global	International	Transnational
Configuration of assets and capabilities	Decentralized and nationally self-sufficient	Centralized and globally scaled	Sources of core competences centralized, others decentralized	Dispersed, interdependent, and specialized
Role of overseas operations	Sensing and exploiting local opportunities	Implementing parent company strategies	Adapting and leveraging parent company competencies	Differentiated contributions by national units to integrated worldwide operations
Development and diffusion of knowledge	Knowledge developed and retained within each unit	Knowledge developed and retained at the centre	Knowledge developed at the centre and transferred to overseas units	Knowledge developed jointly and shared worldwide

Source: Bartlett and Ghoshal (1998: 75).

the assets and capabilities of the firm are both dispersed and centralized from the centre, thus making subsidiaries dependent on the centre for knowledge. But, understanding how transnational professional service firms develop and manage knowledge between subsidiaries through expertise, skills and practice physically embodied in labour, one can look closely at the recent work on the characteristics and work process of professional service firms.

Professional knowledge, including expertise, is the 'core resource and is both the input and output' of the production process (Nachum 2000: 4). Sectors like architecture, accounting, consultancy, investment banking and securities, and legal services depend entirely on the professional knowledge, expertise, reputation and trust of their staff to offer clients and customers bespoke, non-routine solutions, which require close delivery and predominantly co-location (Greenwood and Lachman 1996). Moreover, Nachum's (2000) detailed analysis of the intangible, labour intensive, tailor-made characteristics of professional services illustrates very clearly why delivery is reliant upon close interaction with the client. Accordingly, for professional services, expert staff–client interaction is the key process for knowledge transfer and ultimately the retention of market share, the acquisition of new clients, and expansion into new geographical markets through foreign direct investment.

For geographers, the key process by which professional services interact with clients in world cities is through face-to-face contact, with clients, competitors, suppliers and colleagues (intra-firm knowledge transfer) (Amin and Thrift 1992; Thrift 2000; Jones 2003; Beaverstock 2004). Face-to-face relationships are reciprocal processes which not only share and disseminate tailor-made, non-routine and often 'one-off' solutions to clients, but also act as the medium for interaction with teams of actors where deal-making often requires specialist input from a range of professional services, e.g. lawyers, bankers and accountants. The bottom line, therefore, in professional services is that firms can only deliver their products and aura of professionalism, 'blue-chip' reputation and trust to clients, and interact successfully with colleagues intra-firm and with suppliers inter-firm, through physical relations, which are of course supplemented by the virtual transmission of explicit knowledge. For transnational firms with foreign direct investment in the form of international offices in world city locations, 'services are delivered by local and expatriate staff alike, as in many circumstances, clients require PSFs to stretch knowledge onto a global scale and cross-border' (Beaverstock 2004: 161).

International mobility, therefore, is a key factor of production in the professional service industry. For example, in my recent analysis of global professional labour markets in the accounting industry (Beaverstock forthcoming), mobility was a key facet of the firm's globalization strategies to service existing and reach new clients in world city locations because clients demand a service on site in international financial centres Moreover, I have posited (ibid.) a framework that illustrates the dimension of international mobility in transnational professional service firms (Table 4.2) (by drawing upon Edstrom and Galbraith's (1977) seminal work on transfers within multinational corporations). In short, professional staff are assigned to international locations to fill vacancies, develop as managers and develop the strategy of the organization, and these moves can be highly frequent, between many different locations, mainly fee-earning, and occur throughout a career path.

INTERNATIONALIZATION AND EXPERTISE IN INVESTMENT BANKING

Investment banks have entered world city markets through international office networks because the recipe for global success is intertwined with five major factors – which all have one thing in common: the requirement to be physically located in the market place. First, investment banks are reliant upon the capabilities of their expert staff, whether in a bulge bracket firm or niche boutique, to generate fee income from intrinsic client relations in global markets and product innovation. Investment bankers have to work in teams to be successful, drawing upon individual expertise and generic skills from a range of personnel in global market and product divisions that transcends location (Wrigley *et al.* 2003). As expert knowledge is embodied and cannot be easily reproduced without socialization, investment bankers must be tied to the physicality of the market place where face-to-face contact becomes an essential factor of production in an industry where the market is multi-locational and the product range dispersed between geographical markets. Second, investment banks are client-focused and 'if these banks were not satisfying their customers, they would not be in business' (Economist Intelligence Unit 1999: 21). Investment banks must be in close proximity, co-location, to their clients, who are invariably the major institutional and corporate entities of international financial centres, and their suppliers, especially 'Magic Circle' corporate

Table 4.2 Dimensions of transfer policies in transnational professional service firms

Reasons for transfers			
Dimensions	Fill positions	Develop managers	Develop organisation
Relative numbers	Many	Many	Many
Specialities transferred	Fee-earning	Fee-earning	Fee-earning
Location of host	All countries	All countries	All countries
Direction of flow	Between subsidiaries and between HQ and subsidiaries	Between subsidiaries and between HQ office and subsidiaries	Between subsidiaries and between HQ office and subsidiaries
Age of assignee	Throughout career	Young to middle	Throughout career
Frequency	Many moves	Several moves	Many moves
Nationality of assignee	All nationalities	All nationalities	All nationalities
Personnel information	Extensive lists of candidates monitored by personnel in all offices	Extensive lists of candidates monitored by personnel in all offices	Extensive lists of candidates monitored by personnel in all offices
Power of personnel department	Strong	Strong	Strong
Strategic placement and distribution	Extensive	Extensive	Extensive

Source: Beaverstock (forthcoming)

law firms (Corporation of London 2003). Co-location and interdependency are an essential geographical agglomeration economy for the investment banking industry. Third, as investment banks are market-makers, where reputation and prowess attract the clientele, product innovation can only be efficiently developed in-house, and in co-location with competitors, where traded and untraded path dependencies nurture knowledge spillovers and bind innovation with client relations and supplier networks (Benveniste *et al.* 2003). Fourth, for a global investment bank to claim that they offer customers integrated seamless services, they must have geographical coverage in the principal financial markets and be able to supply products to transnational national clients with a 'global' focus (Berger *et al.* 2003). As Freeman & Co (2003: 16), the investment banking 'think-tank', argue, 'distribution prowess is essential to ensure the "right" products are sold and held by the "right" investors based on the mutual needs of issuers and investors.' In all of these factors, the overriding internationalization driver of the investment banking industry comes down to the common denominator of providing an international office in a world city location for expert staff to engage with the client base of the bank.

Global investment banking is dominated by the organizational activities of the 'bulge bracket' US and European firms who have global geographical coverage (Economist Intelligence Unit 1999). In 1998, the top ten banks accounted for approximately 75 per cent of total global fee income, with Goldman Sachs and Merrill Lynch accounting for over a quarter of international business (International Financial Services London 2000). In 2001, these two banks remained the leading fee earners accounting for 17 per cent of global fee income (Table 4.3). But, given that investment banking rests on the performance of expert staff and close client relations, headcounts (number of experts, fee-earners) are of critical importance in the firm, more than office distribution and number (Beaverstock and Smith 1996). Expert staff are the prime asset of the investment bank and therefore, at an organizational level, are the key attributes for internationalization through local hires and mobility. Investment banking experts present the firm with its competitive advantage in the *global* market. It is the 'skill and commitment of individuals that often makes the difference between success and failure, between trading profits and losses, and between clients returning or taking their business elsewhere' (Seifert *et al.* 2000: 8). What is of critical importance in the rest of this chapter is to illustrate how investment banks deploy expert staff within and between world city office networks, which acts as a significant organization strategy for the bank to successfully penetrate new markets, but moreover as one actant that produces the space of flows and makes world city networks in contemporary globalization (see Beaverstock *et al.* 2002).

MAKING MACRO-WORLD CITY NETWORKS: THE FIRM AND ORGANIZATIONAL MOBILITY

The study reported in this section of the chapter was collected from secondary sources (e.g. firm websites, Annual Reports) and London-based interviews with

Table 4.3 Largest investment banks, 1998–2001

1998	Fee income	% share	2001	Fee income	% share
1. Goldman Sachs	1.5	13.9	1. Merrill Lynch	4.0	9.6
2. Merrill Lynch	1.4	12.3	2. Goldman Sachs	3.1	7.5
3. Morgan Stanley Dean Witter	1.1	9.8	3. Credit Suisse First Boston	2.8	6.8
4. Salomon Smith/ Barney/Citigroup	1.0	9.2	4. Citigroup	2.6	6.3
5. Credit Suisse	0.8	7.3	5. Morgan Stanley	2.5	6.0
6. JP Morgan	0.7	6.5	6. JP MorganChase	2.0	4.9
7. Chase Man.	0.6	5.7	7. UBS Warburg	2.0	4.7
8. Lehman Bros.	0.6	5.1	8. Lehman Bros.	1.4	3.4
9. Deutsche Bank	0.4	4.0	9. Deutsche Bank	1.4	3.4
10. Warburg Dillon Read/UBS	0.4	3.8	10. Bank of America	0.9	2.1

Source: ISFL (2001, 2002).

CEOs responsible for international human resources in 10 global investment banks in 1999/2000.[1] Six of the 10 banks headquartered their global investment banking from London, two from Paris and two from New York, with global corporate finance, fixed income CEOs etc. sitting in these locations, supported by regional management in the main geographical markets (New York, Singapore and/or Tokyo). Two processes responsible for making inter-city relations will be highlighted: the firm and organizational strategy, and the reproductive nature of flow that creates uneven patterns between world cities.

Organizational labour mobility

Investment banking is an industry which seeks profitability and market share through the efficient delivery of products and solutions to institutional, corporate and private clients on a global scale (Economist Intelligence Unit 1999). Interviewed bank websites and other publicity (e.g. Annual Reports) constantly praised the professional credentials of their 'client-focused' and 'client-facing' research, sales, trading and management teams across each bank's entire global product portfolio and geographical markets. Investment bank websites oozed the rhetoric of being able to offer clients 'global–local knowledge', 'seamless services and execution', 'deep, enriched relations', 'innovative solutions', all bounded by trust, reputation, 'ethical behaviour' and 'integrated delivery' and, most importantly, provided by 'expert', 'dedicated staff' and 'global leaders', '*around the world*' or '*worldwide*' (emphasis added). Investment banking is the archetypal professional service industry where the competitive advantage is derived from the knowledge and expertise of the staff and the ways in which they execute their business in deep and trusted client relationships on a world stage. It was therefore no surprise to find at interview with all banks that international mobility is a key organizational strategy for the banks not only to client-face around the globe and ensure that global products are delivered with 'pools' of expert and technical staff available *in situ* during the business cycle but also, very importantly, to deliver 'global' products and 'global' solutions to clients on a global scale. Invariably, as the business of investment banking is rarely tied to one particular location, staff must possess 'global expertise'; from the other side of the fence, for investment banks to claim that they are 'global' they must have staff with global experience and credentials to service clients within and between world city financial markets and management services.

Investment banks transfer knowledge and expertise throughout their international office networks by physically moving staff between world city locations. International mobility facilitates skill exchange between locations and is a surrogate organizational strategy, which reproduces the 'global' branding of the bank. Investment banks require staff with global experience and expertise to project the global branding of the bank and provide competencies to a transnational institutional or corporate client. Within these ten global investment banks, research findings revealed that international mobility made world city networks though four major organizational strategies:

1 Strategic, revenue-earning skill transfers. Banks assigned professional staff
  – whether in front or back office research, sales, trading and IT systems
  functions across all global products (financial markets – equities, foreign
  exchange; corporate finance; fixed income; asset and wealth management) in
  order to check the upside of the business cycle as growth in different world
  city locations created conditions for immediate labour market demand. In
  occupations like trading or research as the 'technical skills would be, if you
  like, universal and it is cheaper to recruit from the local market . . . the only
  reason to expatriate someone is a skill shortage' (IB1). In more client-facing
  functions, like sales, mobility enhances 'succession planning' and 'deepens
  client relationships' (IB3) because of the exchange of technical skills between
  world city markets. For example, IB5, like all other banks, have 'a network
  of people' involved in identical jobs sitting in Tokyo, London and New York;
  if client demand created skill deficiencies in a particular node in the network,
  staff were 'brought in' to fill that skills gap. All banks noted 'severe skill
  shortages' in Tokyo and certain booming emerging market locations (e.g.
  Mumbai). In the emerging market locations, foreign staff were required to
  'develop product lines first matured in New York and London' (IB4). All
  banks noted that international mobility activated through skill transfers would
  account for at least '60 per cent' (IB7) of all corporate postings.

2 Strategic management skill transfers. CEOs and other senior staff were sent
  abroad to manage new offices, departments or product lines; 'there will
  always be a demand for management expatriates either for political reasons
  or genuine skills' (IB1). In IB3, each of the investment bank's international
  offices is headed by a French national, seconded from the Parisian HQ and
  '75 per cent' of IB4's expatriate managing CEOs are American, assigned from
  the bank's New York HQ. In new markets like the emerging markets or newly
  acquired businesses, all the banks sent CEOs to head up the new product line
  because they had the managerial skills, deep technical expertise and 'global
  organizational contacts' (IB10) to succeed. For example, all of IB9's heads of
  global markets and products in Asia, including Tokyo, are expatriates from
  London (British and American). In addition, it was also common for these
  banks to second an exceptionally qualified foreign national who had been
  working in either London or New York to act as the managing CEO of the new
  office or client-venture, supported by colleagues of other nationalities. All
  banks noted that skill transfers activated though the requirement to manage
  new or existing offices, global markets or products would have accounted for
  at least '10–20 per cent' (IB7) of all corporate postings.

3 Strategic training and skill acquisition. All banks circulate more junior
  professional staff to their key international offices, London, New York and
  Frankfurt especially, for training requirements and skill acquisition. Those
  banks with investment banking global HQs outside London regularly sent
  professionals to the London office for rotational training. For example, IB3,
  the French HQ bank, sent 40 staff to London from Paris, and both US banks
  (IB4 and IB5) sent rotation staff from New York. As IB4 comments, 'we have

a rotation program out of New York and they will rotate four times in two years.'

4   International cadre. Interestingly, IB3, IB6, IB7 and IB9 all employ so-called 'Third-Country Nationals', who effectively spend their entire career with the bank working outside their country of domicile. Such workers move frequently between international offices undertaking specific managerial positions associated mainly with non-revenue, back office functions (e.g. IT systems management, human resources, auditing). In all banks, such workers cross all the functional divisions of the banking group, so for example in IB3, who have an estimated 500 'International Cadres', only a small fraction would be 'corporate and investment bankers' as the rest would be involved with retail banking, investment management and private banking.

Mobility, distribution and world city networks

As we have discovered, these investment banks use the physical international mobility of their expert staff, of all nationalities, to transfer knowledge and skill competencies between world city office networks. These workers are actual flows that connect world cities in networks of intra-firm and firm–client relationships. An examination of the world cities of next residence from London and last residence before coming into London reveals a strong circuit of relations between certain world cities in contemporary financial globalization. Table 4.4 and Figures 4.1 and 4.2 illustrate very precisely the dominance of the London and Frankfurt–Tokyo–Singapore–New York–Paris–Hong Kong space of flows, making a world city network of outflows and inflows. The problems of retrieving standard datasets from investment banks does make it almost impossible to quantify the magnitude of flows in the network. But complete evidence from IB1, IB3 and IB10 does quantify the scale, however small, of flows between London and the banks' HQ (if outside London – e.g. Paris) and top world city locations: New York, in particular, where markets are larger and client demand more sophisticated and demanding.

Qualitative evidence from the interviews revealed three major characteristics of the geographies of these people flows in the world city network. First, London and New York were the cities which received and sent out the most expert staff in the bank's entire international office network because these two cities were the location of the bank's HQ or largest office (by staff, turnover and revenue). As these two bankers commented, 'London–New York is by far the biggest path between any two destinations. New York is our world headquarters, Europe is our next biggest region. London is our second largest office' (IB4), and 'London is our largest exporter. Corporate financiers especially move back and forth between London and New York' (IB7). Accordingly, in the majority of banks interviewed, all international mobility in the global investment banking product divisions was originated from London or New York with no evidence of mobility between other offices in the network (e.g. expert staff in Tokyo would not be sent to Singapore; instead, transfers would be initiated through New York or London). The rationale for this distribution has already been highlighted: in London or New York sits the bank's global head of markets (equities, foreign exchange) and products, who orchestrates human resources across the bank's network accordingly; 'we have a global business manager sitting in London responsible for global equities and he's got that advantage of overseeing the business from a global perspective' (IB5). Also, banks identified that it was common practice to fill skills shortages in London from New York, which would involve the import of skills from a whole range of nationalities. Second, more senior, expert staff were sent to emerging market world cities, for example Mumbai, Moscow, Kiev, Buenos Aires, São Paulo, and these workers tended to be employed on traditional expatriate packages to reflect their esteem in the receiving bank and managing functions. In terms of flow, traffic was 'significantly less between emerging markets' (B10), but longer-term (often more than three years), whereas flows between London and New York tended to be contracted as adjusted local packages (i.e. local hires). Expert staff moving to Singapore and Tokyo especially still received enhanced packages, but the shift contractually was to hire on local packages. Third, in all banks, the range of nationalities of expert staff on the move was phenomenal. All nationalities were brought into the banks' investment banking HQs in London, New York, Paris and Amsterdam for all the corporate rationales discussed earlier. Given that each of the bank's global market divisions (e.g. equities) and product lines were segmented into defined geographical regions and markets, trading desks, research teams, analysts etc. were composed of a multitude of nationalities. Admittedly, managing CEOs of the bank's ownership nationality still tended to head up international offices, markets and product divisions outside the HQ, but nationality was not an issue for other revenue-earning capacity.

One final and very important observation to make about flow within these banks is the daily routine of hyper-mobility for investment banking staff. Investment banking is now a multi-location working practice. All banks expected staff in all global markets and product divisions to be client-led and flexible when it came to cross-border working patterns. Three major patterns of world city hyper-mobility can be identified. First, high-frequency short-haul business travel between all European financial centres to meet clients and suppliers (e.g. lawyers), interface with colleagues and senior staff in European offices and HQ (if in Paris, Amsterdam, Frankfurt) and attend team meetings involving cross-border activities (e.g. corporate finance) and for training. Second, high-frequency long-haul business travel, especially between London and New York; as one bank commented, 'we see a large number of backwards and forwards movements between New York and London all the time. That is just normal' (IB1). Banks also commented on the frequency of expert staff making long-haul business trips to East Asian cities like Singapore and then travelling to clients throughout East Asia using Singapore as the hub. Third, business commuting, particularly evident with more senior staff who commute to and from Paris, Frankfurt, Amsterdam, Zurich with overnight stays during the week and time spent back in the 'home' office, usually for very short periods of time when an expatriate or local hire contract is not appropriate (e.g. short-term corporate finance projects involving staff and clients in multiple locations).

Table 4.4 International mobility and world city relations in global investment banks, 1999

Bank	Out of London (time-scale)	Into London (time-scale)		World city of next/last residence
IB1	12 (2–3yrs)	60 (2–3yrs)	Next	Frankfurt (12 employees)
			Last	Frankfurt (40), Hong Kong (16), Singapore (2), Johannesburg (2)
IB2	150 (3–5yrs)	50 (2–3yrs)	Next	Tokyo, Singapore, Hong Kong, New York, Paris, Frankfurt
	50 (1–2yrs)		Last	Tokyo, Singapore, Hong Kong, New York, Paris, Frankfurt
IB3[1]	100 (3yrs)	90 (3yrs)	Next	Paris (50); New York (25), Toronto (25)
			Last	Paris (90)
IB4[2]	N.A. (2–3yrs)	N.A. (2–3yrs)	Next	New York, Paris, Rome, Frankfurt, Madrid, Geneva, Hong Kong, Tokyo, Sydney, Singapore
			Last	New York (70%), other offices (30%)
IB5	250 (3yrs)	300 (3yrs)	Next	New York, Hong Kong, Singapore, Tokyo
			Last	From across the international office network
IB6[3]	20	60	Next	Hong Kong, New York, Singapore
			Last	Amsterdam (60)

IB7[4]	51 (2–3yrs)	10 (2–3yrs)	Next	New York, Tokyo, Hong Kong, Singapore, Paris, Miami
			Last	New York, Rome, Tokyo
IB8	48 (2–3yrs)	6 (2–3yrs)	Next	New York (12), Frankfurt (2), Paris (7), Tokyo (14), Hong Kong (10), Seoul (1), Singapore (1), Mumbai (1)
			Last	Frankfurt (6)
IB9[5]	N.A. (2–3yrs)	N.A.	Next	Tokyo, Hong Kong, Singapore, New York
			Last	N.A.
IB10	46	60	Next	Paris (20), New York (20), Kiev, Geneva, Johannesburg, Chicago, Hong Kong, Tokyo (all 1)
			Last	Paris (46), the remainder are from other international offices

Source: fieldwork

Notes
1 There are 500 'International Cadres'.
2 There are 870 staff on temporary 'International Assignments'.
3 The bank has 200 'Global Career Bankers' and 600 'International Transferees'.
4 The bank also has 20 'Third-Country-Nationals'.
5 The bank has 250 'Standard Expatriates' and 450 'International Assignees'.
N.A., no information.

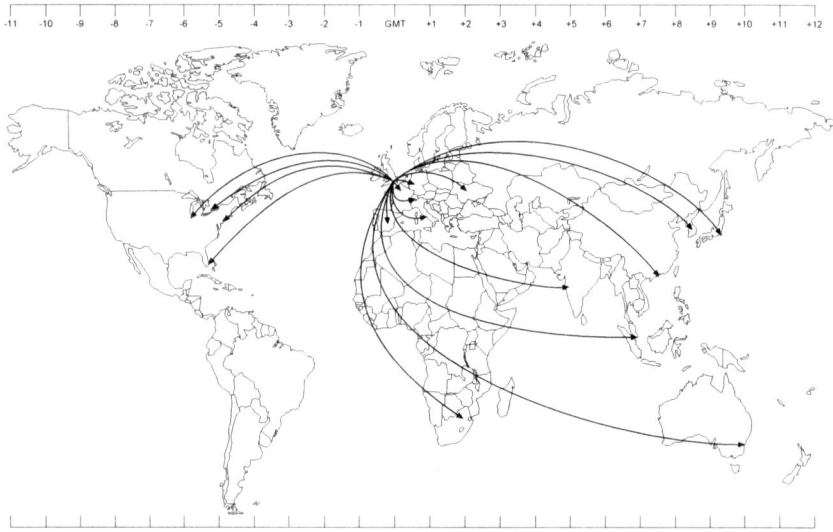

Figure 4.1 World city destinations of expert staff from London (> 1 year)

Making micro-world city networks: career path mobility and transnational working practices

Location is a mantra of upward career-path mobility and for those working in world city international financial centres the City of London, Midtown and Wall Street, New York City, or 'The Golden Shoe', Singapore, are the 'prized locations'. In this section of the chapter, I wish to illustrate how the international career paths and working practices of 10 British investment bankers working in

Figure 4.2 World cities of last residence of expert staff into London (> 1 year)

New York City from the late 1990s onwards make world city networks through their 'micro' past world city working experiences and global physical and virtual business networks and ties.[2]

The 'micro' international career path of the individual makes a significant contribution to the making of world city networks. Changing workplaces, firms, occupations and world city international financial centres are now endemic in the world of finance, banking and professional services. Of these British investment bankers working in New York, three major characteristics stood out in the analysis of their world city career paths (Table 4.5 and Figure 4.3). First, their career paths were dominated by the London–New York dyadic connection, as nine of the group had predominantly only experienced working in London and New York (with three having started working life in UK provincial cities). For eight of these nine bankers, this New York assignment was their first international career-path experience outside London for more than one year. Interestingly, one banker had already experienced two other assignments to New York City with the same employer, NatWest Markets, before the switch of employer in New York City during 1997. Second, those who had experience of inter-firm mobility were predominately with their 'second' employer, with exceptions being respondent 6, who had previously worked for Wang Computers and Instinet in London before moving to a US investment bank in London and then being sent to New York in 1998. Third, incidents of international mobility were linked invariably to training rotations (as in respondent 1) or just seniority and extensive career path experiences (respondent 4). Both these bankers had worked and lived in other world cities for more than six months (e.g. Bahrain, Frankfurt, Paris, Tokyo, Zurich), with either current or past employers. International career paths concretized an array of transnational working and business relations in different cities and employers, which in sum presented the bankers with a world city intellectual infrastructure and network of contacts and social relations, which were portable and transcended space and time.

Once in post in New York, each of the banker's working practices produced unlimited transnational flows of information, practice and social relations between world cities. Of significance in the production of world city networks were the bankers' cross-world city business connections and ties with intra-firm colleagues, clients and suppliers through physical business travel and virtual communication. All of the respondents were involved in business travel, either on a regular basis back to the bank's London office (all respondents) or global headquarters cities (e.g. respondent 9 to Madrid, 5 to Toronto) every six or four months, or for specific business requirements (e.g. respondent 4 to visit clients in Miami). But it was through the virtual media of daily email and telephone communication, and regular video and call conferencing intra-firm that these bankers' produced micro-world city networks. On a regular daily basis, aggregated, these 10 investment bankers were connected in real time to colleagues in European, Latin American, East Asian and Chinese world cities relaying and receiving knowledge, information and intelligence, with all connecting to London (Figure 4.4). The following short biographical account of the banker's career paths and transnational work-

Table 4.5 World city career paths of investment bankers in New York City

Respondent	World city career path
1	London (1991–99, US investment bank) to multiple European short-term postings (including Paris, Frankfurt, Zurich of not more than 6 months) and back to London in between to New York (1999–, same bank)
2	Glasgow (1991–95, Big Four accounting firm) to London (1996–98, German investment bank) to New York (1998–, same bank)
3	Nottingham (1986–87, Lloyds TSB) to Leicester (1987–89, same bank) to London (1989–95, same bank) to London (1995–95, Bank of America; 1995–95, Industrial Bank of Japan; 1995–98, US investment bank) to New York (1998-, same bank)
4	London (1980–1, merchant banking division of a UK Retail Bank) to Zurich (1981–82, same bank) to Bahrain (1982–83, same bank) to Tokyo (1984–88, same bank) to London (1988–97, same bank) to New York (1997-, same bank)
5	London (1976–85, NatWest bank) to New York (1985–89, Nat West Markets) to London (1989–92, same bank) to New York (1993–97, same bank; 1997-, Canadian investment bank)
6	London (1985–93, Wang Computers; 1993–94, Instinet; 1995–98, US investment bank) to New York (1998, same bank)
7	Leeds (1989–93, investment banking division of UK retail bank) to London (1993–97, same bank) to New York (1997-, same bank)
8	Edinburgh (1971–97, merchant banking division of UK retail bank) to New York (1998, same bank)
9	London (1983–85, NatWest Markets) to London (1985–97, Spanish investment bank) to New York (1997-, same bank)
10	London (1996–97, Ford Motor Company) to London (1997–9, German investment bank) to New York (1999-, same bank)

Source: fieldwork

ing practices illustrates very deeply why it is the knowledge-rich workforce of banking, finance and professional service firms that is at the heart of world city network formation and reproduction.

Respondent 1 is a Vice-President in Emerging Markets Trading and Sales in the bank's global headquartered Manhattan office. He joined the US global investment bank on graduation from university in 1991 and in his first four years working for the bank had been assigned to Paris, Frankfurt and Zurich on six-month postings for training purposes (interrupted by six-month stints back in the London office). During that same period, he was involved in many extended business trips to the New York office for both training and client-led business in the US bond market especially. In Manhattan (– 5 hours GMT), his job focuses on the Latin American equity markets, requiring continuous daily dialogue with his equivalents in Buenos Aires and São Paulo especially (both – 4 hours GMT). He starts his working day early before the Brazilian and Argentinean markets open to catch up on the end of day Asian market trading from email and voice mail from his counterpart in Singapore (– 8 hours GMT) and liaises with his London (GMT) colleague (who is responsible for eastern Europe, Africa and the Mid-

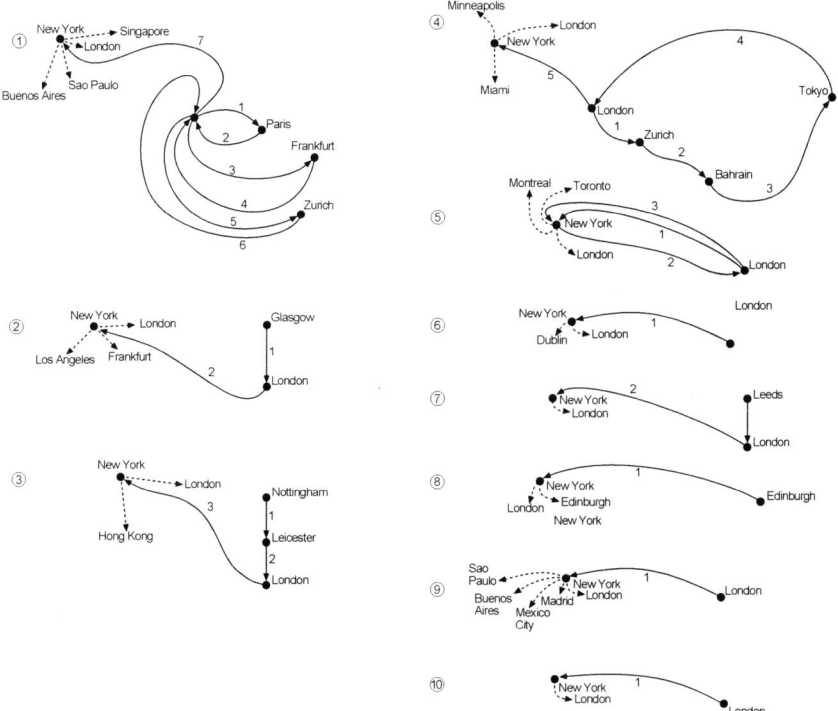

Figure 4.3 World city career paths

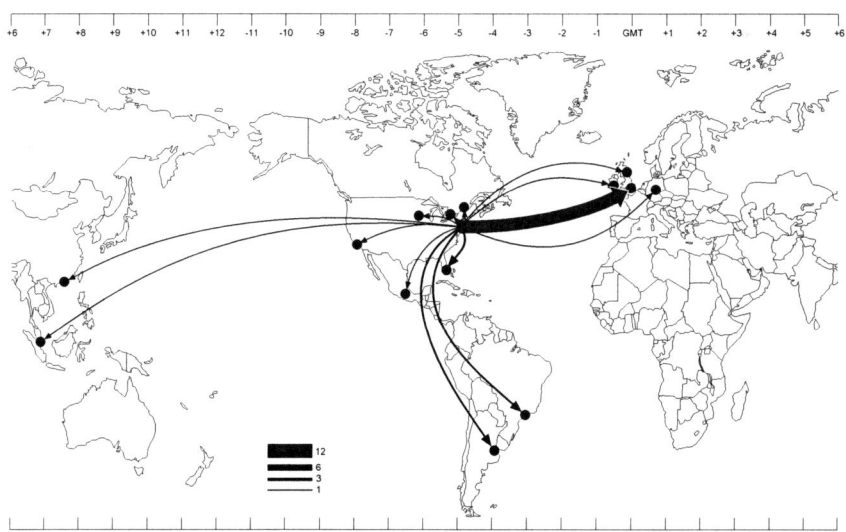

Figure 4.4 Daily connections to world cities

dle East) for late morning/early afternoon market performance. For this banker, daily virtual connections with Buenos Aires, São Paulo, London and Singapore are part and parcel of his working routine because he requires first-hand 'local expertise', 'real-time' knowledge and interpretation of equity market data in different markets to enable him to function in New York. Respondent 1 described his working relationship with his counterparts in these four cities as 'very closely related', involving occasional week-long business trips to Argentina and Brazil, and very occasional longer trips back to London (at least twice a year). Accordingly, for this specific banker, his regular world city relational space is a criss-cross of flows, connections and practice between New York, Buenos Aires, São Paulo, London and Singapore.

CONCLUDING REMARKS AND THINKING 'MICRO' WORLD CITY NETWORKS FROM BELOW . . .

In this chapter I have illustrated that international mobility makes inter-city relations and, more importantly, that the firm is a significant actant in producing the fluidity of cities in globalization and ultimately the making of world city networks. This process-led approach focusing on the action of firms and individuals, conceptualized explicitly in an organizational discourse, enables two major concluding remarks to be posited, levelled at both actors (firms) and recipients (cities) in a network world economy. First, firms are key actants in perpetuating inter-city relations, whether physical or virtual. In a digital global economy the locational decision-making of firms builds the asset base of cities and provides the nodes for communication in the space of flows. Such complex organizational processes are discussed by Beaverstock *et al.* (2002) and their world city network credentials are enhanced in the advanced producer service sector of the economy because competitive advantage in the city is produced by the professionalism, expertise and skills of the workforce, in proximity and co-location with clients, suppliers and competitors. In the empirical study of global investment banks and investment bankers it becomes very clear that these actors can only interconnect with their market through physical relationships in the world city. Firms use international mobility as a globalization strategy and individuals use the opportunity of mobility to enhance world city career paths; both together make inter-city relations. Moreover, incidents of what I have termed hyper-mobility, that is, high-frequency business travel (short- and long-haul) and commuting, are continuous inter-city flows that sustain and feed the network. At the root of this working pattern is the requirement to interact with clients, suppliers and competitors not only through virtual means (email etc.) but, importantly, through face-to-face contact. Knowledge can only be circulated and embedded within the corporate economies of world city networks through socialization and mobility (Beaverstock 2005, forthcoming). In sum, the interconnected high-value service economies of cities in globalization are founded upon highly professional and knowledge-rich expert staff, who are world city hyper-mobile, world city flexible in their working patterns and world city 'street-wise'; who become the 'NYLONers' or 'LONFURT-ers' of the network.

Second, we must remember that understanding cities in globalization and the world city network is not an even process over time and space. The space of flows is a contradiction. Some constituents of the network are more connected than others, some are in the fast lane and some in the slow, or some have just stopped altogether; or some are simply cities not in globalization (see Short 2004). In this research, the empirical findings unearthed a world city network characterized by physical and virtual relations between a New York–London dyad and intensity between European cities (London–Paris–Frankfurt–Zurich–Geneva–Amsterdam); New York and Buenos Aires, São Paulo; London and Asian cities (Singapore and Tokyo); and hyper-mobility especially between London and New York and between London and Frankfurt.

Finally, I would like to argue that world city networks should be theorized as micro-network systems. To borrow phraseology from the transnationalism literatures, unlike Taylor's (2004) world city network 'from above', I would fervently argue that, in order to understand the complex processes of cities in globalization, one should conceptualize world city networks 'from below'. Individual action, whether physical or virtual, is the key agent in understanding world city networks 'from below.' We must not forget that the space of flows is composed of multiple scalars of unequal and uneven connections and flows. In a similar vein to Korff's (1986) first critique of Friedmann's (1986) infamous hierarchy, in order to add value to the conceptual understanding of cities in globalization and the world city network, I would like to throw down the gauntlet and suggest that future research should focus on micro-processes, whether economic, social or cultural, because such actions are the fundamental grass-roots of understanding cities in globalization, and world city networks and relations. We live in a micro-network society composed of multifarious and infinite connections and flows between cities which generate a multitude of different micro-world city networks operating simultaneously and at different speeds in the time-spaces of contemporary globalization.

ACKNOWLEDGEMENTS

I would like to thank the United Kingdom's Economic and Social Research Council Transnational Communities programme for funding this research (Award L214425200I) and the research assistantship of Dr Richard Bostock and Dr James Boardwell. A longer version of the global investment banking case study material is available in Beaverstock (2006)/

NOTES

1 All CEOs interviewed and bank sources are anonymized to protect confidentiality.
2 Between April and May 2000, 10 British investment bankers were interviewed in

New York City who had disembarked from London on inter-company formal second-ment or transfer arrangements for two years or more to the bank's Midtown or Wall Street office in the mid- to late 1990s. Only middle and senior grades of investment banker were interviewed (at Vice-President or above (Senior VP or Executive VP)). These interviewees were a supplement to the 30 highly-skilled expatriate workers interviewed as part of my study of professional service inter-company transfers from London to New York City during the same period (Beaverstock 2005). This research was undertaken before the events of 11 September 2001.

REFERENCES

Aharoni, Y. and Nachum, L. (2000) *Globalisation of Services: Some Implications for Theory and Practice*, London: Routledge.

Amin, A. and Thrift, N. (1992) 'Neo-Marshallian nodes in global networks', *International Journal of Urban and Regional Research*, 16: 571–87.

Bartlett, C. and Ghoshal, S. (1998) *Managing across Borders: The Transnational Solution*, 3rd edition, London: Random House.

Beaverstock, J.V. (2002) 'Transnational elite communities in global cities: connectivities, flows and networks', in A. Meurer and J. Vogt (eds) *Stadt und Region – Dynamik von Lebenswelten. Tagungsbericht und wissenschaftliche Abhandlungen*, Leipzig: Mayr, Deutsche Gesellschaft für Geographie, pp. 87–97.

Beaverstock, J.V. (2004) 'Managing across borders: transnational knowledge management and expatriation in legal firms', *Journal of Economic Geography*, 4: 157–79.

Beaverstock, J.V. (2005) 'Transnational elites in the city: British inter-company transferees in New York city's financial district', *Journal of Ethnic and Migration Studies*, 31: 245–68.

Beaverstock, J.V. (2006) 'Connecting world cities: organisational labour motility in tran-sitional banking', in A.F. Koekoek, J.M. van der Hammen, T.A. Velena and M. Verbeet (eds) *Cities in Globalization*, Utrecht: NGS 339, pp.34–53.

Beaverstock, J.V. (forthcoming) 'Transnational work: global professional labour markets in professional service accounting firms', in J. Bryson and P. Daniels (eds) *The Hand-book of Service Industries*, London: Edward Elgar.

Beaverstock, J.V. and Smith, J. (1996) 'Lending jobs to global cities: skilled international labour migration, investment banking and the City of London', *Urban Studies*, 33: 1377–94.

Beaverstock, J.V., Doel. M.A., Hubbard, P.J. and Taylor, P.J. (2002) 'Attending to the world: competition, cooperation and connectivity in the world city network', *Global Networks*, 2(2): 96–116.

Beaverstock, J.V., Smith, R.G. and Taylor, P.J. (2000) 'World city network: a new metage-ography?', *Annals of the Association of American Geographers*, 90(1): 123–34.

Benveniste, M.L., Ljungqvist, A., Wilhelm, W. and Yu, X. (2003) 'Evidence of information spillovers in the production of investment banking services', *The Journal of Finance*, 58(2): 577–608.

Berger, A., Dai, Q., Ongena, S. and Smith, D. (2003) 'To what extent will the banking industry be globalized? A study of bank nationality and reach in 20 European nations', *Journal of Banking and Finance*, 27: 383–415.

Castells, M. (2000) *The Rise of the Network Society*, 2nd edition, Oxford: Blackwell.

Corporation of London (2003) *Financial Services Clustering and its Significance for London*, London: The Corporation of London.

Edstrom, A. and Galbraith, J. (1977) 'Transfer of managers as a coordination and control strategy in multinational corporations', *Administrative Science Quarterly*, 22: 248–63.

Economist Intelligence Unit (1999) *Global investment banking strategy*, London: EIU.

Freeman & Co. (2003) 'Investment banking. Credit: The rite of passage for investment banks?' June, www.freeman-co.com.

Friedmann, J. (1986) 'The world city hypothesis', *Development and Change*, 17: 69–83.

Gardener, E. and Molyneux, P. (eds.) (1996) *Investment Banking*, London: Euromoney.

Greenwood, R. and Lachman, R. (1996) 'Change as an underlying thing in professional service organisations: an introduction', *Organisation Studies*, 17: 563–72.

International Financial Services London (2000, 2001, 2002) *Banking*, London: IFSL.

Jones, A. (2003) *Management Consultancy and Banking in an Era of Globalization*, Basingstoke: Palgrave.

Korff, R. (1986) 'The world city hypothesis: a critique', *Development and Change*, 18: 483–95.

Lowendahl, B. (1997) *Strategic Management of Professional Service Firms*, Copenhagen: Copenhagen Business School Press.

Nachum, L. (2000) *The Origins of the International Competitiveness of Firms*, Cheltenham: Edward Elgar.

Nohria, N. and Ghoshal, S. (1997) *The Differentiated Network: Organizing Multinational Corporations for Value Creation*. San Francisco: Jossey-Bass.

Sassen, S. (2001) *The Global City*, Princeton, NJ: Princeton University Press.

Sassen, S. (2000) *Cities in a World Economy*, London: Sage.

Seifert, W.G., Achleitner, A., Mattern, F., Streit, C.C. and Voth, H. (2000) *European Capital Markets*, London: Macmillan.

Short, J.R. (2004) 'Black holes and loose connections in a global urban network', *The Professional Geographer*, 56: 295–302.

Taylor, P.J. (2004) *World City Network: A Global Urban Analysis*, London: Routledge.

Thrift, N. (2000) 'Performing cultures in the new economy', *Annals of the Association of American Geographers*, 90: 674–92.

Wrigley, N., Currah, A. and Wood, S. (2003) 'Commentary – investment bank analysts and knowledge in economic geography', *Environment and Planning A*, 35: 381–7.

5 World cities and the internationalization of design services

Paul L. Knox

INTRODUCTION

Professional practice in architecture has long had an international component and a cosmopolitan outlook. The international reach of architecture as a profession was established well before World War II by the commissions of leading practitioners such as Albert Kahn and Le Corbusier (Scully 1988; Frampton 1992). It was consolidated by Philip Johnson and Henry Russell Hitchcock's promotion of the idea of an 'International Style'; by the international migration of Walter Gropius, Mies van der Rohe, and others; by the publication of the Athens Charter developed by CIAM (the Congrés Internationaux d'Architecture Moderne); and by the colonial practices of British, French, and Italian architects. After World War II it was further propagated by the commissions of successive generations of leading practitioners; and, more prosaically, by some large US architecture and engineering (A&E) firms, such as CRS, whose commissions derived from the US government's foreign aid projects (many of them focused on infrastructure projects and 'tied' to the participation of US firms) and from the neocolonial investments of US corporations.

THE INTERNATIONALIZATION OF DESIGN SERVICES

But until recently most practices have been organized around a local, regional, or national framework. Globalization has changed all that. Enabled by digital and telecommunications technologies, by advanced international business services, and by the emergence of clients with transnational operations and a cosmopolitan sensibility, the portfolio of many architecture firms has an international component and the scope of operations of many of the largest firms is now truly global, with multiple international offices covering several continents. Professional publications now contain regular features and updates on international projects and practice, and one publication – *World Architecture* – now specializes in the topic. Meanwhile, the World Trade Organization has attempted to codify international trade in design services through the General Agreement on Trade in Services, signed by WTO members in 1993 and made operative in 1995, and imports and exports of design services have increased dramatically (Tombesi 2003).

According to the American Institute of Architects, mid-sized firms (10 to 20 professionals) in the US are experiencing a significant loss of their market share to large firms. The dominance of large firms has been accelerated by buyouts and takeovers as they seek diversification and market share, and seek to capture more of the profits through vertical integration. Large firms are increasingly engaged in business partnerships with clients: creating business plans, putting together finance packages, doing life-cycle cost analyses, and so on. Strategic alliances have also developed among large firms. Meanwhile, international billings are rapidly increasing. One-third of all large firms had some international work, and one-third of these firms had overseas offices in 2003. Large firms are being drawn increasingly into international work by their transnational clients, and several international practices (as opposed to practices with international projects) have emerged – e.g. KPF, Ove Arup, Foster, Llewelyn-Davies.

Nevertheless, there are several kinds of barriers to international practice: legal and professional barriers (licensing requirements, liability laws, professional standards, procedural requirements, nationality and residency requirements); economic barriers (increased overhead costs, travel, compliance with legal requirements, compliance with technical requirements, different business practices, financing, organization of para-professionals, corruption); technical barriers (regulatory environment, availability of materials, craftsmanship); and cultural barriers (aesthetics, belief systems (e.g. feng shui), ethical standards, business etiquette).

Responses to legal and professional barriers include the propagation of neo-liberal economic policies, support for the General Agreement on Trade in Services (GATS) and for the extension of WTO standards to professional services, including architecture, support for supranational agreements (EU, NAFTA, etc), support for the UIA (International Union of Architects) charter in Beijing (1999: international standards for education, training, and practice in architecture), and support for bilateral accreditation agreements (e.g. the National Council of Architectural Registration Boards in the United States and the Royal Institute of British Architects in the United Kingdom).

Responses to economic barriers involve relationships with foreign affiliates and licensing agreements. Responses to technical barriers include a trend toward performance-based alternatives (in which codes do not limit design, provided that an agreed-upon level of performance can be demonstrated); and establishing local affiliates and alliances. Responses to cultural barriers also include establishing local affiliates and alliances, as well as hiring specialized consultants and agents and US-educated foreign nationals.

GLOBALIZATION, WORLD CITIES, AND DESIGN SERVICES

The relationship of architecture to world cities has several dimensions. Intuitively, the relationship between city skylines and world city status might be expected to be strong; but in fact the relationship between urban form and world city status is highly regional in character (Figure 5.1). Another important dimension concerns the role of architects and other design professionals as key actors in the cultural

Figure 5.1 Top 25 skylines, 2004

milieus of global cities. Design professionals are critical to path creation and path reproduction in the culture industries, and 'star' architects and innovative architecture firms especially help to create and sustain 'atmosphere' in these highly agglomerated services (Capello 2001; Phelps and Ozawa 2003; Kloosterman and Stegmeijer 2004).

Architecture firms themselves have been slow to go global, but the sustained economic boom of the 1990s saw many larger firms extend their operations through office networks that are international in scope. By 2003 the largest firms, such as Nikken Sekkei, Hellmuth, Obata + Kassabaum (HOK), and Gensler, had dozens of overseas projects scattered across every continent except Antarctica. Even smaller firms, like Architectural Resources (Cambridge, MA, USA), Wilson Mason (London, UK), Theofanis Bobotis and Associates (Athens, Greece), and Hayball Leonard Stent (Southbank, Australia), each with 20 or fewer fee-earning architects on staff, had international projects on their books.

The economic logic of globalization had induced an international outlook that is manifest in two practice developments. First, many US- and European-based firms had begun to take advantage of international outsourcing, drawing on pools of skilled but inexpensive labor in South Asia and the Pacific Rim, mostly for specific projects. Total hourly per-employee cost savings through outsourcing for US firms average about 45 percent. A survey in 2004 by Forrester Research projected that 32,000 US jobs would be outsourced by 2005; 83,000 by 2010; and 184,000 by 2015. Currently, around 6 percent of US firms outsource tasks offshore and another 8 percent think that outsourcing some activities overseas is in the long-term interest of the firm (Tombesi *et al.* 2003).

Second, with an international clientele, firms developed global office networks to serve an increasingly complex market. There has been very little systematic empirical research on either of these processes. In this chapter the focus is on the office network development of a sample of the largest architecture firms in 2002, focusing on firms with *global* strategies (defining a global strategy as a network of offices that includes locations on at least three continents). The development of these firms has been reminiscent of the globalization of advanced business service practices analyzed by Peter Taylor and colleagues at GaWC, whereby a relatively small number of firms have emerged to dominate worldwide markets with networks of offices that are located in 'world cities' (Knox and Taylor 1995; Short and Kim 1999; Sassen 2001, 2002; Taylor 2004).

Global strategies

The pursuit of global markets for architectural services goes well beyond the set-ting up of project-based offices for overseas clients. As reflected in the extensive literature on international business management, strategic intent in the global marketplace involves a range of possibilities and varies from industry to industry (Bartmess and Cerny 1993; Hitt *et al.* 1995; Solvell and Zander 1995; Lovelock and Yip 1996; Dicken 2003). Faced with a globalizing economy, firms must ad-dress how best to complement competitive advantages that have been developed in home markets with additional advantages that can be gained from worldwide operations. For advanced business and professional services, the relevant options include mergers, strategic alliances, joint ventures, and the creation of overseas regional and national offices of various kinds, all of which have contributed to the contemporary development of a world city network.

There are reasons to think that architectural firms will have globalized dif-ferently from advanced financial and business service firms. First of all, despite their growth in recent years, most major architectural firms remain smaller than the equivalent advanced business service firms. Obviously this makes provision of a global office network more difficult. Reinforcing this is the project-based nature of architectural practice. Although advanced business services are to some degree project-based – an advertising campaign, a complex legal merger case – architectural work is inherently more 'lumpy' over time with potentially pro-found implications for global strategy. Nevertheless, if we look at the way in which leading architectural firms project themselves on their websites we find that they use globalization discourses that closely mimic prior globalizers in financial and business services. Four themes regularly appear when firms describe their global strategies.

1. They have been very client-led. For instance, the US company *EDAW* is explicit that its location strategy is a response to the global nature of their clients' business (see Table 5.1 for details of firms). Similarly, the Japanese company *Kajima Design* recorded a significant increase in international operations in the 1980s and subsequently reorganized by setting up 'regional subsidiary companies' in response to a growing customer base across the world. This story is repeated time and again. *HOK Group* argues that geographical expansion to serve clients has been important. For *Aedas*, it is

Table 5.1 Firms cited in the text

Firm	Headquarters	Principal Services
Aedas http://www.aedas.com/		Architecture, interior design, landscape architecture, land surveying
Bovis Lend Lease http://www.bovis.com/	Sydney	Construction management, design, consultants, real estate, etc.
CH2M Hill http://www.ch2m.com/corporate_2004/	Denver	Facility and infrastructure planning, design, construction
Dar Al-Handasah http://www.dargroup.com/generic/index.cfm?s=1	Cairo	Architecture, interior design, urban design, facilities planning, etc.
EDAW http://www.edaw.com/	San Francisco	Landscape architecture, urban design, design, planning
A. Epstein & Sons http://www.epstein-isi.com/	Chicago	Architecture, engineering, construction, interior design, graphic design
Gensler http://www.gensler.com/	San Francisco	Architecture, interior design, urban design, planning, consulting
Halcrow Group http://www.halcrow.com/	London	Infrastructure, design, planning, management, etc
HLW International http://www.hlw.com/	New York	Architecture, engineering, interior design, planning

Firm	Location	Services
HOK Group http://www.hok.com/	St Louis	Architecture, design, engineering, interior design, planning
Kajima Design http://www.kajima.co.jp/welcome.html	Tokyo	Architecture, engineering, construction
Khatib & Alami http://www.khatibalami.com/	Beirut	Architecture, planning, engineering, construction management
NBBJ http://www.nbbj.com/	Seattle	Architecture, branding, interior design, landscape architecture, graphic design
Peddle Thorp http://www.pta.com.au/	Melbourne	Architecture, planning, interior design
TY Lin International http://www.tylin.com/	San Francisco	Planning, design/build, construction engineering, program management
Wimberly A T & G http://www.watg.com/	Honolulu	Architecture, planning, interior design
Woods Bagott http://www.woodsbagot.com/	Adelaide	Architecture, interior design, urban design, landscape architecture

stated that increasing numbers of their clients need a more global approach because they are competing in a world arena. *CH2M Hill* explains the practice rationale behind global strategy: their 200 offices around the world are in place because 'personal relationships are important for results.'

2. Seamless service enriches solutions for clients. Going global is not seen as simply an aggregative process. Rather, a new vision of a global holistic practice is presented. Thus *EDAW* claims to be a 'firm without walls' because its offices work together as a seamless whole. For *HOK Group* sophisticated technology allows their employees to work as one 'virtual office.' *Gensler* embarked on a 'new identity program' in 1995 (including name change) to reflect their expanded office network enabling every client to benefit from the firm's 'innovative design and technical proficiency' regardless of location. Similarly, *NBBJ* operates through sharing resources across geographical locations. The promise behind these visions is, as *Aedas* describe their 'global perspective,' to provide a consistent level of delivery in different geographical locations.

3. Global advantages are coupled with local knowledge. Worldwide services do not translate into neglect of the local; with *Aedas* their global perspective is accompanied by 'locally driven solutions.' They claim to provide clients with 'all that is expected from a global player in terms of expertise, international design flair, security and client service but with the invaluable additional benefit of real local knowledge.' More prosaically, *Woods Bagot* use their office network to 'service global clients regionally' and *Bovis Lend Lease* offer both global and local capabilities by combining 'local knowledge with extensive global resources.' *A. Epstein & Sons* have coined the phrase 'think globally and create locally' to describe their strategy. Clearly there is a global–local nexus at the heart of architectural global strategies.

4. There are contrasting global strategies along a generalist/specialist divide. Certain firms combine their global presence with offering a comprehensive package of services. For instance, *Kajima Design* offices are billed as 'full-service architectural organizations.' *NBBJ* claims to have become one of the world leaders in design by creating 'process design,' which maps a success path for each project. The classic case is *Bovis Lend Lease* which offers the capability to manage the entire property life cycle from planning through to development and implementation. Other firms, however have sought out niches in the world market for architectural services: *Peddle Thorp* in tourism, leisure, and entertainment; *Woods Bagot* in health, education, transport, and retail; *Halcrow Group* in transport, water, and property sectors; *HLW International* in media; *TY Lin International* in transportation infrastructure; and *Wimberly A T & G* in hotels, resorts, leisure, and entertainment. Put simply, there are many strategies behind establishing a global practice.

This preliminary look at global architectural strategy certainly suggests that these firms will not differ greatly in their global location practices from advanced business service firms. Thus we can expect architectural practice to broadly adhere to the world city network of global service centers but with some strategic location specificities reflecting the particular route to globalization of architectural practice.

ARCHITECTURAL FIRMS IN WORLD CITIES

The analysis reported in this section was undertaken with Peter J. Taylor, following data collection following data collection and methodology developed in GaWC for analyses of the global organization of advanced business service firms (Taylor *et al.* 2002; Knox and Taylor 2005). This involves constructing a simple 'business service values' matrix, arraying the offices of selected global firms across a large number of important cities worldwide. The 'business service values' indicate how important a particular city is deemed to be within the overall office network of a firm. This is coded as numeric scores ranging from 0 (no importance, i.e. no office in that city) to 5 (the most important city in the office network, i.e. housing the firm's headquarters or decision-making center). For this study, a similar matrix arraying architectural firms across cities in terms of 'architectural practice values' has been constructed. The architectural practice values matrix comprises the office networks of a sample of 21 architectural firms from a total of 68 that were identified as having global office networks. The offices of the 21 firms in our sample were spread across 65 cities worldwide.

A basic geography of global practices

To measure the importance of each of the 65 cities for global architectural practice, we simply sum the firms' practice values for each city. This exercise shows London to be by far the leading city with a total of 51, followed by New York and San Francisco on 26. There are nine cities with scores of over 20 and these are listed in Table 5.2. We can think of these as the premier architectural practice cities. Note that they broadly conform to the geography of global advanced business services: eight of the nine are from western Europe, USA and Pacific Asia. However, although predictably including the 'big three cities' from Pacific Asia, the distribution between Europe and the USA is unexpectedly very lop-sided: London is joined by no other European city whereas there are four US cities listed. The other surprise is Melbourne. Although not an insubstantial global service center, it is by no means in the top rank globally and is not even first ranked for business services within Australia.

This top stratum of architectural global cities can only be understood if we view them in context with the remaining 56 cities. Using practice value sums, the other cities are divided into four other strata and these are mapped in Figure 5.2. Three things immediately emerge from this global distribution. First, the importance of the USA and Pacific Asia with respect to western Europe is confirmed:

Table 5.2 Cities with the highest architectural practice scores

City	Architectural practice score
London	51
New York	26
San Francisco	26
Singapore	25
Los Angeles	24
Washington DC	23
Hong Kong	22
Melbourne	22
Tokyo	22

numerous western European cities are depicted but, after London, they all only figure in the bottom two strata. Second, Melbourne in the top stratum is not a geographical aberration; other Australian cities appear as important cities of architectural practice. Third, Middle Eastern cities are relatively well represented in the middle strata, with more cities at this level than western Europe! Finally a few cities appear from other world regions, but the numbers are quite small and there is no representation at all in sub-Saharan Africa and central Asia.

This basic geography of architectural practice provides several strong hints about the nature of this particular dimension of globalization. Most obviously it appears to be a market-led process that concentrates global architectural practice in leading cities in four regions: USA, Pacific Asia, Australia, and the Middle East. These regional markets are very different: US cities represent a long-term tradition of large-scale building renewal and development; Pacific Asia has been the boom region of contemporary globalization with development focused upon its major cities; Australia possibly represents a smaller version of the US cities process; and the Middle East is a region of concentrated wealth, creating political clients who are trying to boost their local cities. The latter two interpretations do not correspond with findings on global service centers: neither Australia nor the Middle East is as important in this larger globalization and therefore we keep an open mind about what is going on in these two cases of architectural practice concentration. But the demotion of western European cities from constituting a leading global services region to playing a minor role in global architectural practices does seem to be clear-cut. For instance, Paris always appears in the top five world cities for global advanced business services; here it scores a paltry 7 for the architectural practice sum. While London is clearly 'the global place to be' for global architectural practice, it may be that London's preeminence casts a 'shadow effect' on other western European cities: it has been demonstrated that New York has this effect for advanced business services in the United States (Taylor and Lang 2003). However, this can only remain a hypothesis; all we know from Figure 5.2 is that London is the outstanding city of global architectural practice in western Europe with no rivals whatsoever.

Figure 5.2 Global practice scores, 2003

Patterns of global practice

Using principal components analysis to identify distinctive patterns of office net-works among the sample of large architecture firms reveals four components that between them account for 48.7 percent of the total variance. Although this is a very parsimonious solution, it covers a little less than half the variation in the data. Naturally we would have liked to have captured more variance but it appears to be the nature of this particular aspect of globalization that there are many unique features to the global strategies of architectural firms. Perhaps this reflects the fact that the globalization of architectural practice is in its early stages.

The four components divide into two pairs in terms of their importance. Two 'major' components account for 16 percent and 15.4 percent of the total variance respectively, and two 'minor' components account for 8.9 percent and 8.4 percent respectively. The key statistics for each pair are presented in Tables 5.3 and 5.4. The geography is reasonably clear-cut: there is a Global City Arena, an Austral-Asian Arena, a Middle Eastern Arena and a US Domestic Arena. We will describe them in order of importance.

Global arenas of architectural practice

The *Global City Arena* (Table 5.3) is based upon the top four US world cities (New York, Los Angeles, Chicago, and San Francisco) plus London and Tokyo. The two leading global cities, London and New York, have the highest scores. This unequivocally reflects a practice location strategy that concentrates offices in the world's top cities. It is primarily the product of leading US architectural firms. For instance HOK is the largest US architectural practice (second largest worldwide); originating in St Louis, it has followed its clients to become global.

Table 5.3 The two major arenas of global architectural firms

COMPONENT I (16.0%)		COMPONENT II (15.4%)	
Global City Arena		*Austral-Asian Pacific Arena*	
Firm loadings	*City scores*	*Firm loadings*	*City scores*
Gensler	LONDON	0.838	MELBOURNE
0.805	3.42	Peddle Thorp	3.37
HLW Int.	NEW YORK	0.775	LONDON
0.682	3.16	Woods Bagot	2.83
NBBJ	LOS ANGELES	0.708	SINGAPORE
0.615	3.00	Aedas	2.66
RTKL Associates	TOKYO	0.548	SYDNEY
0.606	2.50	Bovis Lend Lease	2.47
HOK Group	CHICAGO	0.594	HONG KONG
0.604	2.33	EDAW	2.22
A Epstein &SII	SN. FRANCISCO	0.573	KUALA LUMPUR
0.549	1.44	CH2M Hill	2.04
Kajima Design	SEATTLE	0.455	BRISBANE
0.501	1.33	Halcrow Group	1.73
EDAW			BEIJING
0.361			1.51

Source: Knox and Taylor (2005).

Note
Firms with loadings above 0.3 are shown in order to identify the important actors (firms) that have created a given component. Cities with scores above 1.0 are shown in order to indicate the spatial configuration of each common strategy.

Gensler perhaps shows the typical expansion pattern: from its San Francisco base, it opened its New York office in 1979, its London office in 1988, and its Tokyo office in 1993. In 1995 it promoted a new global identity and changed its name in the process. HLW of New York made its name originally with skyscraper design in the 1920s, from 1985 to 2000 there was a dramatic change in its client base, and today it assists developers globally with urban design and high-rise develop-ment. NBBJ of Seattle originally became the leading northwest practice but, in the 1990s, they became world leaders with their 'process design' approach, which projected them as new global players in the market. RTKL started in Annapolis and Epstein in Chicago, and both now have a worldwide presence. Kajima of Tokyo is the only non-US firm listed in Table 5.3; its strategy is obviously similar to those described above, for instance, by setting up subsidiary design companies in London and New York as part of its global strategy. The end result of these

Table 5.4 The two minor arenas of global architectural firms

COMPONENT III (8.9%)		COMPONENT IV (8.4%)	
Middle Eastern Arena		US Domestic Arena	
Firm loadings	City scores	Firm loadings	City scores
Dar Al-Handasah	BEIRUT	T.Y. Lin Int.	SAN FRANCISCO
0.863	3.36	0.698	4.87
Khatib & Alami	CAIRO	Ellerbe Becket	WASHINGTON
0.759	2.94	0.624	3.21
Kajima Design	DUBAI	EDAW	MINNEAPOLIS
0.341	2.69	0.504	2.36
	ABU DHABI	Gensler	SEATTLE
	2.32	0.323	1.34
	NEW YORK		DENVER
	2.25		1.17
	LONDON		DUBAI
	1.93		1.08
	DOHA		RALEIGH
	1.62		1.03
	SHARJAH		TAMPA
	1.46		1.00
	KUALA LUMPUR		
	1.11		

Source: Knox and Taylor (2005).

Note
Firms with loadings above 0.3 are shown in order to identify the important actors (firms) that have created a given component. Cities with scores above 1.0 are shown in order to indicate the spatial configuration of each common strategy.

various global strategies has been to create a very clear-cut architectural arena centered on the top strata of world cities.

The Austral-Asian Arena (Table 5.3) is another clearly defined arena, this time regional rather than by city status. Of the eight cities listed, four are from Pacific Asia and three are from Australia, with London making up the numbers. This relates to two features from Table 5.2: first, the 'extra-regional' appearance of London in this component reflects the city's dominance in the organization of global architectural practice; second, the position of Melbourne at the top of the list confirms its surprising appearance as a leading city for global architectural practice. To a large degree this global practice arena has been created by Australian architectural firms taking advantage of growing markets in the Pacific Asian boom region. The Sydney firm Peddle Thorp, for instance, grew initially as a leading Australian practice but with the end of the post World War II boom it turned its attention to successful Asian economies for new opportunities to sustain growth. In 1972 it collaborated with a leading local firm to satisfy the demand for architectural expertise in a rapidly modernizing Singapore. Later it used the same strategy in Kuala Lumpur and Hong Kong. The Adelaide firm, Woods Bagot, has a similar record of expansion from Australia to Pacific Asia. Aedas is a British company that has focused its growth in Pacific Asia and Australia. Bovis Lend Lease exhibits a mix of the previous processes: Bovis, a British firm with Asian and Australian business, was taken over by an Australian firm, Lend Lease, in 1999 to create a truly global enterprise but with a geographical bias to Australia and Pacific Asia. Australian operations began early, in 1951; Singapore was entered in 1973, by 1981 it was in Malaysia where one of its projects was the Petronas Twin Towers. EDAW and CH2H Hill are US firms with major trans-Pacific expansions. The end result of these related expansion strategies is a major architectural global arena within a clearly defined world macro-region.

Neither of these major components has much affinity to the advanced business service fields found in components analyses of global business service firms (Taylor 2004). There is no case where all the top cities come together as in component I above: it would seem that, for architectural services, global cities have a specific 'global city' market and attraction. With advanced business services, Australian cities contribute to 'old Commonwealth' patterns, with little or no linkage to Asian cities to the north: it would seem that Australian architectural firms have taken advantage of booming Pacific Asian cities and carved out a market for themselves in a way that Australian business service firms were never able to achieve. The latter is impressive when it is noted that Tokyo and its firms do not figure in this practice arena. Tokyo and Kajima are part of the Global City arena and not part of their local geographical practice arena.

The *Middle Eastern Arena* (Table 5.4) is another geographical practice arena, with six out of its nine cities from the Middle East, including the top four. On this occasion New York joins with London as an extra-regional global city presence. The other city in the list, Kuala Lumpur, has religious and cultural links to the Middle East. The regional expansion process operating here is of Mediterranean Arab practices taking advantages of the oil wealth and resultant opportunities in the Arabian Peninsula. Both Dar Al-Handasah and Khatib & Alami are Beirut firms with global strategies on a Middle East foundation. The former first expanded throughout the Middle East in the 1970s and main design headquarters were established in London and Cairo. Its main expansion came with meeting the construction needs of 1980s modernization in the Middle East. This company now has 50 offices worldwide; Khatib & Alami remain more concentrated in the Middle East. The end result of these global strategies is the creation of a very distinctive geographical practice arena.

The *US Domestic Arena* (Table 5.4) is the most geographically concentrated of all practice arenas. Note that the USA's three global cities, New York, Los Angeles, and Chicago, are not listed. This is parallel to Tokyo's absence from the Austral-Asian arena: the main cities of the region/country are part of the Global City arena rather than their local geographical arena. Note also that this is the only arena in which London does not appear. This leaves San Francisco and Washington as the cities that dominate this arena. The global strategies described here all have a strong US foundation. The firms are all American, the two with lower loadings also appearing in the Global City component. The end result of these strategies is the maintenance of a domestic practice arena within globalization.

These two minor arenas differ in their relation to previously defined advanced business service fields. The Middle Eastern arena is unique to architectural practice, no doubt reflecting the different nature of the clientele: as well as corporate clients shared with other services, architecture can still find rich patrons in traditional states. In contrast, the US arena is similar to the organization of global business service networks, where US cities also appear as a separate grouping (Taylor and Lang 2003). In this case, for both architecture and advanced business services, the US market is so large within the global economy that it facilitates a safe global strategy based upon keeping a large domestic portfolio.

The absence of western Europe from this identification of global arenas of architectural practice provides a stark contrast with global analyses of advanced business services, where European cities (including Amsterdam, Berlin, Frankfurt, Milan, Paris, and Zürich) dominate several professional fields. Using an exploratory approach to principal components analysis involving searching through extractions of different numbers of components still failed to yield a discernable European component: it is clear that the globalization of architectural practice has not built a western European arena (Taylor *et al.* 2002). Clearly continental Europe has not been a region of extensive construction in the recent past as compared to Pacific Asia, the USA, and even the Middle East. It seems that long-established modern commercial cities collectively have not generated key markets for architectural services under conditions of contemporary globalization. Note that this includes US cities from the northeast, such as Boston, which are conspicuous by their absence from the US Domestic Arena. This does not mean that the metropolises of western Europe and northeastern North America are insignificant in the broader view of international architectural practice. Indeed, the density of long-established firms in these regions means that they account, collectively, for a significant number of international commissions. In Europe, relatively compact national territories mean that there is a great deal of international practice that is intra-European. The contrast with advanced business services is that, beyond London, no city has become a nodal center for *global* networks of architectural practice.

It is clear that architectural practice now has a significant dimension that is truly global and that is somewhat distinctive in comparison with patterns of globalization associated with advanced business services. Although the 'arenas' of practice that emerge from multivariate analysis suggest that the globalization of architectural practice is at a relatively early stage, a definite spatial configuration is emerging as a result of the international strategies of large firms.

REFERENCES

Bartmess, A. and Cerny, K. (1993) 'Building competitive advantage through a global network of capabilities', *California Management Review*, 35: 78–103.

Capello, R. (2001) 'Milan: dynamic urbanization economies versus milieu economies', in J. Simmie (ed.) *Innovative Cities*, 97–128, London: Spon Press.

Dicken, P. (2003) *Global Shift: Reshaping the Global Economic Map in the 21st Century*, 4th edition, New York: Guilford.

Frampton, K. (1992) *Modern Architecture: A Critical History*, London: Thames and Hudson.

Hitt, M.A., Tyler, B.B., Hardee, C., and Park, D. (1995) 'Understanding strategic intent in the global marketplace', *Academy of Management Executive*, 9: 12–19.

Kloosterman, R. and Stegmeijer, E. (2004) 'Delerious Rotterdam: the formation of an innovative cluster of architecture firms', paper presented at the City Futures International Conference, Chicago, July 2004.

Knox, P.L. and Taylor, P.J. (eds) (1995) *World Cities in a World-System*, Cambridge: Cambridge University Press.

Knox, P.L. and Taylor, P.J. (2005) 'Toward a geography of the globalization of architecture networks', *Journal of Architectural Education*, 58 (3): 23–32.

Lovelock, C.H., and Yip, G. (1996) 'Developing global strategies for service businesses', *California Management Review*, 38: 64–86.

Phelps, N. and Ozawa, T. (2003) 'Contrasts in agglomeration: proto-industrial, industrial, and post-industrial forms compared', *Progress in Human Geography*, 27 (5): 583–604.

Sassen, S. (2001) *The Global City*, 2nd edition, Princeton, NJ: Princeton University Press.

Sassen, S. (ed.) (2002) *Global Networks, Linked Cities*, New York: Routledge.

Scully, V. (1988) *American Architecture and Urbanism*, New York: Henry Holt.

Short, J.R. and Kim, Y.-H. (1999) *Globalization and the City*, London: Longman.

Solvell, O. and Zander, I. (1995) 'Organization of the dynamic multinational enterprise: the home-based and the heterarchical MNE', *International Studies of Management and Organization*, 25: 17–38.

Taylor, P.J. (2004) *World City Network: A Global Urban Analysis*, London: Routledge.

Taylor, P.J., Catalano, G. and Walker, D.R.F. (2002) 'Measurement of the world city network', *Urban Studies*, 39(13): 2367–76.

Taylor, P.J. and Lang, R.E. (2003) *American Cities in the World City Network*, Washington, DC: The Brookings Institution.

Tombesi, P. (2003) 'Super Market', *Harvard Design Magazine*, 17: 26–31.

Tombesi, P., Dave, B. and Scriver, P. (2003) 'Routine production or symbolic analysis? India and the globalization of architectural services', *Journal of Architecture*, 8: 63–94.

Part II Inter-city relations in networks and systems

6 Urban network development under conditions of uncertainty

G. A. (Bert) van der Knaap

INTRODUCTION

During the last two decades there has been a growing interest in the role and nature of the dynamics of urban systems (Hall 2002). This interest was fuelled by the internationalization and globalization debate and is closely related to world-systems theory (see Wallerstein 1984; Chase-Dunn 1989). Given the adoption of the systems approach we can observe a shift from a focus on single cities to cities as part of city systems. Thus cities are being perceived as operating in network structures that in turn are also subject to spatial and temporal change. Related to this discussion, Friedmann (1986) formulated the 'World City Hypothesis'. This was taken up by Sassen (1991), who focused on the highest ranking world cities. Although both authors discuss and examine the role of cities as nodes operating in a global arena, in her writings Sassen is more concerned than Friedmann with the impact that rapidly growing international relations have on the internal structure of a single city. She examines the social and economic consequences of these processes in 'global' cities like New York, Tokyo or London. In contrast to Sassen's approach Friedmann is focused on the external relations of 'world' cities, their relative position vis-à-vis each other in either a network structure or a hierarchy.

Although in the literature reference is regularly made to the fact that cities are part of a network of cities, there is not much theoretical concern and not much empirical evidence for understanding the role of cities in such networks. Usually cities are being analysed as singular nodes operating in isolation instead of as part of a larger whole (see also Friedmann 1995). It is against this background that Taylor (2004: 38) observes that there is 'a massive weakness in the evidential pattern of the world city literature, with attribute data (375) far exceeding relational data (50) . . . only 6 percent directly informs [us]'.

A similar observation can be made when we consider the relative position of cities ordered in a hierarchy. Ranking cities does not produce by itself a hierarchy that may provide insights about the interaction of cities with each other. As Taylor (2004: 39) continues: 'Defining a hierarchy requires more than producing attribute measures, . . . there has to be some notion of control up and down different levels'. These observations raise a number of questions related to the role and construction of hierarchies and of networks in urban analysis, as well as the meaning and interpretation of the role of (spatial) scales. Thus issues of scales, hierarchies and networks are closely related, not only at a given point in time, but also from a dynamic perspective. Scales, hierarchies and networks may change as a consequence both of external forces and of their internal dynamics (see Dicken *et al.* 2001). The rate of change may differ for each of these aspects and in this way new spatial structures can be generated.

Networks appear in a variety of forms and it is not surprising that this has generated a rich literature addressing these forms and their dynamics. The network concepts related to this appear to be rather ambiguous and not uniquely associated with a particular form or process. This creates a conceptual confusion and generates a number of uncertainties in theoretical and applied network analysis, which has also implications for the choice of the research methodology. For example, when we want to study, under which conditions and from what perspective can stability be observed in the dynamic systems under observation?

In this chapter we shall first deal with empirical and conceptual developments of city systems. In the next two sections the concept of scales in relation to hierarchies will be discussed and illustrated with empirical examples from the Netherlands and from South East Asia. These examples will also be used to discuss the above issues of uncertainty in a theoretical and applied sense. This will be followed by a discussion of issues relating to the dynamic properties of city systems in which both scales and hierarchies play a role.

City systems and systems of cities

The title of this section pays tribute to a well known publication by Alan Pred (1976) in which he explicitly included ideas of general systems theory (Bertalanffy 1951) into the heart of the study of cities as central places. Cities and city regions were no longer seen as isolated islands with a periphery of smaller places around them. Similar cities in size and function were linked together as one organic whole. This observation was in line with another debate, in which it was emphasized that the crucial factor in geography is spatial interaction (Ullman 1954) This gave rise to numerous studies modelling flows in a central place system or within a city region (Berry 1964a,b). The different spatial scales in this system were usually interpreted as separate levels which were not linked together. Thus on the one hand we could observe a continued interest in spatial patterns; on the other hand there was a growing interest in spatial flows. These two approaches remained separate for a long time and were only linked in applied case studies in the field of urban and regional planning and land-use models (see Lowry 1965).

During the next two decades not much advance was made from a theoretical and conceptual point of view. This period is often referred to as the modern period in which the service economy was developing. The central place paradigm and its related concepts provided at that time the dominant conceptual framework for regional analysis. The emphasis was on the incorporation of quantitative models in the then dominant spatial paradigms. As these models were heavily reliant on classical and neoclassical approaches in economics they were based on equilibrium approaches and on the reduction of complexity by making simplifying assumptions. As the foundations of these models were put forward in the 1930s (see Christaller 1966 (1933)) and 1940s (see Lösch 1954) it became increasingly difficult to understand the present-day urban patterns within the context provided by these theories. This is the case not only for theories relating to the external structure of the city, but also for those referring to the internal structure. Burgess's (1925) North American model of a monocentric city based on Chicago still functions as a role model in a large number of intra-urban studies.

The real world, however, has been changing rapidly under the influence of a large number of processes, such as globalization and the micro-electronics revolution (Forester 1980) and is presently referred to as the network society. Castells (1996) observes that 'we are moving from a space of places to a space of flows'. In this way he emphasizes the increasing importance of human spatial interaction and material flows in society. What has been happening during the 1980s and 1990s is a reintegration of the 'space of places' with the 'space of flows' in which new spatial configurations have been created that are different from those in the past and that are generating a whole set of new and different spatial structures. The classical models of urban structure have lost much of their relevance in this period of post-modernity (see also Hill and Fujita 2003). This has created important changes both in the internal structure of a city as well as in the nature of its linkages. We can observe:

- An increased fragmentation of space in a functional and spatial sense changing the spatial pattern from rather homogeneous structures into distributed heterogeneous structures. At the same time there is a shift from a mono-centred city to a multi-centred urban space.

  An important question here is what the forces of spatial reintegration are and how this process takes place.
- Spatial fragmentation has simultaneously occurred on different spatial scales. At the same time related activities are happening on different scales, thereby linking functions and scales together.

  An important question here is how we conceptualize these linked activities from a spatial perspective.
- The reduced meaning of Cartesian geometry.

  This is caused by the increased use of electronic communication and related long-distance control. As a consequence of this, faraway places have come nearer and, vice versa, nearby places are sometimes at a considerable distance. This changing role of distance as a factor of spatial friction has stimulated a debate about the 'death of distance', but distance is no longer physical distance alone; the concept also includes functional and cultural distance at the same time, thereby changing the concept of distance into a multi-dimensional concept operating on different scales at the same time.

• The lost meaning of boundaries.

In a network society in which related activities are occurring at different scales at the same time the concept of a daily urban system has lost much of its integrating ability as increasingly more activities are taking place at locations outside its boundary. The 24-hour city has become edgeless in this sense. This is creating a number of serious problems in urban governance and the management of the urban system at large.

The direction of these four changes and their implications for the spatial organization of society are by no means clear. Edward Soja (2000) attempted to structure the direction of change by formulating six visions or discourses on post-modern urbanization, which are based on his analysis of the changing urban structure of Los Angeles. These discourses can be ordered in three related pairs. In the first pair he emphasizes the urbanization process by discussing the *Postfordist Industrial Metropolis*, in which the instrumental and self-regulating economy plays an important role. In the discourse about the *Cosmopolis,* the impact of global and local processes is stressed. The second pair of discourses is focused on the spatial implications of urban form. *Exopolis* emphasizes the changing role of the rural and urban fringe and in this context also the changed meaning of core–periphery issues. In the *Fractal City* he deals with the growing problems of increasing social inequalities between citizens at different locations within the city. The third pair of discourses is focused on the consequences for urban regulation. Emerging issues about the role of and changes in public and private space are dealt with in the concept of *Carceral Archepolis,* whereas *Sim City* emphasizes the distinction between real and imagined urban space.

These emerging different urban spaces are to a large extent the result of processes of both economic globalization and local economic restructuring. The possible outcomes as indicated above lead to a number of contradictions when traditional conceptual models are being used. Camagni (2001) argues that what we need is a combination of perspectives in which a cognitive logic is linked with a spatial logic, taking into account symbolic and functional approaches and combining these with territorial and network approaches (see Figure 6.1).

Important here in this context is that the urban system is viewed not as a static structure, but as a system in flux, characterized by interdependencies and uncertainties about the nature and the direction of change. Mechanisms reducing uncertainty at one particular level are counteracted by dynamic forces at the upper and lower levels in the system. The urban system is creating a milieu in which two counteracting forces are operating at the same time; one is directed at the reduction of uncertainties through evolutionary processes and the other is creating dynamic uncertainties through the innovation processes. The first type of uncertainty reduction is emphasized by Camagni (1991) and this will in his view eventually lead to spatial stable, often equilibrium searching, outcomes.

In the discussion above reference has already been made to the fact that forces of globalization and localization have changed the role of distance as a structuring factor in the spatial organization of an urban system.

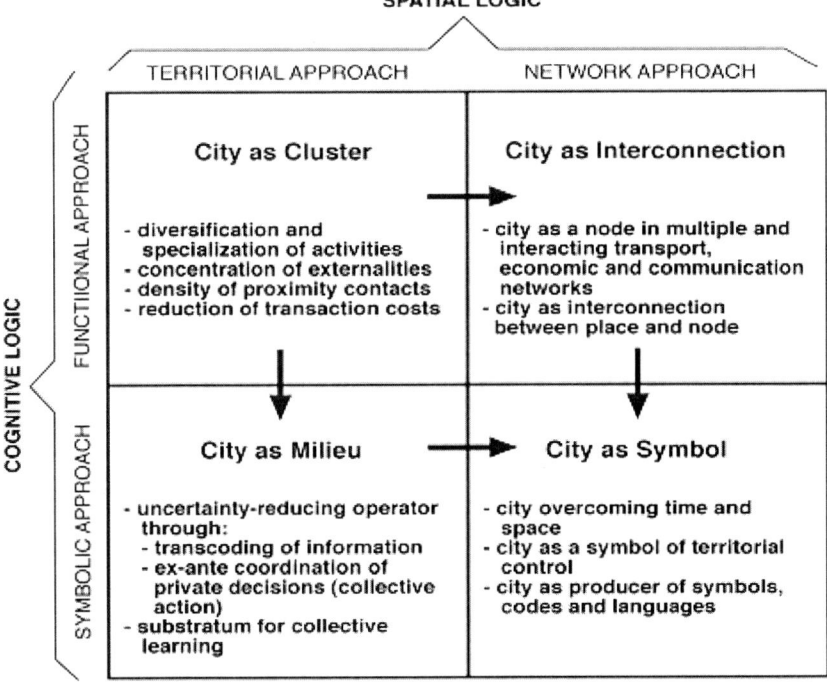

Figure 6.1 The roles of global cities, a theoretical taxonomy (Camagni 2001)

We can observe the occurrence of deconcentration processes in a large number of cities, caused by forces of localization and spatial fragmentation, while at the same time a concentration of higher-order economic activities is taking place in a limited number of cities (globalization). Thus processes of concentration and deconcentration are occurring at the same time, but at different places (levels) in the *hierarchy* and at different places (nodes) in the *network*. A consequence of these joint processes is a spatial debundling and rebundling of economic activities leading to bundles with a different composition of goods or activities from before at different locations in the system (see also Figure 6.2).

Thus we observe an increased social and economic differentiation between cities, which can be characterized by different combinations of bundles of goods: *the service complex* related to knowledge-intensive production, *the financial complex* related to financial markets and flow of capital and *the managerial complex* based upon clusters of head offices of multinational firms predominantly engaged in the manufacturing of goods.

Ideally, in the classical central place system, similar cities in the system – I am referring here to factors such as city size and level in the hierarchy – have similar bundles of goods and activities (see Table 6.1). Often cities are ranked on the basis of these criteria (see Hall 2002).

In a network structure, however, the multiple hierarchies that arise are often incomplete and cities also participate in different hierarchies at the same time. The

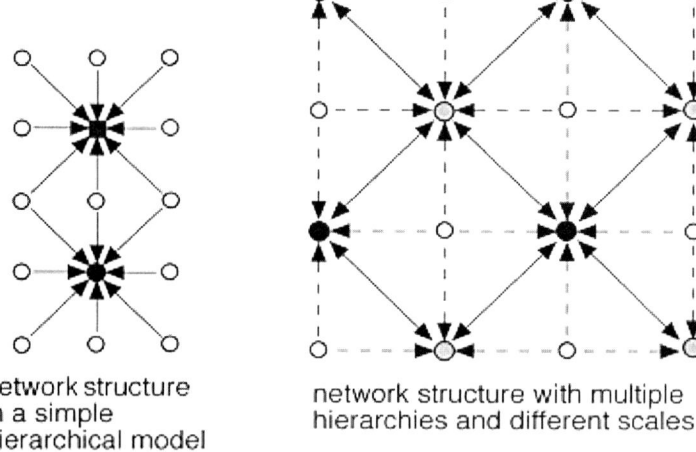

network structure
in a simple
hierarchical model

network structure with multiple
hierarchies and different scales

Figure 6.2 Inter-urban linkages with multiple hierarchies

effect of these changes for a city is an increased dependency on a network of cities
rather than on a limited number of cities. At the same time the scale (spatial reach)
of the network is increasing, creating global networks. We are dealing here with
networks related to different scales whereby the spatial reach becomes related to a
variety of scales operating from local and regional networks to global networks.

Table 6.1 Central place systems versus network systems

Central place system	*Network system, with multiple hierarchies*
System features	
centrality	nodality
size dependency	size neutrality
primacy	flexibility
Economic features	
homogeneous goods and services	heterogeneous goods and services
vertical accessibility	horizontal accessibility
mainly one-way concentrated flows	two-way flows and diffuse structure
transport costs	information costs
perfect competition over space	imperfect competition with price discrimination
Spatial features	
centralised and concentrated	decentralised and dispersed
separation between city and country side	integration of city and country side
hierarchies	networking
homogeneous land use	mixed or mosaic land use, fragmented space

Source: adapted by the author from Lahti (1990: 107) and Batten (1995: 320).

The spatial outcome of this change, from a hierarchical structure towards a network structure, results in an increased heterogeneity and an increased fragmentation of space. This pattern is not represented by a static equilibrium outcome of the process, but represents a spatial structure that is permanently in flux sometimes moving towards an equilibrium situation and at another point in time moving away from it.

Given this perspective cities can be considered as physically integrated places with space and time as constraints (see Graham and Healey 1999). City systems thus have become fluid structures which can be characterized by a mixture of hierarchical dominance and network organization and each of these may reflect an equilibrium outcome at a given point in time. In a central place system, city size and economic functions are fixed and these create a structural inflexibility compared with cities in a network structure, where city size is a neutral feature, not directly a fixing function. Such types of cities can enjoy network externalities and they can possess functions which are unrelated to their actual size, as these are based on borrowed size from the network at large (see Rusten 2000).

The spatial process of urban economic and social restructuring referred to above is a continuous dynamic process that generates different types of uncertainties far away from equilibrium. Not only the type of uncertainty is changing (see also van Asselt 2000); the sources of uncertainty may also vary during the course of the process. Here we can identify at least three different sources of uncertainty:

1. Representational uncertainty occurs when different conceptual models may apply to the particular case under analysis. This has consequences for the selection of the type of hierarchy or the relevant level(s) of aggregation.
2. Spatio-temporal uncertainty refers here to the choice of a particular network structure. The emerging network structure may vary. This is dependent upon the nature and the frequencies of interactions between actors and thereby between cities in the system. This variation is caused by differences in the particular properties of the linkages and related to this by the different capacities of the nodes. Thus linkages and nodes are heterogeneous within a network.
3. Dynamic uncertainty is related to the state of the economic process under analysis. On the firm level this type of uncertainty can be caused by changes in the product life cycle, which may generate different types of linkages or changes in the position in the value chain caused by changes in the external environment of the firm.

The linkage structure of supplier relations may serve as an example of both spatio-temporal and dynamic uncertainty (see Figure 6.3). This figure is based upon a survey among service firms in the Netherlands and it displays the linkages of each firm in the survey, agglomerated by municipality, with other service firms within the country and also abroad. These linkages indicate the intensity of their supplier relations, as measured by the frequencies of occurrence during the course of a year. The network structure of the producer services revealed by this contact

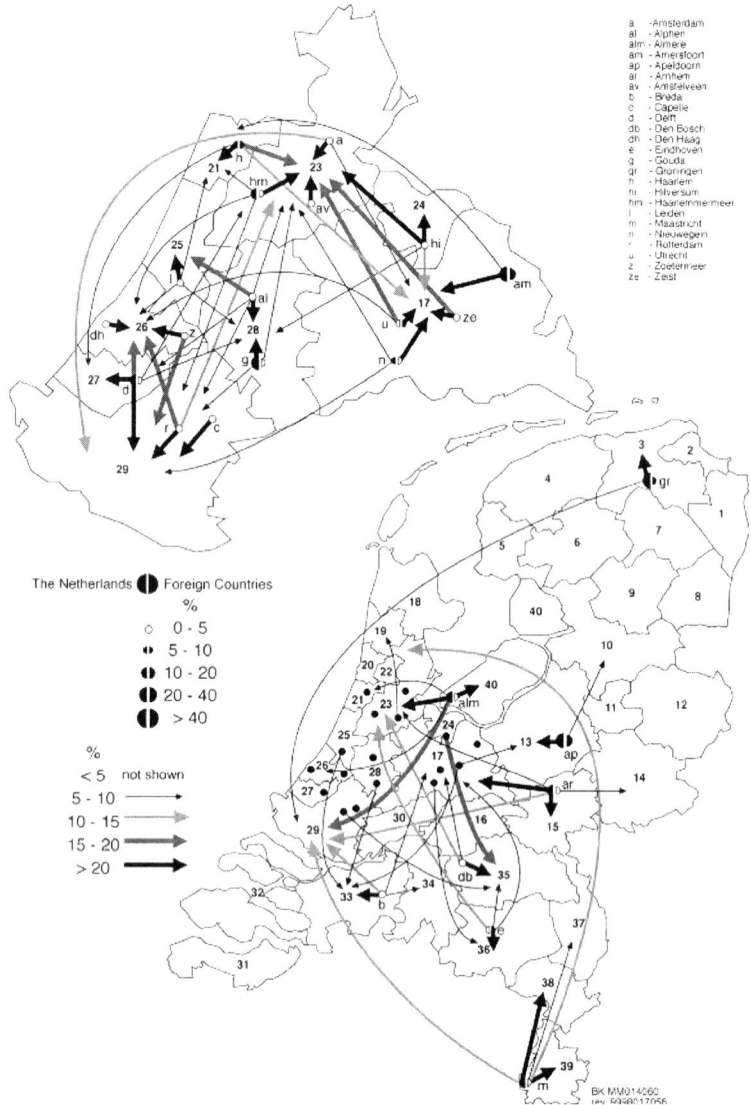

Figure 6.3 Direction and size of the supplier relations (Municipalities versus Corops)

pattern (see also van der Knaap and Wall 2003) indicates a considerable degree of spatial concentration within the western part of the country and only a limited number of contacts with firms in the northern and southern regions.

This can be partly explained by the fact that the growth of the producer service firms in size and number started in the western region. This region is characterized by a complex urban structure in which cities display a considerable amount of complementarity related to the specific economic features of the cities concerned.

A limited number of linkages, however, does not imply that the linkage itself is not of economic importance for the firm(s) concerned. Here it may be relevant to distinguish between strong and weak links (Granovetter 1973). Weak links may indicate a spatial flexibility and economic dynamism, whereas strong links may refer to a spatial fixity caused by a process of path dependency and thereby reinforcing the existing spatial structure. Changes, for example in the life cycle of a firm and the additions of new nodes (cities) as a result of the emergence of new centres of activities in other regions may lead to shifts in the observed network structure over time (see also Riccaboni and Pammoli 2002). This is more likely to happen under conditions where weak ties are prevalent and the resulting new or adjusted structure is being enabled by the dynamic uncertainties in the urban system.

The evolution of business networks is related not only to the life cycle of a firm, or to a leading industrial sector, but also to the phase of the economic cycle (van Duijn 1983). In periods of expansion and rapid economic growth there is less emphasis on the creation of networks than in other periods (see Table 6.2). For example, during a recession, mature firms seek to strengthen their market position through mergers and acquisitions, thereby expanding their network structure temporarily or more permanently.

Van Duijn (1980) pointed also at the role that government plays in the process of network formation during the different phases of the long wave cycle, through regulatory legislation; in particular through incentives during a downturn of the economy.

This is especially the case for sunset industries like textiles or iron and steel (see also Boussemart and de Bandt 1993), but also for emerging innovative sunrise industries, where government instruments are directed, amongst other things, to stimulate growth through cooperation and the development of forward linkages.

The above discussion raises a number of questions related to the functions performed by the network, the role of scale in this context and, related to this,

Table 6.2 The evolution of networks and economic development

Stage in economic development	Life cycle of a leading sector	Intensity of network development	Role of governement in network development
Expansion	Expansion	++	regulatory legislation
Rapid growth	Expansion	+	little government intervention
Recession	Maturity	+++	isolated support for weak classes of industry
Depression	Decline	++++	restructuring problems
Recovery	Decline and Innovation	+++	regulatory legislation

Source: after van Duyn (1980: 92).

the relevant type of aggregation. One has to realize that networks are conceptual constructs referring to process and structure at the same time (see Yeung 1994; Brenner 2001). Networks cannot be observed directly. Even if they could, they cannot be observed in their entirety as the boundaries of a network are not defined (see also the discussion above related to the edgeless city). What we can observe are the building blocks for creating a network, namely a series of dyadic combinations between nodes. The network structure constructed on this basis is to a large extent a function of its relation with other geographical and economic concepts, such as place, locality and space, or the type of economic production, such as the nature of the product and the organizational characteristics of the firm.

LINKING CITIES

In the literature dealing with urban network structures a considerable terminological confusion has been created around the content of concepts of an urban system. Various terminologies exist to address these, e.g. polynuclear cities, network cities and urban networks. Often these terms are being used interchangeably although there are substantive differences between them. When these distinctions are not properly explained representational uncertainty will occur (see Table 6.3).

The spatial contents of these concepts vary considerably on a large number of structural aspects, such as the scale of operation, related to this the mesh of the network, the spatial form of the network and the structural differences between the nodes. In a polynuclear network the central city is still the dominant player and exhibits therefore more hierarchical (vertical) linkages than horizontal, network type linkages. This difference in the orientation of the linkages has consequences both for the relation between the central city and the other cities in the network and for the shape (topology) of the network. Related to this topological variation is the variation in spatial scale. These differences in the structure of the network also have consequences for the functional characteristics of each of the cities and for the population size of the urban system as a whole. When considered in this

Table 6.3 The basic features of polynuclear cities, network cities and urban networks

	Polynuclear city	*Network city*	*Urban networks*
Scale	Local	Regional	National
Mesh	5–10 km	10–30 km	30–100 km
Network type(topology)	Star shape	Closing of circuits	Networks at different scales
Role of the central city	Dominant	Mutual dependent	Dominant at higher levels and mutual dependent at the same level
Population size (average)	1.5 million inhabitants	3 million inhabitants	6 million inhabitants

Source: van der Knaap (2002, Table 5.3).

way the three concepts, namely polynuclear cities, network cities and urban networks, are thus nested concepts, each of which has a relevance of its own.

Cities are linked not by a single network, but by a large variety of networks, each of which has been aggregated in different ways related to the particular characteristics of the linkage structure concerned, namely physical networks, institutional networks and functional networks. Conversely, cities can link a multiplicity of networks together as they are positioned on these networks. They may however occupy different positions on each of them. When they occupy similar or nearly similar positions on each of these networks this may reinforce their position in each of these networks and this may lead to agglomeration advantages. Such cities may in addition to this enjoy considerable network externalities.

Physical networks are transport networks supporting the transportation of goods, people and information. In general the nodes on each of these networks will differ considerably from each other because of the different functions they perform. However, the networks supporting the movement of people and information are more similar to each other than the networks of goods. This is caused by the interchangeability of the functions they perform. Major destination centres, like London, Paris, Frankfurt and Amsterdam, are also major Internet hubs with multiple hubs supporting at least 2.5 Gb/s in communication access. The performance of these nodes is dependent upon the capacity, intensity and the quality of the linkages concerned, as well as on their own characteristics.

Usually, institutional networks are operating at a smaller scale as they are often the expression of intra-national or international governmental cooperation. The latter type can be observed on a continental level and below, reflecting trade agreements, such as Mercosur and ASEAN, or forms of closer economic cooperation, such as the European Union. On a national level we are dealing with forms of regional cooperation, such as intermunicipal, to achieve a more efficient spatial level of organization for administrative or planning purposes.

The different roles functional networks perform at different levels of aggregation are of particular interest in the context of network analysis. Functional networks refer to business relations within or between firms creating intra-firm or inter-firm networks, but also to extra-firm networks which link firms and other types of organizations, such as governmental organizations (see Yeung 1994). The role of these networks is wide-ranging as they cover a whole spectrum of organizational possibilities and related forms between markets and hierarchies (see Williamson 1985). Thus as governance structures they link horizontal (inter-firm) networks to vertical (intra-firm) networks. These types of relations between firms thus occur at the micro level and they create urban or regional networks at the intermediate (meso) level.

LINKING REGIONS

Networks, as has been discussed earlier, are intermediate structures that exist temporarily at a particular moment in space and time. The stability of the observed network structure will vary with both the level of aggregation and the time frame under consideration. As the level of aggregation increases one may assume that over both space and time the network structure will display an increasing amount of stability. A high level of aggregation, however, does not automatically imply that there is a strong degree of spatial stability.

This can be illustrated with the analysis of the development of trade networks in the electronic industry in South East Asia. The economies in this region did change rapidly over the last three decades, both internally and with respect to the external relations of the countries concerned. These changes are reflected in the structure of the trade relations with respect to imports during the period 1985–95 (see Figure 6.4) in a number of ways. This regards the volume of trade and the orientation of the trade relations. The former is reflected in the increasing size of the nodes (regions) and the latter in the number and size of the linkages. In 1985 there exists a strong pattern of triangular trade between Hong Kong, Taiwan and Singapore, while the latter also displays strong links with Malaysia. Over the next decade this dominant pattern gradually disappears as other linkages are added to the system of trade relations leading to overall changes in the pattern. First of all the number of linkages is increasing while at the same time there is an emerging differentiation between the nodes. Next to Korea, Malaysia and China are grow- ing rapidly, relative to the other regions, and this results in a spatial and functional differentiation leading to three different types of regions (nodes). Although the overall economic growth in the region is remarkable the differentiation between the nodes is being strengthened during the last period (to 1995).

In terms of outward linkages the role of Japan in the total structure is decreas- ing. Internally we can observe a further increase in the linkage structure between all regions and a diminishing dominance of the three regions of the early triangle: Hong Kong, Taiwan and Singapore.

On a sub-regional level the dynamics and change in network structure are re- flected by the changing relations between two main nodes of the interregional sys- tem of trade relations, namely Hong Kong and Singapore, during the two decades 1980–2000 (see Figure 6.5). Although these two city regions keep a dominant position throughout the period, their trade relations and thus the pattern of the linkage structure are in continuous flux over a series of five-year periods. This is especially the case for Singapore and its relations with Malaysia, Thailand, the Philippines and Indonesia, which change every time period. Thus, in the two examples presented here, we can observe different dynamics in the network struc- ture over space and time. This is of particular interest in the present discussion as the economic development in this region is often presented in the metaphor of the 'Flying Geese' model (see Akamatsu 1962; Ozawa 1995), in which Japan is the lead goose.

This model is in essence a four-level hierarchical model in which Japan is the dominant actor and three levels below the top level are related to a particular stage of economic development. A crucial issue here is to understand the nature of the relationship between the levels and how these are related to the hierarchy. From the examples given here, it is evident that both hierarchical (vertical) and horizontal networks play a role at the same time.

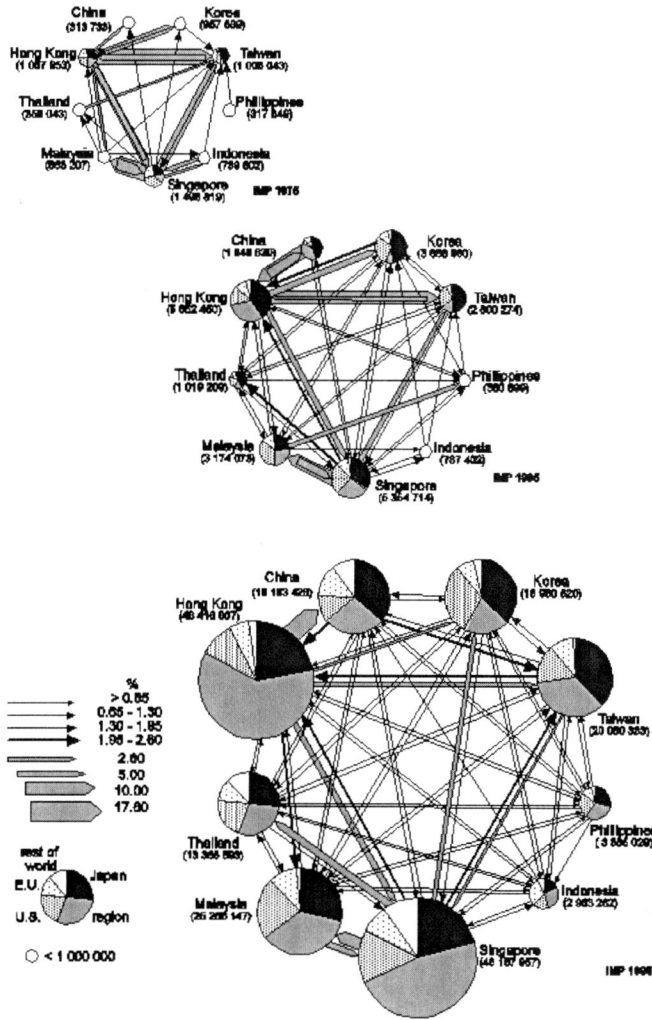

Figure 6.4 Asian region import trade structure (1985–95)

SCALES AND HIERARCHY

Hierarchy theory links three levels of observations to each other (O'Neill 1988): the level of interest with the next higher and the next lower level. At each level the process can be observed, albeit at a different speed. The lower levels display a faster speed than the upper levels. Thus the spatial variations that could be observed within the sub-regional networks of Singapore and Hong Kong are not necessarily in contradiction with the slower dynamics at the higher level for the region as a whole. When these results are compared with the analysis of the evolution of the networks of the micro-electronic industry in South East Asia we may note that different types of networks are revealed.

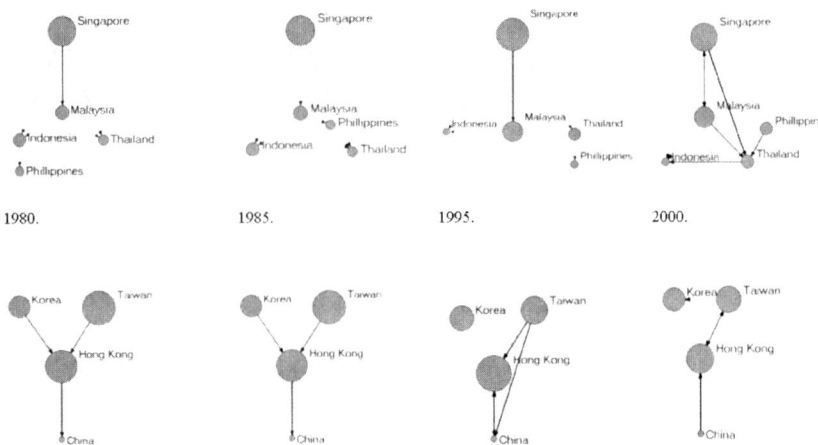

Figure 6.5 Sub-regional nodal trade relations within the micro-electronic industry (1980–2000). (Note: the sizes of the nodes in these figures are only based on the relative size of the total world trade of participant countries in the sub-regional relation; comparison of the absolute nodal size amongst the Singapore oriented and the Hong Kong oriented node is not possible)

There does not exist a fixed number of scales or fixed levels of aggregation. However, scales at different levels in the hierarchy cannot be selected at random. Scales are linked together according to the internal logic of the process under analysis. Randles and Dicken (2004) use the structure of musical chords as an analogy to explain this coherence. Some set of scales, like chords, are mutually interdependent and others are not. The latter combinations are internally inconsistent and do not fit the problem under analysis. Thus the nature of the problem and the related scales of relevance are also important in the determination of the type of hierarchy involved.

A related conceptual problem is the choice of the theoretical framework to understand and explain the development and use of linkage structures in an urban context. It has been demonstrated by a number of authors (see also van Oort 2002) that urban theories tend to be scale-dependent. The case of theories dealing with the issue of urban agglomeration advantages provides a good example here. The focus of these theories is on the economic characteristics of the cities as nodes in the systems and the way in which linkages are being generated and maintained. They indicate the conditions under which spillover effects may occur in an urban context. Thus, dependent upon the level of aggregation and the theoretical position chosen, we may or may not be in a position to observe the existence of spillover effects.

In the above discussion it has been emphasized that scales (spatial levels) and hierarchies are linked and, in addition to this, that economic processes on each of these levels are operating at different speeds. Both deterministic and probabilistic perspectives fall short of allowing us to understand the nature and the direction of the various processes. The interplay between the levels creates a situation of systemic complexity that cannot properly be understood by considering each level in isolation. This comes close to what Portugali (2000: 49) describes as a process of self-organization, that is 'The phenomenon by which a system organises its internal structures independent of external causes'.

Order in these systems is achieved by the process of circular causality described above. Temporal stability then will be achieved through the interaction of macroscopic behaviour (slow speed) at a high level of aggregation and microscopic behaviour (high speed) at a lower but related level. The associated spatial structures exhibit similar features in terms of their propensity to change. The inter-urban network structures are likely to change at a much slower rate than the intra-urban network structures operative at the neighbourhood level. This is also the case when we look at economic organizational dynamics at the sectoral level. Take, for example, the case of changes in the micro-electronics industry. In this industry we can observe a process of continuous readjustment. This is reflected both in the related generalized commodity chain and at the firm level in the continuous reorganization of outsourcing and supplier relations. These adjustments occur in an interdependent way within the context of the internal organization of the production chain and the related locational aspects of the individual establishments (see also Dicken *et al.* 2001).

As these types of economic systems are in a situation of continuous flux they tend to be far away from equilibrium, and when dynamic stability is being reached different network patterns have been generated at different levels. At the same time there are also different patterns of temporal stability at the same level, as the outcome of the process of circular causality between microscopic and macroscopic behaviour is not predictable. Thus we may argue that two forms of chaos exist (see also Cheng and Masser 2003 and Portugali 2000), one at the local level, microscopic chaos, which displays a considerable amount of probabilistic indetermination, and one at the macro level, exhibiting macroscopic chaos with a considerable amount of random deterministic behaviour. This leads in turn to the conclusion that the structure of the networks is also codetermined by the next higher and lower levels in the system (see O'Neill 1988). To understand the process of network formation we also need to understand the nature of the linkages between the levels and the way in which these linkages influence the spatial outcome at a particular level.

CONCLUSIONS

The above approach to network analysis generates a whole series of new questions. It has been observed earlier that networks are structure and process at the same. An interesting research question in the above context, then, is to study the nature and features of this relation. Under which conditions can we learn about the process aspects and what are the consequences for the spatial outcome? Conversely, given the deterministic structure at the macro level, to what extent and in what way does this influence the process at the micro level? As the structure of the network is also determined by the process of circular causality between the various levels in the system, questions can be raised about the influence these levels exert on the process of network formation and its related spatial structure.

It has been argued that to understand the properties of networks a distinction should be made into three types of uncertainty: representational, spatio-temporal and dynamic uncertainty. These three types of uncertainty are related in a variety of ways. At different levels of aggregation (scales) different conceptual models play an ambiguous role, and each of them emphasizes a different aspect of the phenomenon under analysis. Some of these can be observed at one particular level and not in others. In some conceptualizations the concept of a hierarchy is emphasized, in others it is the concept of a network, stressing horizontal relations between cities. An important attribute of a network is its spatial and temporal stability.

Also, stability or change of the network appears to be a scale-dependent property. Both are related to dyadic orientation of the linkages and properties of the nodes; this raises questions about the role of the concepts themselves in the analysis of the spatial dynamics and the stability of the patterns that emerge.

It is not a big step given the preceding discussion to conclude that a network is an intermediate structure that derives its stability from the interaction between the levels internally as well as from the influence exerted by external factors. The latter may also have an impact on the direction of change. The resulting networks thus may have a variety of forms determined by the state of the process.

Thus an important research issue is to identify the factors that determine the relative stability of a network and its related form and explain under which conditions this can be (re)produced. The idea of (re)production should not be taken in its literal meaning, as each network is a unique outcome of the process, but should be interpreted in terms of its structural features and characteristics; thus in this case we are referring to a class of network types.

REFERENCES

Akamatsu, K. (1962) 'A historical pattern of economic growth in developing countries', *The Developing Economies*, preliminary issue, 1: 3–25.

van Asselt, M. (2000) 'Perspectives on uncertainty and risk: the PRIMA approach to decision-support', Ph.D. thesis, University of Maastricht.

Batten, D. (1995) 'Network cities: creative urban agglomerations for the 21st century', *Urban Studies*, 32: 313–27.

Berry, B.J.L. (1964a) 'Approaches to regional analysis: a synthesis', *Annals of the Association of American Geographers*, 54 (1): 2–11.

Berry, B.J.L. (1964b) 'Cities as systems within systems of cities', *Papers and Proceedings of the Regional Science Association*, 13: 147–64.

von Bertalanffy, L. (1951) 'General systems theory: a new approach to the unity of science', *Human Biology*, 23: 303–61.

Boussemart, B. and de Bandt, J. (1993) 'The textile industry: widely varying structures', in

H.W. de Jong (ed.) *The Structure of European Industry*, Dordrecht: Kluwer Academic Publishers, pp. 203–35.

Brenner, N. (2001) 'Limits to scale, methodological reflections on scalar structuration', *Progress in Human Geography*, 25 (4): 591–614.

Burgess, E.W. (1925) 'The growth of the city: an introduction to a research project', in R.E. Park and E.W. Burgess (eds) *The City*, Chicago: University of Chicago Press.

Camagni, R. (1991) *Innovation Networks*, London: Belhaven Press.

Camagni, R. (2001) 'The economic role and spatial contradictions of global city-regions: the functional, cognitive, and evolutionary context', in A.J. Scott (ed.) *Global City-Regions, Trends, Theory, Policy*, Oxford: Oxford University Press.

Castells, M. (1996) *The Information Age: Economy, Society, and Culture. Vol. 1: The Rise of the Network Society*, Oxford: Blackwell.

Chase-Dunn, C. (1989) *Global Formations: Structures of the World-Economy*, Cambridge, MA: Basil Blackwell.

Cheng, J. and Masser, I. (2003) 'Modelling urban growth patterns: a multi-scale perspective', *Environment and Planning A*, 35: 679–704.

Christaller, W. (1966 (1933)) *Central Places in Southern Germany*, translated by C.W. Baskin, Englewood Cliffs, NJ: Prentice-Hall.

Dicken, P., Olds, K. and Yeung, H.W. (2001) 'Chains and networks, territories and scales: towards a relational framework for analyzing the global economy', *Global Networks*, 1 (2): 89–112.

Forester, T (1980) *The Microelectronics Revolution*, Oxford: Basil Blackwell.

Friedmann, J. (1986) 'The world city hypothesis', *Development and Change*, 4: 12–50.

Friedmann, J. (1995) 'Where we stand: a decade of world city research', in P.L. Knox and P.J. Taylor (eds) *World Cities in a World-System*, Cambridge: Cambridge University Press, pp. 21–47.

Graham, S.D.N. and Healy, P. (1999) 'Relational concepts for space and place: issues for planning theory and practice', *European Planning Studies*, 7 (5): 623–46.

Granovetter, M. (1973) 'The strength of weak ties', *American Journal of Sociology*, 78: 1360–80.

Hall, P. (2002) 'Christaller for a global age: redrawing the urban hierarchy', in A. Mayr, M. Meurer and J. Vogt (eds) *Stadt und Region: Dynamik von Lebenswelten*, Leipzig: Deutsche Gesellschaft für Geographie, pp. 110–28.

Hill, R.C. and Fujita, K. (2003) ' The nested city: introduction', *Urban Studies*, 40: 207–17.

van der Knaap, G.A. (2002) *Urban Activity Space: Reflections about Urban–Rural Change [Stedelijke Bewegingsruimte: over veranderingen in stad en land]*, Scientific Council for Government Policy (WRR), vol. 113, The Hague: Staatsuitgeverij.

van der Knaap, G.A. and Wall, R. (2003) 'Linking scales and urban network development', in H. van Dijk (ed.), *The European Metropolis* (e-book) Rotterdam: Erasmus University, p. 35.

Lahti, P. (1990) 'Technological change and the urban future: some aspects and questions', in: H. ter Heide (ed.), *Technological Change and Spatial Policy*, Geographical Studies no. 112, Utrecht: Department of Geography.

Lösch, A. (1954) *The Economics of Location*, translated by W. Storper, New Haven, CT: Yale University Press.

Lowry, I. S. (1965) 'A short course in model design', *Journal of the American Institute of Planners*, 31: 158–66.

O'Neill, R.V. (1988) 'Hierarchy theory and global change, in T. Rosswall *et al.* (eds), *Scales and Global Change*, Chichester: John Wiley, pp. 29–45.

van Oort, F. (2002) 'Agglomeration, economic growth and innovation', Ph.D. thesis, Tinbergen Institute, Rotterdam, Research Series, 260.

Ozawa, T. (1995) 'Structural upgrading and concatenated integration: the vicissitudes of the Pax Americana in tandem industrialization of the Pacific Rim', in D.F. Simon (ed.) *Corporate Strategies in the Pacific Rim: Global Versus Regional Trends*, London: Routledge, pp. 215–46.

Portugali, J. (2000) *Self Organisation and the City,* Berlin: Springer Verlag.

Pred, A. (1976) *City Systems in Advanced Economies, Past Growth, Present Processes and Future Development Options*, London: Hutchinson Press.

Randles, S. and Dicken, P. (2004) '"Scale" and the instituted construction of the urban: contrasting the case of Lyon and Manchester', *Environment and Planning A,* 36: 2011–32.

Riccaboni, M. and Pammoli, F. (2002) 'On firm growth in networks', *Research Policy,* 31: 1405–16.

Rusten, G. (2000) 'Geography of outsourcing: business provisions among firms in Norway', *Tijdschrift voor Economische en sociale Geografie*, 91: 122–35.

Sassen, S. (1991) *The Global City: New York, London, Tokyo*, Princeton, NJ: Princeton University Press.

Soja, E.W. (2000) *Postmetropolis: Critical Studies of Cities and Regions*, London: Blackwell Publishing.

Taylor, P.J. (2004) *World City Network: A Global Urban Analysis*, London: Routledge.

Ullman, E.L. (1954) 'Geography as spatial interaction', *Proceedings of the Western Committee on Regional Economic Analysis*, Berkeley, CA: University of California Press, pp. 63–71.

Van Duijn, J.J. (1983) *The Long Wave in Economic Life,* London: Allen and Unwin.

Wallerstein, I. (1984) *The Politics of the World-Economy*, Cambridge: University Press.

Williamson, O.E. (1985) *The Economic Institutions of Capitalism,* New York: Free Press.

Yeung, H.W. (1994) 'Critical reviews of geographical perspectives on business organisations and the organisation of production: towards a network approach', *Progress in Human Geography*, 18 (4): 460–90.

7 City networks as tools for competitiveness and sustainability

Roberto Camagni

INTRODUCTION

The paradigm of city networks, complementary to the traditional one of urban hierarchy (as epitomised by the 'central-place' model) and initially proposed by the southern European tradition of spatial analysis (Dematteis 1985, 1990; Camagni 1993a), has gained interest and support in other scientific and policy contexts, particularly in regional and urban geography (Castells 1996; Malecki 1997; Nijkamp 2003). Currently we speak about 'city networks', '*reti urbane*' and '*réseaux de villes*' to define new organisational forms of the urban structure and new tools for urban policies. Recently the concept was supported by the EU spatial strategy document, the ESDP – European Spatial Development Perspective, particularly in its Noordwijk (1997) and Glasgow (1998) drafts, but also in its final draft approved in Potsdam by the ministers responsible for spatial planning (EU 1999), in the specification given by this author, namely that of 'complementarity' and 'synergy' networks.

For some people these are only new catchphrases. In my opinion, we are in fact confronted with a new paradigm in spatial sciences, providing that some precise conditions are met:

- its exact meaning is thoroughly defined;
- its economic rationale is justified;
- the novel features of its empirical content are clearly identified and distinguished from more traditional spatial facts and processes that can be easily interpreted through existing spatial paradigms.

In fact, if we examine the third condition, the concept of spatial networks is sometimes merely used as a substitute for 'interaction': an exchange of goods, services, information and contacts among places and nodes, at the local as well as at the global scale. In this case, the traditional paradigm (and related models) of spatial interaction can be easily utilised, unless one could demonstrate that the probability of such exchanges is mainly independent of distance and the size of nodes.

By the same token, the term 'network' is sometimes used to interpret relations and flows that take place within an urban hierarchy among centres of different hierarchical level. Also in this case, we would not need a new concept to identify well-known phenomena, unless we refer to interactions taking place among centres belonging to the same hierarchical level (which are not considered in standard central place models such as the one elaborated by Christaller).

While interaction is supported by research work interested in the 'space of flows', other studies focusing on a description of the 'space of places' often use the network concept as a synonym (or an explanation) for polycentrism, a merely geographical and descriptive concept, not adding much in interpretive terms.

Finally, the same term is sometimes used to identify social linkages that characterise local communities or associations kept together by ethnic, linguistic, civic or even criminal goals. While in this case the micro-foundations of these phenomena are studied by other disciplines, such as sociology or political science, their spatial effects may well be interpreted through other existing concepts, like that of social capital (Coleman 1990; Putnam 1993; Bowles and Gintis 2002; Glaeser *et al.* 2002) or local milieu (Aydalot 1986; Camagni 1991a; Maillat *et al.* 1993).

The issue is mainly terminological, but nevertheless central to our discourse: it appears preferable to use the term 'network' for selective and formalised linkages among well defined economic actors and spatial units, transcending proximity relationships and taking place at a trans-territorial level. On the other hand, the term 'milieu' may be used to encompass more informal and 'atmospheric' relationships, taking place inside local territories thanks to cultural proximity and social cohesion.

Once agreed on what precedes, we have to admit that there are spatial phenomena that cannot be interpreted through the usual tools – spatial interaction, urban hierarchy, social capital – for which the new concepts, namely urban milieux and city networks, could be of use. The reference here is to spatial relations taking place at two different spatial levels:

- inside the city or the metropolitan area, in the form of cooperation agreements and coalitions of local actors, often developed through 'strategic planning' processes;
- among cities of similar size and hierarchical level, in the form of cooperation networks for selected and targeted goals, generally inside regional areas but in some case also at longer distances. In some cases the centres involved perform similar tasks and functions inside the urban system; sometimes on the other hand we observe a spontaneous or organised division of labour among centres in a regional context.

In both cases, these linkages and relations give rise to a *network surplus* as a consequence of synergies and cooperation.[1]

In a scientific context where the economy is increasingly seen as a system or web of links between individuals, firms and institutions, with links depending on experience and evolving through learning processes (Malecki and Oinas 1999), the relevant theoretical building blocks on which the network concept or paradigm may be constructed are the following:

- *recognition of cooperation as an original organisational and behavioural form* for individuals and companies, intermediate between hierarchy and resort to the market, *in a condition of complexity and uncertainty*, following the well-known works of Coase and Williamson (Coase 1988; Williamson 1985, 2002);
- *recognition of cities as collective actors*, competing between each other and developing their own development strategy. This statement finds its economic rationale in various recent theoretical reflections, referring to the possibility of correctly defining territorial competitiveness, and consequent collective goals of local communities pursued through intentional development strategies (Camagni 2002).

As a consequence, it is possible to say that networking – intended in a micro-economic sense as cooperation among individuals, firms and institutions concerning collective action, public/private partnerships and the supply of public goods – may become a scientific paradigm for interpreting the macroscopic spatial behaviour of collective agents like cities, competing and cooperating in the global arena where locations of internationally mobile factors (professionals, corporations, institutions) are negotiated and large territorial projects are decided.

The paper is organised as follows. The next two sections following this introduction deal with the theoretical building blocks on which the city network paradigm will be built: the first is devoted to the micro-economic analysis of firm cooperation networks in terms of institutional economics and game theory, conceiving cooperation as a tool to cope with uncertainty and complexity; the second section deals with the interpretation of cities as economic actors, looking at the way cities engage in inter-regional trade and competition and in the consequent necessity for them to define shared development strategies. The third section then tries to define 'city networks' in theoretical terms, building on competition and cooperation relationships. The fourth section presents some empirical analyses and cases where the concept of city network is, implicitly or explicitly, utilised for spatial policies.

THE FIRST DIMENSION: NETWORKS OF ACTORS INSIDE THE CITY AND THE CONCEPT OF LOCAL MILIEU

Modern economic and territorial systems are characterised by intrinsic complexity: in their internal structure, in their internal and external relationships, in the forces that determine their development, in the structure and evolution of preferences of actors and in governance systems. This complexity has relevant effects not only on the evolution, predictability and control of their development path, but also on the decision-making process of the single territorial actors, firms and individuals.

The hypotheses and axioms of perfect information and absence of transaction costs, usual in mainstream economics, become less and less acceptable in the interpretation of the real economies. It is therefore necessary to pass from the abstract 'theory of choice' of economics textbooks to a 'theory of contracts' based on the observation of the constraints that the real market imposes on the single actors (Williamson 2002). The same champions of the axiomatic approach in economics recognise that 'instead of single transparent axioms there looms the likelihood of psychological, sociological and historical postulates' (Hahn 1991: 47).

The cognitive context is one of bounded rationality. In the traditional field of choice theory, this means to abandon the maximisation criterion for a satisfying criterion *à la* Simon, but in the field of contract theory it means more stringently that we have to accept the incompleteness of all complex contracts, the possibility that the pay-off conditions of any transaction could not be clear *ex ante* to all partners, the possibility of opportunistic and free-rider behaviour; all elements which clash with the hypotheses of the traditional model of exchange and production. The necessity 'to embed transactions in more protective governance structures' (Williamson 2002: 439) becomes clear, in the same way as the necessity to dispose of operators and institutions, of a private (trust, cooperation) and public nature, that could force order into the incentive structure, reduce conflicts and allow the realisation of mutual advantages from the exchange.

For some simple typologies of contracts, the pure market is a sufficient governance tool – if supported by efficient institutions and legal frameworks, as the experience of 'transition' countries has widely shown, in negative terms. For the more complicated typologies of contracts, at the other extreme, the most efficient governance structure is direct control through hierarchy; for most intermediate cases, a cooperative form of governance is requested.

In this last case, social capital and milieu linkages in the local economy prove crucial: in fact, they work as facilitators of cooperative behaviours and reducers of uncertainty. In all cases, the existence of clear rules of the game, of reliable juridical institutions in charge of the control over the execution of contracts, of shared codes of economic behaviour, appears fundamental.

The concept of the *milieu innovateur* was created in order to understand innovative processes involving small enterprises, and interprets spatial development as an effect of synergies developing over limited territorial areas (Aydalot 1986; Camagni 1991a). It is defined as a set of relations which bring together and integrate a local production system, a system of actors and representations and an industrial culture, and which generates a localised dynamic process of collective learning.

A fundamental element of a milieu is geographical proximity, which results in a reduction in production and transaction costs. But geographical proximity is also accompanied by socio-cultural proximity – which must inevitably be present for a milieu to exist. We can define it as the presence of shared models of behaviour, mutual trust, common language and representations, and common moral and cognitive codes. Geographical proximity and socio-cultural proximity mean a high probability of interaction and synergy between economic agents, density of informal repeated contracts, absence of opportunistic behaviours, high division of labour and cooperation within the milieu. This is what we have called local relational capital (Figure 7.1), and it is composed of cooperative attitudes, trust, cohesion and a sense of belonging to a community.

The role of the local milieu in terms of economic theory is linked to three types of cognitive outcomes, supporting and completing the normal mechanisms of information circulation and coordination of agents performed through the market: reduction of uncertainty in decision-making and innovative processes; *ex ante* coordination among economic actors, facilitating collective action; and collective learning, a process occurring within the local labour market and industrial atmosphere, enhancing competencies, knowledge and professionalities. These outcomes, particularly the first two, are functions enabling well-known cases of 'market failure' to be overcome.

A. The function of *reducing uncertainty in innovative processes*. Local relational space is seen as a means of reducing uncertainty, since – thanks to geographic and cultural proximity – collecting, evaluating and particularly transcoding information, selecting decisional routines, controlling and coordinating competition (all functions usually performed by research and development

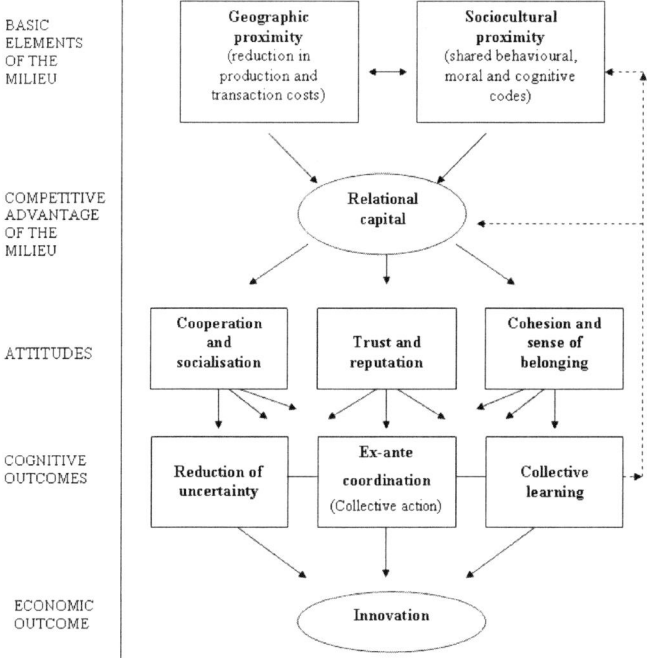

Figure 7.1 Basic elements and functions of a local milieu

or strategic planning teams in large enterprises) are carried out collectively within the social context of the local milieu (Camagni 1991b).[2]

B. A function favouring *ex ante coordination* of local actors and realisation of 'collective actions', due to the presence and development of conventions, norms of behaviour, shared codes of social inclusion and exclusion, mutual trust. Once again it is evident how the characteristics of a 'pure' market, and the way information is assumed to operate within it, mean that it is not possible for complementary investment decisions to be taken concurrently, since the profitability of each decision is dependent on the concurrent decisions of other subjects (Richardson 1960; Camagni 2002); the presence of a milieu realises this important governance function of helping coordination of simultaneous decisions of different actors.

C. A function favouring *collective learning*, which in the local milieu, and particularly in the local labour market, finds a permanent environment where it can be incorporated (Camagni and Capello 2002). Thanks to the high mobility of skilled labour and professionals inside the local milieu, and thanks also to the wide cooperation agreements and synergies taking place, the learning processes that are at the base of innovation adoption and diffusion manifest themselves in a collective and socialised way, inside the local territory but outside the single firms.

In sum: cooperation – technological, commercial and financial – appears as an original economic paradigm in an era of continuing innovation and rapid techno-logical change, where the market 'fails' in the orientation of dynamic and innova-tive behaviour. The market does not deliver sufficiently correct and timely signals (Camagni 1991b). The new behavioural form has two major advantages. It avoids the high transaction costs that are inevitable when crucial inputs are demanded through the market, and it reduces the high costs involved in adopting a strategy of internal development of a new technology or competence.

Cooperation also has some definite costs, however, which should be taken into account by the firm, and by theories attempting to interpret the new behavioural pattern. Some of these costs, or risks, refer to the relationships among partners – possible opportunistic behaviour and coordination costs, asymmetry in infor-mation and bargaining power, which affect the allocation of advantages among partners (Camagni 1993b) – while others refer to the limited appropriability of the advantage of the cooperative game and the possibility of free-riding, all elements that impinge on the probability of a cooperative game being initiated at all.

In general, the choice whether to engage in a cooperative or collective action will depend on whether the benefits of cooperative action (Bc), reduced by regula-tion and coordination costs (Cr, Cc) exceed the benefits of individual action (Bi) reduced by transaction costs (Ct):

$$Bc - (Cr + Cc) > (Bi - Ct)$$

Regulation costs depend on the possibility of free-riding, the principal barrier to cooperative behaviour, and the consequent existence of incomplete markets: some goods will never be produced, in spite of there being a demand, because it is impossible to exclude people from their free use ('public goods'), and this will lower collective welfare. Rational individual behaviour will involve a defection from cooperative behaviour but, if generalised, results in an irrational collective outcome, namely a lower utility even at the individual level. This is a well-known case of market failure and social dilemma (Arrighetti 2003).

Interestingly enough, particular elements of the territorial context, like the presence of reciprocal trust, reputation and social or 'relational' capital, reduce the regulation costs and the risks of defection; the same elements at the same time reduce the temptation to defect, as they provide a signal of huge and certain social sanctions against opportunistic behaviour.

The second type of cooperation cost regards coordination costs that arise because the outcome (pay-off) of the cooperative action is uncertain, depending widely on the characteristics and the content of the action itself. Only in textbook examples may they be assumed to be perfectly known beforehand. The costs refer to the elaboration of a common scheme/strategy, in the presence of information asymmetries, and to uncertainty about the cooperative attitudes of the different partners. The same territorial elements that were indicated above, and in particular the presence of relational capital or a milieu, reduce coordination costs and may be seen as favourable preconditions for the building of collective actions.

All these elements have profound effects on the definition of new policy styles and new urban governance tools. In fact, it is possible to contend that a new type of community governance is emerging, namely urban *strategic planning*:[3] promoting and recognising social networks; focusing on shared values and identity and providing opportunities for discussing and coordinating issues; supplying public goods, for example in the self-organisation of neighbourhood amenities; organising mechanisms for cooperatively sharing risks. In most cases the local community is able to effectively carry out collective actions where spontaneous mechanisms or actions by the public administration fail on account of a lack of crucial information regarding the behaviour of various partners, their abilities or needs (Bowles and Gintis 2002).

THE SECOND DIMENSION: CITIES AS COLLECTIVE ACTORS

The second theoretical building block referred to in the introduction is the hypothesis that cities and territories behave like collective actors. This hypothesis is based on the following reflections.

First of all, it is possible to demonstrate the sound theoretical basis of a concept of territorial competition and competitiveness: not just firms but also territories compete with each other. In fact, regions, local territories and cities, because of their intrinsic openness to the movement of both goods and factors, operate in the context of inter-regional trade within a regime of 'absolute advantage' and not within a regime of 'comparative advantage' (Camagni 2002).[4] As a consequence, if their absolute competitiveness is inadequate or declining with respect to the other regions, the spontaneous adjustment mechanisms which in the comparative advantage regime and in the case of nations always assure them a role in the international division of labour – namely devaluation of the currency and wage–price flexibility – either do not exist or are inadequate to re-establish equilibrium. Conditions of weakness, due to inadequacies in production factors, adverse geographic circumstances or poor accessibility, may well result in mass unemployment and, if public transfers of income are not sufficient, emigration and possible abandonment. Local communities may react to this last condition through explicit and intentional strategies and policies.

Second, it is widely accepted that local firms, in their search for competitiveness, rely not only on local externalities (public goods, human capital and social overhead capital), but increasingly on selected external assets and 'specific resources' that cannot be easily obtained via spontaneous market developments. Therefore firms are increasingly engaged in a cooperative process with other local firms, (collective) actors and the public administration for the conception and provision of these resources (Colletis and Pecqueur 1995; Cooke and Morgan 1998).

This cooperation among firms and social actors is facilitated by particular territorial conditions, determined by a particular richness of inter-firm interactions: the milieu effect, as seen before. This condition generates cumulative learning processes, enhancing the innovativeness and the competitiveness of the local territorial system.

The third reflection concerns the fact that firms increasingly use locations as competitive tools, exploiting their global mobility in order to optimise production and distribution costs. Location territories, on the other hand, are not just the passive objects of location decisions by firms, but communities made up of economic subjects which act in their own interest by trying to keep or attract firms. Local workers, subcontracting firms, suppliers of intermediate inputs, services and factors, are all agents which can achieve their goal, not just by competing on prices and wages with other communities (sites), but also by upgrading the quality of their service through direct or indirect tools that involve the community and the local public administration. Locations are in a sense bought and sold on a global market, where demand and supply confront each other.

As a consequence of the above, local territories, cities and milieux compete (and, as we shall see further on, cooperate) with each other, building their own 'competitive' or 'absolute' advantage. In fact, beyond the specific competitive advantages strategically *created* by single firms, an increasing and crucial importance is played by territorial synergies, by cooperation capabilities enhanced by an imaginative and proactive public administration, by externalities provided by local and national governments, and by the specificities historically built by a territorial culture.[5] As is clear, they are all artificial or created advantages, open to the proactive, voluntary action of local communities and their governments.

But if individual firms and individual actors undertake collective activities, facilitated by (and creators of) trust and social capital; and if significant cognitive synergies, readily apparent in the local milieu, result from their various inter-actions; and finally if these actions and these processes draw additional vitality from cooperation with local public administrations, especially in the creation of attractive territorial projects and innovative schemes; then it appears justifiable to go beyond methodological individualism – which regards only single firms and individuals as economic actors – and to argue in favour of the logical validity of a 'collective' concept such as that of *territory*, and to affirm that territories and cities compete among themselves, using the creation of collective strategies as their instrument.

THE STRUCTURE OF THE URBAN SYSTEM: FROM CITY HIERARCHY TO CITY NETWORKS

The need for a new paradigm: the three logics of spatial behaviour

According to textbooks of theoretical geography and urban economics, the ana-lytical model which still best describes the structure of the city system in strictly economic and locational terms is Christaller's and Lösch's central-place model developed in the 1930s and 1940s. After the basic refinements introduced by Isard, Beckmann and McPherson, a huge literature has grown along the same logical foundations and simplifying assumptions with the works of Parr, Beguin, Mulligan and others. The model still remains the most elegant, abstract but con-sistent representation of the hierarchy of urban centres.

Nevertheless, real city systems in advanced countries have significantly de-parted from the abstract Christallerian pattern of a nested hierarchy of centres and markets. The reduction in transport costs and the demand for 'variety' by the consumer have broken the theoretical hypothesis of separated, gravity-type, non-overlapping market areas. 'Location economies' as described by Hoover, and synergy elements operating through horizontal and vertical linkages among firms, have led to the emergence of specialised centres, in contrast to the typical de-specialisation pattern deriving from the theoretical model. High-order functions sometimes locate in small (but specialised) centres where the model's expecta-tions are only for lower-order functions.

This evidence is not at all new, and the deficiencies of the model are often highlighted, but to change the underlying assumptions would mean changing the model itself, and no other set of clearly defined hypotheses has ever replaced the original ones.

Yet, other evidence is at variance with the logic of the model. Urban policies are increasingly addressed towards economic goals: enhancing the efficiency of the local production fabric, attracting new sectors and functions, expanding the markets of local firms through better external transport and communication link-ages. According to the logic of the model, these kinds of goals lack any economic rationale: the location of sectors and the roles of single centres are defined on the sole basis of city size.

Our hypothesis, or theoretical conjecture, is that a new form of inter-urban interaction and consequently a new organisational structure of the entire urban system may be conceived, based on the new paradigm of firm behaviour illustrated above: something that we can call *city networks* (Camagni 1993a).

From a theoretical and abstract point of view, it is possible to identify three logics of spatial behaviour of the firm: we may call them the territorial, competitive and networking logics (Table 7.l).

According to the first logic, the territorial, a firm sells to (and buys from) the geographical space it gravitationally controls. Space is therefore organised into the well-known Löschian honeycomb of market areas, where the friction of space, embodied in transport costs, at the same time differentiates the products of competing firms and represents the strongest entry barrier into the market. The crucial function of the firm is production and its strategy consists in the control of the market area defined around its geographical location.

According to the second logic, the competitive, the market of a firm is not restricted to the local territory, as transport costs do not play a relevant role; the firm may sell anywhere, trying to control the widest *share* of the global market. Competitiveness, differently achieved and interpreted by the different firms, becomes the crucial element in the economic arena, and marketing the crucial function of the firm; the market of each production unit is limited by both its relative economic strength and by consumers' demand for 'variety'. 'Two-way' trade, or the geographical interchange of the same products in two directions, becomes the rule with, for example, people from Turin no longer being obliged to buy only Fiat cars.

According to the third logic, the network, innovation becomes the crucial function of the firm and the control of innovation assets and their time trajectories its main goal. The firm, wherever located, may overcome crucial know-how weaknesses in its internal structure and surrounding 'milieu' by linking up with other firms and by establishing trans-territorial cooperation agreements.

How is it possible to pass from the locational logic of the single firm to the general spatial allocation of activities and functions?

It is well known that, in what we call the territorial logic, agglomeration economies may explain the coexistence of lower-order functions in centres where higher-order functions are already located, and that gravity-type considerations may attract different firms towards the centre of their market areas, where demand density is higher.

In the competitive logic, on the other hand, agglomeration may derive from supply rather than demand considerations: the agglomeration of firms belonging to the same sectors ('district economies') or the same industrial complex (control of component suppliers, *'filières'* of local specialisation) allows higher levels of static and dynamic efficiency to be achieved, giving rise to specialised industrial areas and 'innovative milieux' (Aydalot 1986), made up of vertically or horizontally integrated firms (the long-standing concept of 'localisation economies').

Table 7.1 The three logics of spatial organization

Organizational logics	Territorial	Competitive	Network
Levels & aspects			
Firm:			
Nature	Local market firm	Export firm	Network firm
Crucial function	Production	Marketing	Innovation
Strategy	Control of market areas	Control of market shares	Control of innovation assets and their trajectories
Internal structure	Single unit	Specialised functional units	Functionally integrated units
Entry barriers	Spatial friction	Competitiveness	Continuing innovation
City system:			
Principles	Domination	Competitiveness	Cooperation
Structure	Nested Christallerian hierarchy	Specialisation	City networks
Sectors	Agriculture, government, traditional tertiary activities	Industry: industrial districts and filière of specialisation	Advanced tertiary activities
Efficiency	Scale economies	Vertical/horizontal integration	Network externalities
Policy strategy	None: size determines function	Traditionally: none, as export base determines growth. Nowadays: strengthening of competitive advantage of each centre	Intercity cooperation; intercity transport and communication network provision
Intercity cooperation goals	None (except military or diplomatic goals)	Intercity division of labour	Economic, technological and infrastructure collaboration
Networks of cities	Hierarchical, vertical networks	Complementarity networks	Synergy networks, innovation networks
Single city:			
Nature	Traditional city	Fordist city	Information city
Form	Relative internal homogeneity	Monofunctional zoning	Multifunctional zoning, polycentric city
Policy goals	Power and image	Internal efficiency (clockwork city)	External effectiveness and attractiveness
Symbols	Palace, cathedral, market	Chimney, skyscraper	Airport, trade fair

The third logic is more complicated. In spatial terms it involves the presence of:

- nodes of localised and specialised know-how (in 'poles', 'districts', 'parks', 'valleys', 'corridors' etc.) interlinked through cooperation agreements and financial/technological/marketing alliances; or
- multi-functional nodes interconnecting different economic and spatial networks. In this respect the old concept of 'urbanisation economies' is revitalised here in terms of interaction and synergy among network functions: the city gains a role as a node of interchange and interconnection among a set of worldwide networks of physical and information interactions.

It is widely known that scale economies and generic agglomeration economies are the main efficiency elements that shape the spatial structure of location centres under the first logic. On the other hand, economies of vertical and horizontal integration are the main efficiency elements in the second logic, and 'network externalities' in the third one. In this last respect, the network operates as a 'club good', delivering advantages only to the members of the club, an intermediate structure between private and public goods (Capello 1994).

The three logics of spatial organisation presented here are of course to be considered as theoretical archetypes, and not directly as historical behavioural patterns. In some respects, they have always coexisted, as they apply specifically to different sectoral specificities (to the primary, secondary and tertiary or information sectors respectively). Nevertheless, as these sectors or functions have prevailed in different and successive periods in recent history, the three logics may be assumed, albeit very cautiously, as the leading paradigms of different 'accumulation regimes'.

The city network paradigm

The territorial logic is the basic theoretical underpinning of the Christallerian hierarchy of centres, and nowadays applies well, even if in abstract and simplified terms, to the spatial behaviour of agricultural production and markets (except for 'industrialised' agriculture and 'specialised' agriculture producing diversified products, like special wines etc.), public administration and government functions, private and public service activities – in particular, 'traditional' ones (retail and wholesale trade, health, education) but also modern ones (private consultants, banking and insurance, advertising) and in general activities where the customer bears the transport cost.

The model presents many drawbacks which significantly limit its empirical relevance in modern societies:

(i) Since Christallerian implementations of the central place model overemphasise the role of transport costs, such models cannot be properly used to understand industrial location and markets.

(ii) It neglects input–output relationships, and in particular *horizontal linkages* among specialised firms and, in spatial terms, horizontal linkages among specialised centres of *similar size,* performing different but complementary functions.

(iii) It neglects 'network externalities', or the 'synergetic surplus' that may come to the partners (firms or cities) of a cooperation network.

Under these circumstances, our hypothesis is that a new paradigm of spatial organisation should be considered, the *network paradigm*, which links with the new logics of spatial behaviour we have labelled as the 'competitive' and the cooperative, 'network' logic.

As far as the competitive logic is concerned, it lies at the basis of the well-known phenomenon of industrial districts, specialised by sectors or by *filière*, and, as a result, a host of territorial relationships among centres based on privileged complementarity relations in both production and marketing.

The third logic, the 'network' logic, in turn determines a set of privileged synergetic relationships between centres that cooperate or interact in the same fields or functions, through information, communication or transport networks. In parallel to a previous statement concerning network relationships between firms, a city network may be considered as a 'club good', providing externalities to the partners which cooperate on the basis of horizontal linkages and perform the same functions. Also in this case, networks might be seen as a way of generating (urban) scale economies in a cooperative way, without implying a growth of the single centres, and of distributing the consequent advantage among the partners.

Therefore, in the organisation of the city system two kinds of city networks appear:

A. *Complementarity networks,* made up of specialised and complementary centres, interlinked through a set of input–output and market relationships. Inter-urban division of labour at the same time ensures that there is a sufficiently large market area for each centre and that scale and agglomeration economies are achieved. Good examples of these networks are provided by the specialised cities in Holland's Randstad or in the Veneto region in Italy.

B. *Synergy networks,* made up of similar, cooperating centres. In this case the necessary economies of scale are provided by the network itself, which integrates the market of each single centre. Examples of these networks are the financial cities, whose markets are virtually integrated through advanced telecommunication infrastructures, or tourist cities connected through cultural or historical 'itineraries'.

These considerations may be synthesised in the following definition: city networks are systems of relationships and flows, of a mainly horizontal and non-hierarchical nature, among complementary or similar centres, providing externalities or economies respectively of specialisation/complementarity/spatial division of labour and of synergy/cooperation/innovation.

Immaterial relationships and physical flows of people and goods are highly complementary, as is widely known. In similar ways, city networks rely on both physical integration – through transport and communication networks – and virtual, economic integration – information exchange, financial transactions, cooperation in multiple fields, including provision of high-order public services in culture and education.

Thanks to the existence of complementarity networks among centres, high-order functions may be supplied by medium-sized centres, provided that they remain specialised in a few sectors/*filières*, and that the entire regional/national market is assured to each of them by a spontaneous or agreed division of labour among centres. Smaller size often means higher quality of life, but presence of high-order functions may imply also higher incomes. By the same token, different small cities organised in tourist itineraries (typical 'synergy networks') may reach sufficient visibility and attractiveness for a sustained economic development.

On the other hand, an apparently opposite tendency could involve 'global city-regions' (Scott 2001), which are typical cases of 'synergy networks'. The advantage and the profitability coming from their mutual interconnection, and therefore from operating on a worldwide scale, may be so huge as to generate a push towards excessive physical size, well beyond the size justified by net economies of scale or implied by any local rank-size rule (Camagni 2001).

What is presented here is mainly a deductive 'conjecture', in search of a corroboration of the underlying theory through proper empirical validation. Many aspects still require further, detailed examination, such as the economic effectiveness and the laws of motion of the new organisational logic and the way in which the new hypothesised network linkages may be observed and measured. The main difficulty in this field is that the nature of the problem requires 'flow indicators' between centres, while mainly 'stock indicators' exist.

DO CITY NETWORKS REALLY EXIST?

Two econometric experiments

Some years ago, an empirical experiment was run in order to detect and identify possible 'city networks' in the real world of spatial interactions, starting from the definition given above (Camagni *et al.* 1994). Flow data were utilised by using telephone calls between 'telephone districts' in northern Italy in 1990 (before the advent of cellular phones).

Reliable data were available and they referred to a totally ubiquitous network, eliminating potentially misleading supply effects linked to the territorial architecture of physical networks – as for example transportation data. A 'network' relationship between couples of centres was hypothesised when actual communication flows significantly exceeded the interaction expected on the basis of a traditional doubly constrained entropy model. By using such a model it is in fact possible to capture all relevant but generic interaction that occurs within a regional or national urban system as a consequence of size and distance, while we are interested in capturing the extra interaction occurring as a consequence of special and selective economic relationships between the centres.

A rectangular matrix (36 × 119) was used for flows between the Lombardy telephone districts and the districts of the entire northern Italian macro-region. The results were encouraging and in many respects unexpected. In particular they showed that city networks, as we define them, do in fact exist, and that:

- they are not ubiquitous but very selective in space;
- they do not replace more traditional, hierarchical forms of spatial organisation; rather, the two organisational forms of the city system appear to be complementary;
- a super-gravitational effect of the capital city, Milan, was apparent on the major northern Italian cities (Figure 7.2: see the white empty bars, indicating flows underestimated by the model, or a wider real interaction), indicating a strong hierarchical structure;
- a similar super-gravitational effect of Milan is exerted on the major cities in Lombardy;
- in the context of the Lombardy region, city networks show up as statistically significant in three spatial conditions, characterised by a high synergetic potential: a) within the metropolitan area of Milan, linking its major sub-centres; b) within some industrial districts featuring distinct sectoral specialisation and economic 'vocation' – Como (silk production), Lecco (mechanical engineering), Seregno (furniture and furniture design), Busto–

Figure 7.2 The super-gravitational effect of Milan in northern Italy. (Note: super-gravitation is expressed by white empty bars, indicating an under-estimate of real interaction by the model. The sea boundary in the Liguria region is not drawn. Source: Camagni *et al*. 1994)

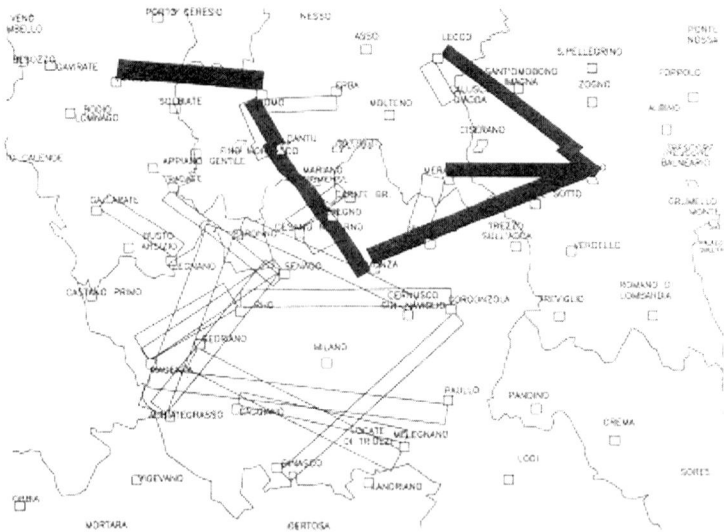

Figure 7.3 City networks in the Milan metropolitan area. (Note: analysis is carried out at the lower territorial level, the 'telephone sector'; empty bars indicate an under-estimate of real interaction by the model, black bars an over-estimate)

Gallarate–Legnano (textiles and light mechanical engineering), Vimercate (high-tech industries) (Figure 7.3); and c) in the eastern part of the region, linking major provincial capital cities (Bergamo, Brescia, Mantova, Cremona and Verona). These last centres are in a sense structuring a strong alternative to the dependency to Milan via a city network, with a prominent role for the largest one, namely Brescia;

• however, city networks did not show up over longer distances, for example along the main Po valley central axis, where they were expected, except for some bilateral links (Turin–Novara, Bergamo–Brescia) and the already mentioned gravitation towards Milan.

A second empirical analysis refers directly to another relevant question: does network behaviour really generate advantages for partner cities? Does a 'network surplus' really exist?

From what was said in the theoretical framework, the main economic rationale for network behaviour is no longer to minimise transport costs and maximise control over non-overlapping market areas. It is rather to exploit scale economies in complementary relationships and synergic effects (external economies) in co-operative activities, achieved through participation in a network; i.e. the *network externality* (or network surplus) element is the main economic advantage explaining network behaviour.

An empirical inquiry trying to measure these advantages was run recently, on the 'Healthy City' network of the World Health Organization (Capello 2000). This network links together cities of different sizes and countries in a joint programme dealing with quality of life and health in urban areas; its existence provides a concrete example of an institutional network aiming at: a) widening information exchange among public managers on policy strategies in specific health-related areas; b) providing an opportunity for cities to launch and run joint policies in specific health-related areas (WHO 1995).

Through a direct survey, three main indicators were constructed: a connectivity indicator (c), an indicator of intensity of use of the network (i) and an indicator of policy performance thanks to the use of the network (p).[6] The following relation among these indicators was assumed:

$$p = f(c, i)$$

The econometric analyses gave the following results, in line with expectations: a) the higher a city's degree of connectivity, the higher its performance in terms of policies implemented thanks to the existence of the network (and vice versa); and b) if a city plays an active role in the network, the resulting advantages increase.

Through a cluster analysis, some distinct classes of behaviour were found, achieving different levels of success:

- A group of cities showed *opportunistic, short-term but effective behaviour, addressed to acquiring legitimacy*. In this cluster the main advantage achieved through the network was political legitimacy for local policies; cities were aware of the advantages achieved, but seemed to play a 'hit and run' strategy; they did not manifest any willingness to develop interaction further and did not exploit strategic advantages from the network (e.g. innovation).
- A second group of cities showed an *explorative behaviour, addressed towards commitment and learning*. These cities participate in the network with serious commitment, make significant investment in the experience and have developed a serious learning process. However, they do not show a particular advantage from the network.
- A third group of cities was characterised by pure *efficiency behaviour, addressed towards information gathering and catching up*. These cities entered the network with simple information-gathering goals, and achieved typical advantages like increased economic efficiency, increased information and economies of scale. Mainly eastern European and non-European cities belonged to this cluster.
- A fourth group of cities showed *strategic behaviour, addressed towards achieving synergies and innovation*. In this cluster, cities gained the greatest and most strategic advantage from the network – the acquisition of know-how – and were the ones showing the greatest number of local success stories, derived from participation in the network.

In summary, the statistical analysis provides evidence that a potential network surplus may be differently exploited by each partner city, according to the different behavioural patterns and strategies that are selected. The achievement of the maximal advantages from the network requires commitment to participation and an open, synergistic attitude.

City networks as policy tools

France was the first country to consider *réseaux de villes* (city networks) explicitly as tools in spatial policies. Since 1989, 27 city networks have established following a bottom-up procedure (Figure 7.4), and after 1991 they were considered as partners of national or regional governments. Through the so called *Contrats de Plan* with the national government, they received financial support for joint territorial projects.

The general goal was, by and large, territorial competitiveness on the European scale. More specific goals, shared by many networks, are:

* scale economies and synergy in tourism and agriculture development;
* joint management of basic services, cooperation in planning;
* development of common high-order projects: university facilities, infrastructure, entertainment parks and facilities;
* increased visibility through territorial marketing;
* lobbying (e.g. on large national infrastructure projects).

Different assessment reports were run by Datar, the French national agency for spatial development planning, in 1998, and by the Club National des Réseaux de Villes (CNRdV, 2001). Both underlined the relevant potential of the city network tool, especially because it allows reaching economies of scale in project design and sufficient territorial size for ambitious schemes, unthinkable if single

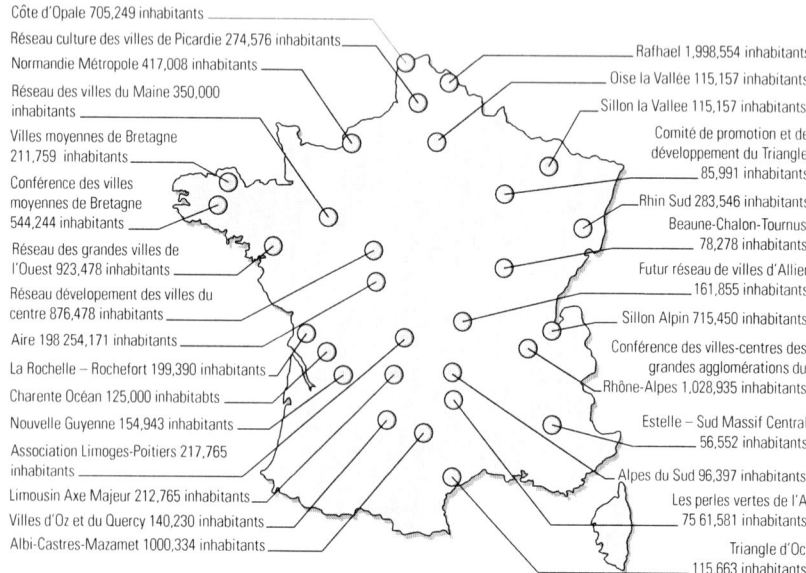

Côte d'Opale 705,249 inhabitants

Réseau culture des villes de Picardie 274,576 inhabitants

Normandie Métropole 417,008 inhabitants

Réseau des villes du Maine 350,000 inhabitants

Villes moyennes de Bretagne 211,759 inhabitants

Conférence des villes moyennes de Bretagne 544,244 inhabitants

Réseau des grandes villes de l'Ouest 923,478 inhabitants

Réseau dévelopement des villes du centre 876,478 inhabitants

Aire 198 254,171 inhabitants

La Rochelle – Rochefort 199,390 inhabitants

Charente Océan 125,000 inhabitabts

Nouvelle Guyenne 154,943 inhabitants

Association Limoges-Poitiers 217,765 inhabitants

Limousin Axe Majeur 212,765 inhabitants

Villes d'Oz et du Quercy 140,230 inhabitants

Albi-Castres-Mazamet 1000,334 inhabitants

Rafhael 1,998,554 inhabitants

Oise la Vallée 115,157 inhabitants

Sillon la Vallee 115,157 inhabitants

Comité de promotion et de développement du Triangle 85,991 inhabitants

Rhin Sud 283,546 inhabitants

Beaune-Chalon-Tournus 78,278 inhabitants

Futur réseau de villes d'Allier 161,855 inhabitants

Sillon Alpin 715,450 inhabitants

Conférence des villes-centres des grandes agglomérations du Rhône-Alpes 1,028,935 inhabitants

Estelle – Sud Massif Central 56,552 inhabitants

Alpes du Sud 96,397 inhabitants

Les perles vertes de l'A 75 61,581 inhabitants

Triangle d'Oc 115,663 inhabitants

Figure 7.4 City networks in France

centres engage alone. Once again, the full exploitation of this potential is linked to cooperation commitment, explicit support by the local policy-makers, and the establishment of common, efficient management structures.

Perhaps Germany is now the country with the most convinced (and convincing) use of the network paradigm (*Städtenetze*) for policy purposes (Hildebrand 2003). Since 1989, 26 city networks have been established, some of them with a private character (Figure 7.5). They are supported as policy networks by the Federal Government and the Länder, but usually with no specific financial support for the specific projects they propose, which follow the general evaluation and approval procedures.

The general goal is competitiveness and strengthening of the German decentralised urban system. The leitmotiv is for sure sustainable territorial development. The specific objectives of the single networks are similar to the French ones, but a major attention is devoted to common physical planning strategies and local transport infrastructure development.

In Italy, in spite of the early development of the concept in theoretical terms, and in spite of the general rhetoric about the usefulness of the tool, city networks were never established explicitly. One of the rare, implicit but interesting, exam-

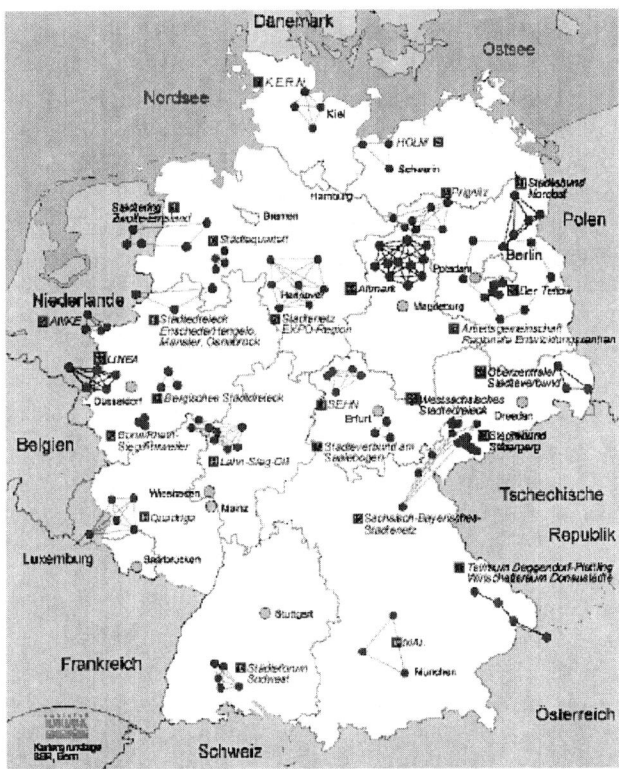

Figure 7.5 City networks in Germany

ples resides in the support of the national government for the establishment of historic and tourist itineraries linking sites of low visibility located along the Roman routes used by the medieval pilgrims.[7] The national programme for the 2000 Jubilee supported the restoration of these sites (monasteries, churches, paths), their receptive capacity, and their networking through investments in information, organisation, transport services, tourist logistics and management, historical research and publishing.

CONCLUSIONS

In this paper, it is shown how the logics that shape the city system are more complicated than the simple 'territorial' and hierarchical logic of the traditional central-place model. A firm's control of the market of outputs, inputs and innovative assets is attained not only through managing a gravitational area, but also and increasingly through cooperative, trans-territorial network relationships.

The new behavioural logic of the firm parallels and partly determines the new organisational logic of the city system, where phenomena of specialisation and networking also appear. Processes of urban complementary specialisation in a regional context allow the single centres to take advantage of the entire regional market in the specialisation sector, reaching relevant economies of scale and hosting functions that, in the traditional hierarchical model of urban system, were reserved to higher-order centres. The spontaneous or partly planned division of labour among the Dutch cities of the Randstad or the one among the cities of the polycentric Veneto region in Italy are good examples. By the same token, centres of similar rank and size, specialised in similar sectors, may cooperate in order to reach a superior critical mass, benefiting from network externalities. Examples range from the case of world financial cities, operating on a unique and integrated market through communication networks and institutional arrangements, to second-order tourist cities, integrated in 'itineraries' through information and organisation networks, reaching in this way a sufficient attractiveness in order to compete with the champion cities.

In this sense, we distinguish between complementarity and synergy networks. Economies of specialisation and division of labour on the one hand and scale economies reached through the network on the other represent the economic rationale of the new spatial paradigm in the two respective cases. Similar goals may lead cities to cooperate in order to establish innovative territorial schemes and projects, like a common airport serving the entire network.

The new concept, and the new logic, of city networks do not replace but complement the traditional hierarchical logic of the city system, which remains the fundamental spatial logic of many sectors (such as consumer services, traditional agriculture and public administration) and is still visible in the territory as the historical organisational form of the spatial structure from times when these sectors were the leading ones. From a terminological point of view, our proposal is to use the term 'city network' only for the new theoretical and empirical realm of selective horizontal linkages among centres of similar size and rank, leaving the usual terms of hierarchy and spatial interaction respectively for vertical, hierarchical linkages and for generic, gravity-type, relationships.

The paradigm of city networks has been enriched by empirical evidence able to measure two important aspects of the city network paradigm. The first concerns the possibility of detecting and empirically identifying 'city networks' themselves in the real world of spatial interactions. In the Lombardy region, not only Christallerian gravity-type relationships were envisaged but also cooperative relationships among cities of similar rank and size. The second important aspect regards the possibility of empirically determining whether (and in which conditions) the network logic, which is cooperative in nature, brings specific advantages to the partners (firms, institutions or cities). A 'network surplus' was in fact evident and partly measured in the case of an international city network, the network of Healthy Cities.

The new pattern of territorial relationships opens up new opportunities for the planning activity, as a city is confronted with expanded alternatives regarding its development path. In particular, through cooperation inside a city network, each centre could be able to develop high order functions – and benefit from the consequent income levels and 'surplus' – without increasing its own size, as would be implied by the constraints of scale economies and indicated by the central-place model. There is scope therefore for intentional city strategies, both at the level of the single centre and at the level of the entire city system.

NOTES

1 For a review of the concept of 'network surplus' see Capello (1994: chapter 2).
2 We can also mention here the function of promoting informal guarantees for the *honouring of incomplete contracts*, which the milieu can provide thanks to its networks of interpersonal relations. In modern economies – where an increasing importance is attached to product differentiation and ill-defined qualitative factors, and uncertainty regarding the properties of new, not yet existing, products is high – incomplete contracts gain importance. As a result, conditions of trust, respectability and reputation on the one hand and sanctions and exclusion for opportunistic behaviour on the other, conditions which local communities of limited size can provide, gain significance and relevance in the economic sphere.
3 In the international literature, (urban) strategic planning is defined as the collective construction of a shared vision of the future for a given territorial area, through processes of participation, discussion and listening. It is an agreement between administrators, actors, citizens and various partners to implement this vision through a strategy and a series of projects, with varying interconnections with each other, which are justified, evaluated and shared. Strategic planning can finally be defined as the coordination of responsibilities assumed by the different actors for the implementation of such projects (Gibelli 1996).
4 This is contrary to conventional wisdom. As Armstrong and Taylor (2000: 123) affirm: 'That trade is based on comparative advantage and not absolute advantage is universally accepted and rarely tested.' In our opinion, this statement, when referred to regions, should not be accepted.
5 As Porter puts it: 'There is growing recognition that company success also has much

to do with things that are outside the company', such as 'supplier relationships and the benefits of partnering' (Porter 2001: 140).

6 The connectivity indicator was a weighted sum of the number of business meetings and the number of Multi-City Action Plans cities were involved in. The intensity of use was formulated as the ratio between the number of cities by which each city was inspired to launch a policy and the connectivity indicator. The indicator of urban performance was built as the ratio between the number of successful policies launched by each city thanks to cooperation within the network and the total number of local successful policies developed by each city.

7 Via Appia (from Rome to Lecce in the south-east), Via Appia Traianea (from Benevento to Bari in the south), Via Flaminia (from Rome to Rimini in the north-east), Via Romea (from Rimini to Venice), Via Francigena (from Rome to the north-west and to Santiago de Campostela through the Piccolo S. Bernardo and Moncenisio Pass).

REFERENCES

Armstrong, H. and Taylor J. (2000) *Regional Economics and Policy*, Oxford: Blackwell.

Arrighetti, A. (2003) *Economia dell'Azione Collettiva*, Parma: Facoltà di Economia, Università di Parma.

Aydalot, Ph. (ed.) (1986) *Milieux innovateurs en Europe*, Paris: GREMI.

Bowles, S. and Gintis, H. (2002) 'Social capital and community governance', *The Economic Journal*, 112: F419–F436.

Camagni, R. (ed.) (1991a) *Innovation Networks: Spatial Perspectives*, London: Belhaven-Pinter.

Camagni, R. (1991b) 'Local milieu, uncertainty and innovation networks: towards a dynamic theory of economic space', in R. Camagni (ed.) *Innovation Networks: Spatial Perspectives*, London: Belhaven-Pinter, pp. 121–44.

Camagni, R. (1993a) 'From city hierarchy to city network: reflections about an emerging paradigm', in T.R. Lakshmanan and P. Nijkamp (eds) *Structure and Change in the Space Economy: Festschrift in Honour of Martin Beckmann*, Berlin: Springer Verlag.

Camagni, R. (1993b) 'Interfirm industrial networks: the costs and benefits of cooperative behaviour', *Journal of Industry Studies*, 1: 1–15.

Camagni, R. (2001) 'The economic role and spatial contradictions of global city-regions: the functional, cognitive and evolutionary context', in A.J. Scott (ed.) *Global City-Regions: Trends, Theory, Policies*, Oxford: Oxford University Press, pp. 96–118.

Camagni, R. (2002) 'On the concept of territorial competitiveness: sound or misleading?', *Urban Studies*, 13: 2395–2412.

Camagni, R. and Capello, R. (2002) 'Milieux innovateurs and collective learning: from concepts to measurement', in Z.J. Acs, H.L.F. de Groot and P. Nijkamp (eds) *The Emergence of the Knowledge Economy*, Berlin: Springer, pp. 159–46.

Camagni, R., Diappi, L. and Stabilini, S. (1994), 'City networks in the Lombardy Region: an analysis in terms of communication flows', *Flux*, 15: 37–50.

Capello, R. (1994) *Spatial Economic Analysis of Telecommunications Network Externalities*, Aldershot: Ashgate.

Capello, R. (2000) 'The city network paradigm: measuring urban network externalities', *Urban Studies*, 37(11): 1925–45.

Castells, M. (1996) *The Rise of the Network Society*, Oxford: Blackwell.

Club National des Réseaux de Villes (2001), 'Réseaux de villes, réseaux de vie: un maillage porteur d'avenir', *La Gazette des Communes, des Départements, des Régions*, November: 228-74.

Coase, R. (1988) *The Firm, the Market, and the Law*, Chicago: University of Chicago Press.

Coleman, J.S. (1990) *Foundations of Social Theory*, Cambridge, MA: Harvard University Press.

Colletis, G. and Pecqueur, B. (1995) 'Politiques technologiques locales et création des ressources spécifiques', in A. Rallet and A. Torre A. (eds) *Economie industrielle et économie spatiale*, Paris: Economica, pp. 445–63.

Cooke, P. and Morgan, K. (1998) *The Associational Economy: Firms, Regions and Innovation*, Oxford: Oxford University Press.

Dematteis, G. (1985) 'Verso strutture urbane reticolari', in G. Bianchi and I. Magnani (eds) *Sviluppo multiregionale: teorie, metodi, problemi*, Milan: Franco Angeli, pp. 121–132.

Dematteis G. (1990) 'Modelli urbani a rete: considerazioni preliminari', in F. Curti and L. Diappi (eds) *Gerarchie e reti di città: tendenze e politiche*, Milan: Franco Angeli, pp. 27–48.

EU (1999) *European Spatial Development Perspective*, Brussels: Committee for Spatial Development.

Gibelli, M.C. (1996) 'Tre famiglie di piani strategici' in F. Curti and M.C. Gibelli (eds) *Pianificazione strategica e gestione urbana*, Florence: Alinea.

Glaeser, E., Laibson, D. and Sacerdote, B. (2002) 'An Economic Approach to Social Capital', *The Economic Journal*, 112: F437–F458.

Hahn, F. (1991) 'The next hundred years', *The Economics Journal*, 101: 47–50.

Hildebrand, A. (2003) 'Las redes de cooperación entre ciudades', paper presented to the Workshop CUIMP on *Estrategias Territoriales*, Barcelona, 15–17 December 2003.

Maillat, D., Quévit, M. and Senn, L. (eds) (1993) *Réseaux d'innovation et Milieux Innovateurs : un Pari pour le Développement Régional*, Neuchatel: Edes.

Malecki, E.J. (1997) 'Entrepreneurs, networks and economic development', *Advances in Entrepreneurship, Firm Emergence & Growth*, 3: 57–118.

Malecki, E.J. and Oinas, P. (eds) (1999) *Making Connections: Technological Learning and Regional Economic Change*, Aldershot: Ashgate.

Nijkamp, P. (2003) 'Entrepreneurship in a modern network economy', *Regional Studies*, 37(4): 395–405.

Porter, M. (2001) 'Regions and the New Economics of Competition', in A. Scott. (ed.) *Global City-Regions: Trends, Theory, Policies*, Oxford: Oxford University Press, pp. 139–57.

Putnam, R.D. (1993) *Making Democracy Work*, Princeton, NJ: Princeton University Press.

Richardson, G.B. (1960) *Information and investment*, Oxford: Oxford University Press.

Scott, A. (ed.) (2001) *Global City-Regions: Trends, Theory, Policies*, Oxford: Oxford University Press.

WHO (1995) *Twenty Steps for Developing a Healthy Cities Project*, Copenhagen: World Health Organization, Regional Office for Europe.

Williamson, O. (1985) *The Economic Institutions of Capitalism*, New York: The Free Press.

Williamson, O. (2002) 'The lens of contract: private ordering', *American Economic Review, Papers and Proceedings*, 92(2): 438–53.

8 Firm linkages, innovation and the evolution of urban systems

Céline Rozenblat and Denise Pumain

INTRODUCTION

Because a city cannot be conceived of as an isolated system but is always part of a *system of cities* (Berry 1964), interaction between cities is an essential component of the dynamics of urban systems. Each town or city is a persistent and relatively autonomous entity whose evolution is influenced or limited by the towns and cities in the same interaction networks (Pred 1977). It has been demonstrated that competition between towns and cities for resources and growth is the main driving force in the dynamics of systems of cities, which also explains the pervasiveness of their structure (Pumain 2000). Most systems of cities link neighbouring towns and cities located on the same regional or national territory. But cities also have relationships with more distant competitors, especially in specialised networks, which is more and more the case in the current context of an increasingly global economy. Therefore, the hierarchical organisation of cities in systems of cities is no longer an inclusive one (if ever it was, as a too strict acceptation of the Christallerian model of embedded hierarchical levels would suggest), since the multiple links connecting towns and cities can go anywhere. However, many of the relationships between urban actors are recurrent or use the same communication channels. Recurrent interaction patterns shape the structure of urban systems. This structure is universally characterised by strong hierarchical differentiation (Pumain 2006). Several orders of magnitude separate the importance of towns and cities, in terms of population, gross product or influence. This importance can be defined by the relative position of a given city in interaction networks (centrality, betweenness), as well as by the urban attributes representing the cumulative effects of that position over time. Interaction flows in turn reflect (or are induced by) this structure of the urban system, since they are generated by the attributes of one city rather than another.

We therefore share P.J. Taylor's (2001) view that there is a need for more studies about networks of cities, particularly studies that draw on databases of inter-urban flows. This paper presents the results of a survey illustrating the gradual emergence of an urban system on the scale of Europe, through the interactions created by firm linkages. A subsidiary owned by a multinational firm in a foreign country is interpreted as a directed interaction between the city where the head-quarters are located and the city where the subsidiary is owned (after the foreign subsidiary is set up or acquired). We examine the position of cities in these owner-ship networks, which provides insights about the main factors in a city's capacity to participate in the formation of a system of cities on the European scale.

CITIES AND THEIR NETWORKS IN EUROPE

The integration of European society and territory is achieved mainly through the development of new linkages between cities, especially the largest. But, even if the familiar process of hierarchical diffusion of innovation in urban systems (Pred 1977) is at work here, these linkages cannot be considered in principle as being limited to only a few cities or capitals. National urban systems as a whole are or will be involved in the expansion of interactions between territories at the inter-national level. However, after a wave of comparative studies of large sets of Eu-ropean towns and cities in the 1980s and 1990s (Brunet 1989; Conti and Spriano 1990; Pumain and Saint-Julien 1996; Cattan *et al.* 1999), there has been a shift in research towards monographs addressing a small number of global cities (Sassen 1991; Hall 1995).

Global cities raised new theoretical questions about the role of cities in the globalisation of the economy (Friedmann 1986). Meanwhile, the changing sta-tus of cities and regions in a context of weakening nation-states was everywhere challenging the forms of their governance (Scott 2001). On account of regional decentralisation and increasingly independent urban management, 'city-regions' began negotiating directly with the actors involved in building global networks. For instance, the deregulation of the airline industry in 1993 gave a stronger role to airlines, airports and their strategic agreements in the organisation of the air transport networks connecting cities (Storper 1997; Graham 1998). Urban devel-opment can no longer be understood (and perhaps never was) without considering the networks and systems to which cities belong. In more and more situations, the economic competitiveness of places is now assessed from the perspective of networks. For example, the European Commission recommends a policy for developing polycentric urban systems to improve the capacity of the European space to be equitable and redistribute key activities (European Communities 2001; ESPON 2003). On another scale, urban managers frequently request assessments of the position and influence of their city in the European urban system, but there is a dearth of comparable urban data.

Relative positions in multinational networks can be defined according to two kinds of complementary information. To describe the power of nodal cities, their importance can be compared using a scale of urban attributes, or by measuring how they are connected by a subset of links or flows in one or several networks. These two methods of measurement are not opposite in significance, since both types of data can be represented either as attributes of the node, or by the intensity of some linkages (there is always a dual representation in networks). Even when the analysis is strictly limited to the relative position of the nodes in terms of their accessibility within a network, this structural analysis can give rise to different rankings of the nodes: a recent study of the worldwide air transport network (Guimera *et al.* 2005) demonstrated that the most connected cities (the nodes with the highest score or number of incident connections) are not necessarily the most central in the network (that is, cities through which most shortest paths go). Apart from physical networks, most studies of the global power of cities, in terms of the importance of their international activities or their actual or estimated linkages, give figures relating to economic activities or multinational firms (Cohen 1981; Rozenblat and Pumain 1993; Taylor and Walker 2001). Another recent study conducted for the public management board of the City of Marseilles, Euroméditerranée, and Datar (the territorial development and regional action division of France's ministry of regional development), compared 180 European cities (urban areas with a population of at least 200,000 in 2000) in the European Union, Norway and Switzerland using a set of 15 synthetic indicators of their global weight or influence (Table 8.1) (Rozenblat and Cicille 2003).

The classification of cities according to their global influence revealed four different principles of spatial ordering and distribution of international functions (Rozenblat and Pumain 2004), satisfactorily validating the theoretical approach previously suggested by Cattan *et al.* (1999) and put forward in ESPON (2003):

1. principle of hierarchical networking;
2. principle of national integration;
3. principle of selective specialisation;
4. principle of regional cross-border integration.

Table 8.1 Fifteen indicators of global influence

Indicators	Source
Population of urban agglomerations 2000	National censuses
Population change (1950–1990)	Geopolis, 1993
Harbour traffic 1999	Journal de la marine marchande, ESPO
Airline passengers 2001	Airports Council International
Airline and railway accessibility 2002	Amadeus Global Travel Distribution
Headquarters of large European firms	Forbes, 2002
Stock exchanges	The Bankers Almanac 2002 (Reed BI)
Tourist overnight stays 2001	National censuses and tourist sites
Fairs and exhibitions 2002–2003	Paris Chamber of Commerce
International conferences 1993–2000	Union des Associations Internationales
Museums 2002	International Council of Museums (ICOM)
Cultural sites and events	Michelin 2001
Students 2001	National or regional institutes
Scientific publications	Institute for Scientific Information (ISI) 2000
Research networks	CORDIS, 2002

Source: Rozenblat and Cicille (2003).

Particularly in the case of network functions, the combination of these principles creates complex effects, exhibiting at the same time high concentrations in a few cities and a wide variety of urban configurations.

For example, the map representing the distribution of multinational firms' headquarters (Figure 8.1) reveals that, besides the dominance of London and Paris, many headquarters are still located in old industrial cities of the Ruhr, and they include new activities such as banking, insurance and new technologies. In the case of banking, concentration, which occurred earlier than in other sectors, increased the size of financial groups while reducing the number of banks. Banking networks are now multinational and use the upper part of the national urban systems for locating their branches (Figure 8.2). The major stock exchanges in Europe are located in only four cities: London, Paris, Frankfurt and Luxembourg. They concentrate 80 per cent of European financial flows (Pagetti 1998). German cities are organised into a polycentric system of eight cities with stock exchanges. Other locations of European stock exchanges linked to Euronext include Lisbon, Porto, Valencia and Bilbao.

Even if the networks of economic activities are largely coordinated through telecommunication systems, physical accessibility still contributes to the development of every international function. As an illustration, we provide an accessibil-

Figure 8.1 Headquarters in European cities

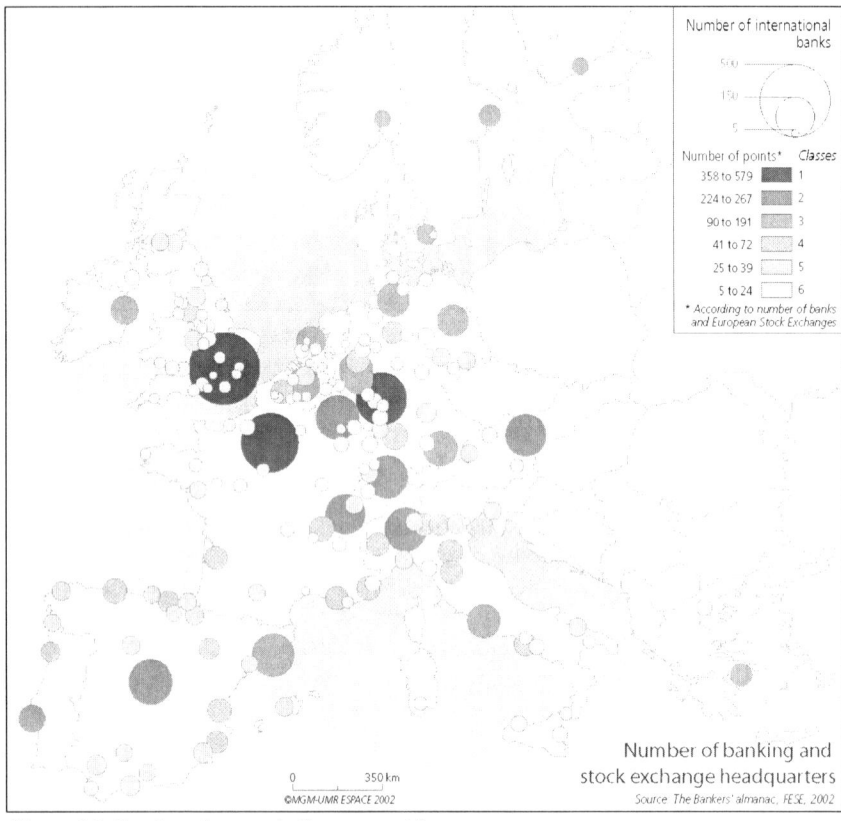

Figure 8.2 Stock exchanges in European cities

ity map, which measures for each of the 180 European cities the number of cities
that can be reached by plane or train in a one-day return trip (Figure 8.3). Despite
the hierarchical ranking produced by the airline connections of large capitals,
many places located in border regions are also highly accessible, thanks to the
high density of neighbouring cities. This is the case in the Rhine Valley in Germany (especially Düsseldorf), in northern Italy with Bologna, and in France with
Lyons. Accessibility is theoretically related to the centrality of cities within all
urban systems, because their potential for trade depends on their position. Hence
good international accessibility is favourable to the development of international
functions. We observed that our indicator is only partly related to a city's size.
Other significant correlations are found with regional urban density and proximity
to international borders. With an equal population size, the number of cities that
are accessible from a European city is fairly highly correlated with the number
of international banks located in that city ($R=0.76$), the number of conferences
held there ($R=0.7$) and, to a lesser extent, the number of headquarters ($R=0.49$),
tourists (0.43) and international fairs (0.41).

According to that study (Rozenblat and Pumain 1993, 2004), there is some

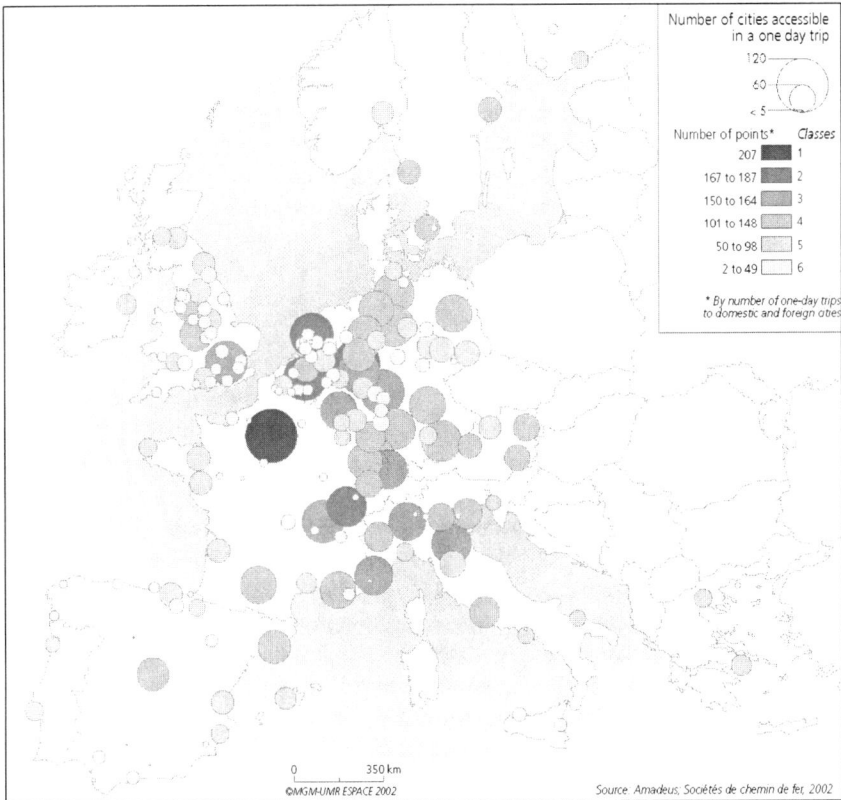

Figure 8.3 Accessibility of European cities

evidence that regional and long-range accessibility, together with urban hierarchy, national divisions and economic specialisation, are the basic structures influencing the integration process of cities in Europe. Their effects are combined differently depending on the nature of each function and the city's location in the European urban system, since the international division of labour creates some specialisation in each city and corresponding specialised inter-urban connections.

AN EXAMPLE OF THE INTER-URBAN CONNECTION PROCESS

We shall examine here how a specific kind of inter-urban connection is created through the location process of multinational firms in European cities. The spatial strategies of firms aim to select locations that supply many kinds of local resources (natural and human resources, knowledge, infrastructure and institutions) to serve their purposes. The multiple locations of a multinational firm reflect complementary choices and may create networks linking different cities. Each city where a firm is located contributes to the competitiveness of that firm's economic network . There is thus an interaction between private global strategies and local attractiveness. The wealth and resources of each territory combine with global economic and political actors' behaviour to form a complex system (Dicken and Malmberg 2001). The associated advantages probably constitute a new kind of external economies, which can be called 'network economies' by analogy with 'agglomeration economies'. In order to understand the interdependencies created between urban developments (Storper 1997) through this process, we have decided to observe directly some effective linkages created by multinational firms between European cities.

Source material: two surveys of foreign firms' networks in Europe

We conducted two surveys (1990 and 1996) of foreign subsidiaries of multinational firms in Europe (Rozenblat 1992; Rozenblat and Pumain 1993; Rozenblat 1997). We collected information on the 300 largest European firms (i.e. headquartered in Europe), about the exact location of all the entities they owned in other European countries, their size (turnover and number of employees), their sector of activity and their function in the firm. The questionnaire, which was sent by post, asked each subsidiary the date it was set up or acquired, its activity and functional role in the group and various size-related criteria for the purposes of comparison. Although most companies simply sent back their annual reports, without answering our questions precisely, we extracted information from the annual reports and requested additional information by letter or telephone where required.

We thus obtained homogeneous and useable information about thousands of subsidiaries, including their addresses, functions, activities and ownership links with sub-subsidiaries. From one-third of the companies at each date (approximately 100 of the largest 300 groups) we collected a sample of approximately 3,000 subsidiaries in 1990 (western Europe) and another sample of more than 4,000 in 1996 (western and central Europe). Based on the links of secondary ownership describing up to five successive levels of branching, we established the architecture of the ownership linkages connecting them.

Firm ownership linkages have rarely been used to describe the structure of city networks, although they are highly significant. The architecture of ownership linkages is a marker of the decision-making and power channels between firms. Financial ownership controls the strategic orientation of the subsidiaries. However, not all firms are organised in the same way. There are differences in the structure of holdings depending on the degree of centralisation of the decision-making process. For example, in 64 per cent of the 81 cases analysed by Francfort *et al.* (1995), the company organisation is highly centralised. In concrete terms, this involves a 'system of control' operated either by means of 'contracts' for subsequent control of results (60 per cent of the cases) or by 'standards, means or objectives' for prior planning of actions (40 per cent) (Mintzberg 1994; Francfort *et al.* 1995). The architecture of ownership linkages thus reflects the hierarchical decision-making process in a majority of groups of companies. According to many specialists, these features are even more important in the case of multinational corporations (Veltz 1998; Zimmermann 1995; Michalet 1997; Mucchielli 1998; Bouinot 2004).

The circulation of decision-making between the many entities of a multinational firm also varies depending on their functions. Headquarters perform fewer and fewer administrative functions to concentrate on strategic functions. These strategic functions, including strategic policy, common principles, oversight and benchmarking tools, are becoming more and more centralised. At the same time, purchasing is increasingly centralised to generate economies of scale. But 'it is rare that production is formally organised on a world level' (Veltz 1998). Production is generally organised into product or geographical divisions, for which the continent appears to be the most relevant scale. Between units on the same continent, coordination is closer and a decision to concentrate on some sites can put them in competition with each other. The marketing function seems to be the link between all the organisational scales, because it is present at all levels. In the case of research and development, the project structure fosters association between researchers, engineers and marketing specialists, without necessarily bringing them spatially closer (Zarifian 1993).

This description of operations within multinational groups demonstrates the decisive role of the legal structure of firms in their concrete organisation. We therefore decided to use the legal structure to describe the economic linkages between cities in the globalisation process (Figure 8.4). Instead of using the GaWC method, which implies that cities are connected to all the other cities where a subsidiary of the same group is located (Figure 8.4b), we use the network of firm ownership in the strict sense to determine which cities are actually connected by ownership linkages (Figure 8.4a). Contrary to the GaWC's research, we do not create more information than we have in our data. The networks linking cities through the location of multinational firms that we show thus represent the minimum but actual inter-firm networks, whereas the GaWC maps show the maximum possible networks.

Examples of firm networks through ownership linkages

In Figure 8.5, the location of the headquarters of three large firms is represented by a triangle. The links between cities are legal ownership relationships (subsidiaries at least 50 per cent owned) either between the headquarters and subsidiaries located in a foreign city (the size of the circle representing the city is proportional to the number of branches) or between a national subsidiary and its sub-subsidiary located in a foreign city. These three examples of firm networks demonstrate that firm ownership linkages between cities in different countries are sometimes very indirect, highlighting the role of intermediate subsidiary companies in the globalisation process. One headquarters can own several subsidiaries, which in turn own several sub-subsidiaries, and so on. A firm's foreign subsidiaries are not necessarily owned or directly controlled by its European headquarters, but by intermediate headquarters. This legal organisation may reflect a decentralised management organisation, based on a regional and/or sector division of labour

a- Empirical observations

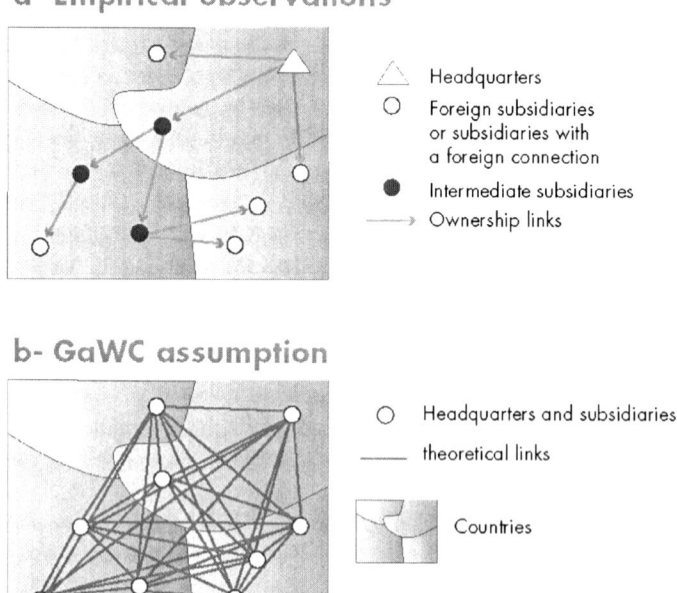

△	Headquarters
○	Foreign subsidiaries or subsidiaries with a foreign connection
●	Intermediate subsidiaries
⟶	Ownership links

b- GaWC assumption

○	Headquarters and subsidiaries
—	theoretical links
	Countries

Rozenblat, UMR ESPACE, 2004

Figure 8.4 Firm linkages

or markets. Several decision-making levels (continental, national, by branch or product) can remain relatively independent from each other (as shown by Veltz 1998; Crozet *et al.* 2004).

This representation of the effective linkages between firms considerably improves on the description that is generally made of the globalisation process using data on foreign direct investment (FDI) (Michalet 1999; Mucchielli 1998). FDI is calculated by national statistics offices from capital flows across international boundaries. These figures do not take into account possible multiple locations of investment within a foreign country. Also, because they do not identify the national origin of headquarters, they do not show the globalisation networks created by firm ownership linkages. FDI data thus produce a fragmented image of the globalisation of companies, failing to show their overall strategies. This image is not without interest, in particular for representing spatial interactions between countries, but is only a partial view of complex multinational economic networks.

Types of intermediate cities as a distinctive characteristic of city networks according to firm ownership linkages

Using 100 firm networks, we calculate the links connecting two cities *i* and *j* according to the number of times that a firm located in city *i* owns a subsidiary in

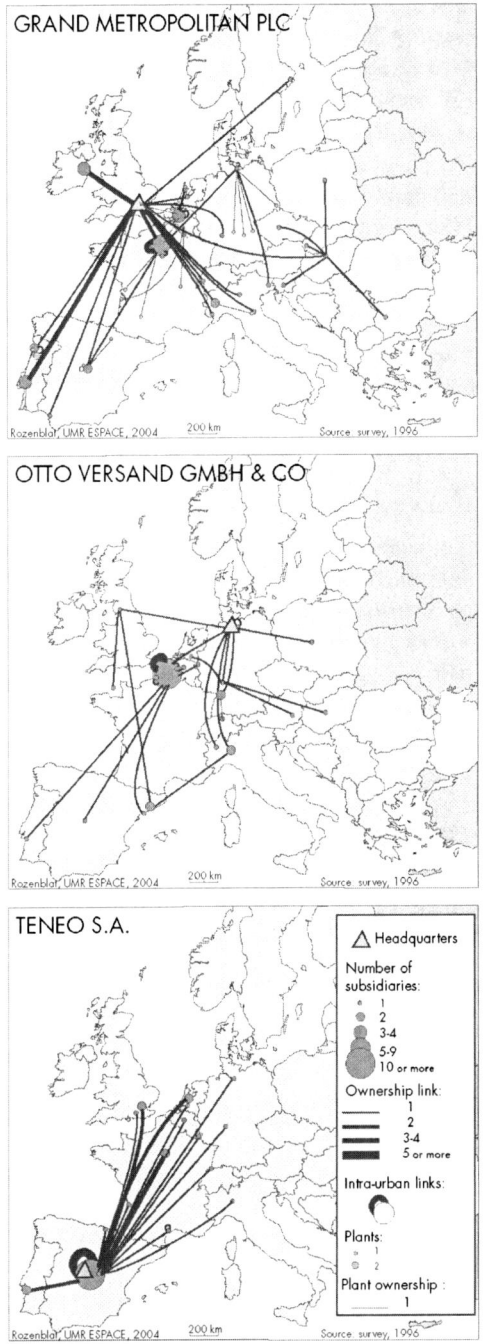

Figure 8.5 Three examples of multinational firms' ownership networks in 1996

city *j* (Figure 8.6). One link of legal ownership represents one link between city *i* where the headquarters of firm *K* (*A*, *B* or *C* on Figure 8.6) is located and city *j* where the subsidiary of firm *K* is located. To construct the graph of connections between cities through firm ownership linkages, we look at the orientation of the links. In order to describe the basic networks connecting cities through multinational firm ownership linkages, a first qualitative distinction can be made according to the relative position of cities where subsidiaries are located. A comparison of the nationalities of the three cities – the city of the intermediate subsidiary, the city of its headquarters, the city of its sub-subsidiary companies – shows up three 'types' of 'intermediate' city (Figure 8.7):

A: 'Bridgehead' cities. These cities host one or several subsidiaries of a firm in a foreign country; they in turn have one or more subsidiaries in the same country. The bridgehead city is used as an entry point for foreign investment: upstream it is connected to foreign cities by links of subordination (e.g. firm control), and downstream it dominates other cities in its own territory through linkages of economic control. When companies are organised by region, these intermediate headquarters generally control the subsidiaries in the country of investment. Most of these intermediate cities that act as entry points for foreign investment are the largest cities in each country, especially those where the urban system is dominated by primate cities, such as London, Paris, Vienna, Milan, Madrid, Barcelona and Brussels. From these central cities, foreign companies thus radiate to other cities in the same country.

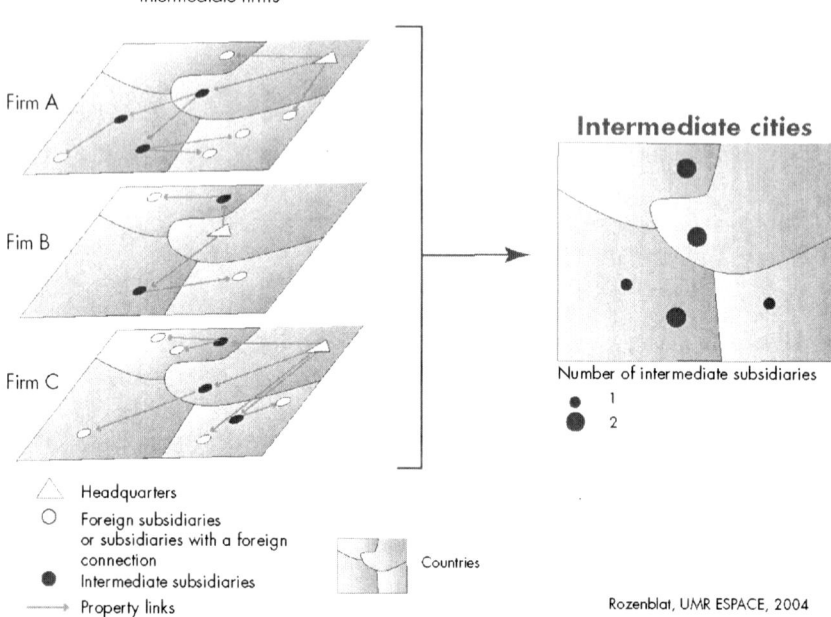

Figure 8.6 From firm linkages to urban networks

A- Bridgehead Cities

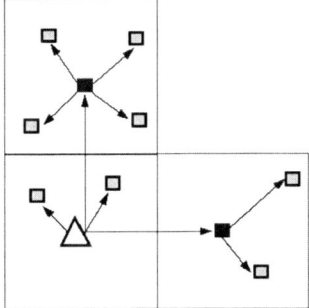

B- Outpost Cities

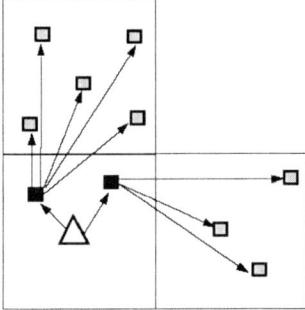

C- Turntable Cities

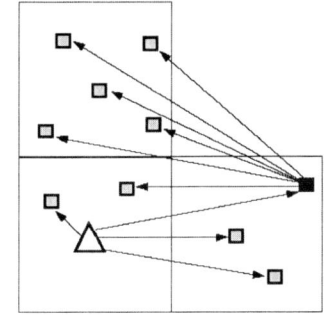

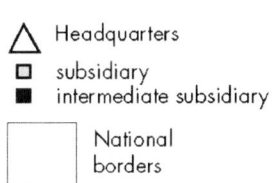

△ Headquarters
□ subsidiary
■ intermediate subsidiary

National
borders

@UMR E.S.P.A.C.E.,Rozenblat, 1998

Figure 8.7 Intermediate cities in international firm ownership linkages

However, this type of nodal function has different spatial configurations. For example, Madrid and Barcelona perform this role with the same intensity, but not with the same territorial extension: Madrid controls sub-subsidiary branches of foreign companies all over Spain, whereas Barcelona's range is limited to Catalonia. From this point of view, London is similar to Barcelona: more than half of the foreign sub-subsidiary companies that it controls are located in its own urban area. In general, the bridgehead function reinforces national or regional urban hierarchies by magnifying the central role of some cities, which are essential in the early stages of capital diffusion. These cities provide companies with the necessary information and knowledge about a whole national territory. These cities are also selected for their accessibility in terms of transport, institutional contacts etc. This process reinforces the need for central administrative offices and infrastructure in these cities, which in turn enhances the attractiveness of the cities.

B: 'Outpost' cities. For each city we measured the number of domestic subsidiaries controlling sub-subsidiary companies abroad. The city is used here as a node for foreign direct investment: upstream it is connected to a city on its own territory where the headquarters of the group are located, and downstream it is directly connected to foreign cities in which it controls subsidiaries. These outpost cities act as springboards towards foreign countries, as is the case for Paris, London and Vienna. In other cases, this situation results from the vertical or horizontal integration of companies whose various headquarters have remained fairly independent. Veba, a company present in both our surveys, is typical of this form of organisation. The company is the result of an industrial merger of petrochemical firms (Huls AG, Raab Karcher and Veba Oel headquartered in Essen, and Stinnes Interoil headquartered in Hamburg), energy companies (Preussenelektrika in Hanover), storage firms (in Essen and Lubeck) and wholesale firms (in Essen and Kaiserslautern). Veba's foreign subsidiaries are controlled from the total subset of these cities. It is well known that similar mergers or acquisitions of companies have occurred since the 1990s, with the same intensity in all countries. Conversely, not all urban systems offer companies the same opportunities for maintaining headquarters simultaneously in several places. Whereas German cities (Essen, Stuttgart, Köln–Bonn and to a lesser degree Frankfurt) and some British cities (such as Birmingham) and Austrian cities (such as Linz) can perform the function of headquarters for foreign subsidiary companies, this is less common in French cities other than Paris. In the territory of France, a study of inter-urban command by the headquarters of multi-entity firms confirms, on another level of company organisation, the strong polarisation of companies' national capacity in Paris (Rozenblat 1998). Other French cities may act as small, secondary centres, because they host the headquarters of large national companies. This is the case, for example, of Clermont-Ferrand (Michelin), Lyons (Renault Trucks), Saint-Etienne (Casino) and Strasbourg (Aventis, Kronenbourg). In recent years, headquarters located in French regional capitals have tended to lose strategic power. Indeed, in the past decade, most strategic functions have moved to the capital, whereas an increasing number of entities dedicated to other functions (mainly production) have moved out of Paris (Jourdan 2004).

C: Multinational 'turntable' cities. For each city we counted the number of foreign subsidiaries that own sub-subsidiary companies abroad. Amsterdam seems to play a specific role as an international node for almost half the groups in our sample (in both 1990 and 1996). Many European groups locate foreign subsidiaries in Amsterdam, and these subsidiaries develop sub-subsidiaries abroad. The function of these intermediate subsidiaries is primarily financial. They benefit from the advantageous tax treatment offered by the Netherlands for both foreign subsidiary companies and companies that own other subsidiaries abroad (Mignolet and Pierre 1998). They also take advantage of Amsterdam's international functions, which are more developed than one might assume from its population size (under 2 million, Rozenblat and Cicille 2003). The international functions of Amsterdam are particularly developed in the sectors of finance and air transport, which are the activities most strongly related to attractiveness for the location of registered offices. The city of Luxembourg also plays a similar role, but with less intensity (12 groups from the sample in 1990 and eight in 1996).

In the hierarchical legal organisation of the companies, the intermediate sub-sidiaries enable cities to maintain positions of both control and dependence at once. Indeed, the cities accommodate the two positions simultaneously because the same subsidiary is at once under the domination of its head office and domi-nant over its sub-subsidiaries. Crozier and Friedberg (1977) have stressed the essential role of nodes in any organisation. These nodes have a dual function (Crozier and Friedberg 1977: 164). On the one hand, for the organisation, they constitute a 'segment of environment', a source of information on the organi-sation's ability to adapt to the environment (past history, institutions, resources, culture etc.) and thus can be considered as 'reducers of uncertainty'. In addition, for the environment, they represent 'the organisation and its interests'. They form part of 'a permanent process of exchange through which an organisation opens, so to speak selectively, into the broader system in which it takes part, and by which it integrates more or less permanent parts in its own system of action for thus being able to adapt it to its own requirements' (1977: 179). For this reason, a group's intermediate subsidiaries are the interfaces between the group and the economic environment and territory. One aim of our approach is to understand the role played by these intermediate subsidiaries between territories, given that their nodal position depends on the 'degree of monopoly available to each partner vis-à-vis the other in both space and time' (Crozier and Friedberg 1977: 172). The positions of the intermediate subsidiaries (and the nodal cities that host them) depend on the nationalities of their headquarters and of their sub-subsidiary com-panies.

Centrality and hierarchical innovation diffusion as characteristics of inter-city linkages and their evolution

The combination of all firms' networks according to their location in cities can give rise to a more general network representing the position of cities within linkages of firm ownership. Because the ownership linkages are oriented, the network is also an oriented graph where the arcs connect a subsidiary-sending city to a subsidiary-receiving city. According to the analysis outlined above, the arcs follow the decision-making path and thus reflect a city's position in the process of European economic integration. In our approach, foreign subsidiaries are considered as being under the direct control of the decisions taken by their headquarters, which can centralise the design of new products, the organisation of production, decisions about training, and industrial and financial strategies. Through a common policy, the headquarters control all the activities of the group's entities and subsidiaries (Dicken 1992). By putting all the foreign subsidiaries under the direct control of the European headquarters, centralisation of decisions within the firm is taken to the extreme. Domestic companies taking part in the globalisation process are not taken into account. This representation is only a partial reflection of the actual operation of groups, since practices may vary from one group to another, and even within the same group, depending on the sector of activity. This step nevertheless makes it possible to map out a system of control–dependence between the cities hosting the headquarters and the cities where the groups' foreign subsidiaries are located. This approach has been used previously by Alan Pred for the United States (1973, 1977), and by Paul Le Fillâtre (1964) for France (headquarters–entity links were identified from the first annual survey of firms by INSEE), in order to represent the spatial configuration of economic control between cities.

On the basis of the relationships of control between headquarters and their subsidiaries, one can define relative measures of centrality among cities. In geography, the centre always represents a place that dominates relationships based on unequal exchanges. The exchanges generated by the relationships of subsidiary ownership are indeed asymmetrical, since they include interactions of dependence that arise from strategic decision-making, financial transfers and sometimes human transfers or trade in goods. Centres and peripheries are thus distributed among the nodes of a network shown in the oriented graph representing the geographical positions of the headquarters of the groups relative to their foreign subsidiaries.

We plotted the graph of inter-city linkages F_{ij} through foreign firms' property by using the following equation:

$$F_{ij} = \Sigma_k PL_{kij}$$

where PL_{kij} represents the ownership link between the headquarters located in city i and the subsidiary located in city j (in another country) for the firm k.

The position of each city on the graph of the relationships of control or dependence defines its degree of centrality. Interpreting the cumulated linkages in terms of 'power over' and 'power to' command other cities, means a capacity or a medium (Friedmann 1978; Allen 1997; Taylor *et al.* 2002). The transfer of companies' power to their space of reception is part of the process through which a centre reproduces the conditions of its centrality and a periphery reproduces the conditions of its peripherality. However, the dynamics of cities, considered on the scale of urban systems, are clearly a different process from the centralisation that describes the concentration of power at the level of the internal organisation of institutions, whether economic or associative (as opposed to decentralisation). Furthermore, the location of the centre depends on the geographical scale that is considered, a fact that is usually overlooked in traditional approaches (Myrdal 1957; Hirschmann 1958; Amin 1973; Wallerstein 1980; Reynaud 1981). Depending on the geographical scale under consideration, the centres and peripheries may change role, since a central position in one network can become peripheral in a network on a higher scale.

Within a relatively short period of time, between 1990 and 1996, this system of enterprise control between cities has reinforced urban interdependency on an increasingly vast territory, while remaining highly selective (Figure 8.8). By a process of hierarchical innovation diffusion, the most central cities have strengthened their positions in the networks of control: this is the case for the London–Paris–Brussels triangle, as well as for German, Dutch and Swiss cities in the centre of Europe. On the geographical periphery, cities such as Lisbon, Stockholm and Budapest have joined the network. Thus, the system of European cities strongly integrated by multinational corporations is reinforced and made more compact. Interactions between these cities are increasing, with a growing number of ownership linkages. They can be considered to form a 'European urban system' representing the major centres connected through European economic integration. The most central cities are above all those for which the ratio of the number of headquarters to the number of foreign subsidiaries is highest (Figure 8.8). From this point of view, London is the main centre in the network of company control. However, Paris is connected to the highest number of cities and therefore has the highest betweenness centrality. In 1990, Brussels, Milan and Madrid formed a first peripheral crown, which was confirmed in 1996. New linkages of economic control have arisen since 1990 in the vicinity of Vienna and from Brussels to Milan. These 'transversal' links, which are developed across the levels in the size hierarchy, are more common in 1996, in the vicinity of Düsseldorf, Essen, The Hague and Zurich. These cities form secondary centres of control by multinational corporations.

The most satisfying interpretation of these observations is that investment in foreign markets by European firms can be seen as a special case of the hierarchical diffusion of an innovation in a multinational urban system. The process of foreign investment, leading to the international expansion of firms, is one of the organisational innovations of the last third of the twentieth century (Dunning 1977, 1992). The strategy of firms is first to enlarge their own market by taking positions on territories outside their country of origin (Dicken 1992), second to

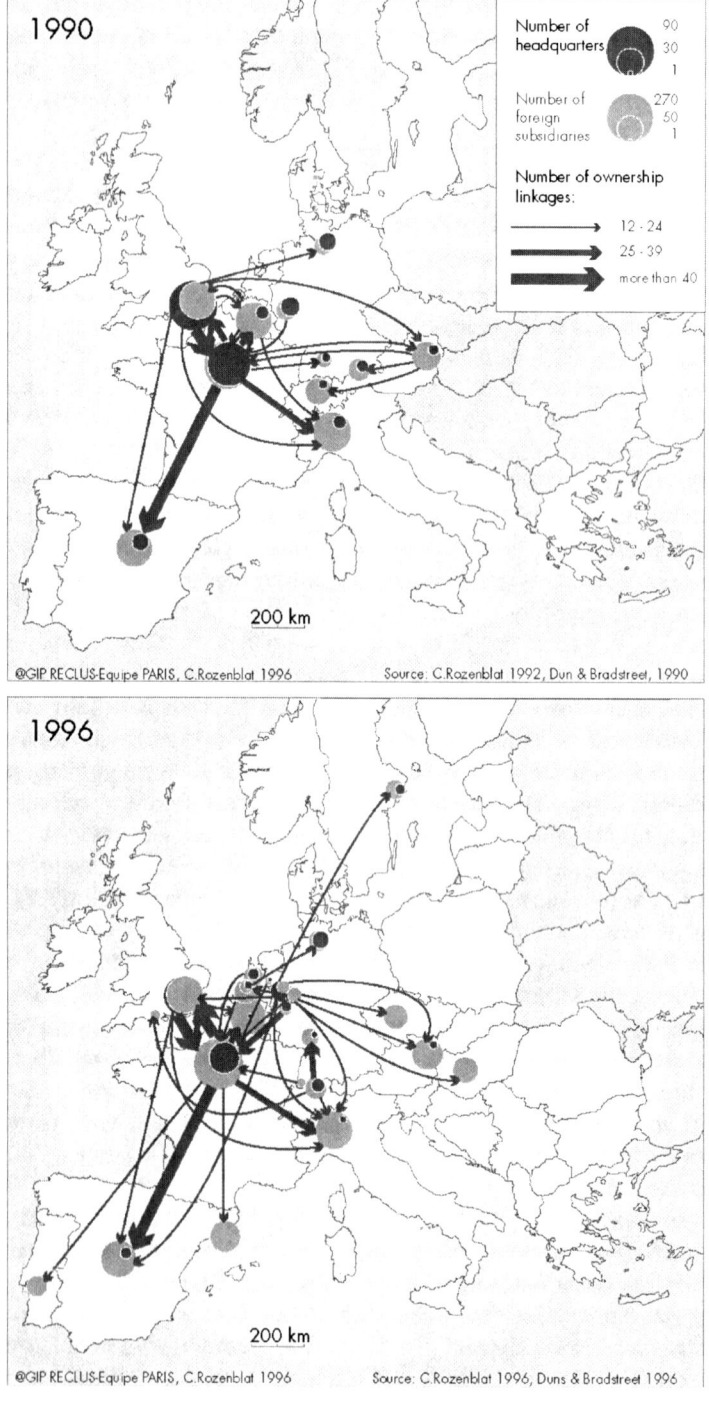

Figure 8.8 Centres in economic networks: main linkages through subsidiary ownership

diversify their production to adapt it to local demand (Krugman 1988), and third to make use of some comparative advantage that the new locations can offer, within the framework of 'flexible capitalism' (Porter 1996, Storper 1997). First they select locations that offer good connections in a whole country and that minimise the risks associated with their investment. Most European capitals and large economic metropolises now meet these conditions. From a city's perspective, that innovation requires a minimum level of international functions in order to provide a favourable business environment and suitable living conditions for foreign firms' employees. Since more than half of European companies' headquarters are concentrated in London and Paris, it is not surprising that these two leading cities, far ahead of any other in size, form the core of the diffusion process. The first tiers of cities reached at a second stage by the 'innovation' process are of two kinds: cities that belong to the core of urban Europe and specialise in international activities, such as Brussels and Luxemburg, and major cities or capitals of the main countries, such as Milan and Madrid. Although the diffusion process follows a classical hierarchical pattern in what could be considered to be an emerging European system of cities, it remains hindered by national boundaries and cities' former functional specialisations.

Different sectors are at different stages of the diffusion process of globalisation. The most advanced sectors in this process have a higher number of foreign subsidiaries (Mucchielli 1998). At the dates of both surveys, chemicals, electrical and electronics sectors account for more than half of the sample of foreign subsidiaries. These activities are also among the most concentrated in the large cities. They tend to spray out of the large cities like all businesses involved in the manufacture of staple consumer goods. This trend is also observed for some chemicals subsidiaries that process farm goods (such as those of Ciba-Geigy), which are naturally located outside urban environments (Rozenblat 1992). Subsidiaries in the construction sector and financial services were the most concentrated in large cities in 1990. The concentration of business services subsidiaries in the largest cities increased between 1990 and 1996, whereas most other types of activity seem to have diffused to smaller cities (Rozenblat 1997). Some urban specialisations appear in the subsidiaries hosted, very often reinforcing existing local specialties: with equal activity and city size, Amsterdam and Zurich are specialised in financial activities and business services. But these specialisations are not very strong and the higher a city is in the urban hierarchy, the less strong its specialisations (Pumain and Saint-Julien 1989; Rozenblat 1992). Thus, the largest cities, which accommodate more subsidiaries, have a more diverse range of activities: their markets (consumer and labour), infrastructure and hosting capacity increase their attractiveness for all types of activity from abroad.

In addition to urban effects, there is a strong national effect that results in significantly different rates of penetration of foreign subsidiaries in different countries. The global impact of the size of cities takes different values depending on their nationality (Rozenblat and Pumain 1993). For instance, with equal population, a Dutch city will tend to attract more foreign subsidiaries than a French city. This can be checked statistically using variance analysis (taking into account the nationality of the cities) combined with regression (taking into account their population). One thus obtains a strong statistical explanation for the number of foreign subsidiaries in a city ($R^2 = 94\%$). This demonstrates that national effects remain highly significant in the diffusion process (approximately 30 per cent of the variance, with city size alone accounting for 64 per cent). The level of functions not only reflects the size that cities have acquired over a long period, mainly in line with the development of the national urban system to which they belong. A country's openness also determines its cities' average capacity to host foreign multinational corporations. According to Mucchielli and Puech (2003), the varying attractiveness of European countries for foreign companies can be attributed mainly to differences in payroll costs. Other authors stress the tax burden (Mignolet and Pierre 1998).

The two geographical levels, urban and national, thus combine to determine location choices in a multilevel approach, also according to regional data analysed by Mucchielli and Puech (2003). Cities' population size and national structural characteristics are still essential for understanding the current development of territories. It is therefore important to keep in mind the spatial pattern of European cities. The hierarchy of cities and their spacing remain two major factors in their integration into global networks. Networks where large cities are regularly spaced, as is the case in eastern Europe, could induce a broad diffusion of global networks in the urban system. However, foreign investment in those countries is riskier than elsewhere in Europe (Michalet 1999). The evolution of the relative concentration of headquarters and subsidiaries, compared with the concentration of urban size, seems to depend on how long the process of multinational integration of cities has been operating in each country. The urban hierarchy can thus be interpreted as a kind of attractor, exerting a feedback effect from the macro geographical system to the behaviour of urban actors. Over the long term, the various hierarchies of the same territory fit together and mutually reinforce each other. The urban hierarchy itself can also undergo transformations due to the concentration of the other functions. These characteristics can be interpreted within the framework of a more comprehensive evolutionary theory of urban systems.

INNOVATION AND EVOLUTIONARY THEORY OF URBAN SYSTEMS

Many uncertainties still hamper our ability to forecast the future of urbanisation. With the advent of new technologies and the political reorganisation of territories, geographers observing globalisation processes are puzzled by the possible next trends in urban concentration, dispersion and relative growth. By applying ideas and models developed by a new field of investigation known as complexity theory, we could learn more about the universe of possible evolution stemming from observed urban dynamics, and perhaps discover some abstract hidden processes that may better explain the similarities appearing in urban structures and evolution, despite the overwhelming diversity of physical, economic, political, social and cultural forms that urban systems take around the world. The concept of urban system is actually a good example of a complex system (Pumain 1998). Urban systems produce self-organised multilevel structures that are evolving via dynamic social processes where non-linearity, discontinuity and irreversibility, as well as permanent adaptation through cooperation and competition, are general rules.

What are the emergent properties of urban systems and how are they produced? We shall focus here on structural features that characterise the macro level of a system of cities, and that emerge from the interactions between towns and cities at the meso level. In this case, interactions occur through multiple exchanges of people, goods and information that circulate continuously from one city to another through a multitude of networks (Allen *et al.* 1999; Massey *et al.* 1999). Since it is hardly possible to represent all of them in a model, we have selected those that are considered responsible for the emergence and maintenance of structural features at the upper level, and chosen to represent them in an abstract way.

The main structural features of urban systems are defined according to the principles that organise their internal diversity, regardless of the period or country in which the system is observed. Such general features are well known. First there is a principle of hierarchical organisation, including important scaling properties for attributes such as size, spacing and level of socio-economic activities of towns and cities. These features were previously formalised by central place theory. Geographers have criticised that theory because it is static and based upon a restricted set of urban activities (mainly services to residents), which does not consider specialisations in non-central functions (for instance industry and tourism). Therefore, a more general framework is needed.

However, this does not mean that all the observations explained statically by central place theory are no longer valid. Hierarchical structure and spacing regularities are still relevant, as shown for instance by the map in Figure 8.9, representing the location and size of towns and cities (urban agglomerations with populations of more than 10,000 defined in a comparable manner) in countries of western and eastern Europe. A simple analysis of this map by the cartographic filtering method illustrates the evidence of the hierarchical and spatial organisation of urban systems, according to research by Céline Rozenblat (1995). In Figure 8.10, the agglomerations in Figure 8.9 were initially connected by a straight line when the distance between them was less than 25 km. The resulting pattern clearly shows three main types of urban system in Europe: a central part with much higher densities, a more contrasting western part (mainly France and Spain) and an eastern part with highly regular spacing. Combined with size, these three types would fit into the typology established by Etienne Juillard ('Rhenish', 'Parisian' and 'peripheral' urban networks, Juillard and Nonn 1976). But the point here is that the general pattern is maintained if the minimum distance is raised to 50 km, or the minimal population size to 100,000 (with separating distances of 100 and 150 km). This result confirms the high consistency of the hierarchical and spatial organisation of urban systems.

The hierarchical organisation is an emergent property that characterises the level of observation of systems of towns and cities. It is produced by the multiple

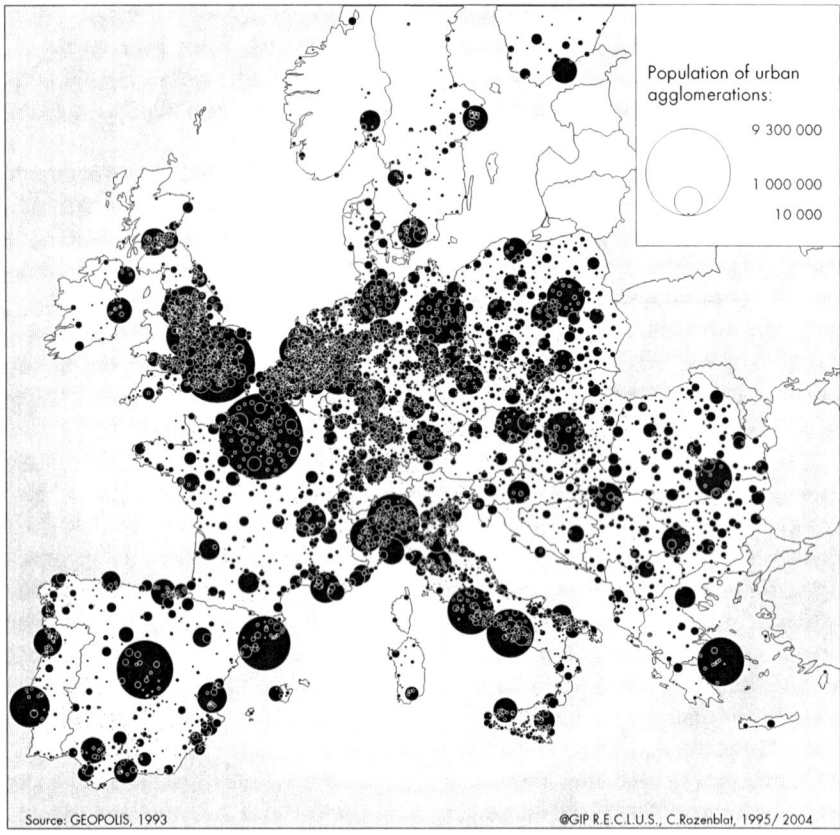

Source: GEOPOLIS, 1993   @GIP R.E.C.L.U.S., C.Rozenblat, 1995/ 2004

Figure 8.9 Population of European cities

interactions occurring between individual towns and cities. The facts that a town
or city keeps its size to a given proportion of other cities' sizes, that the spacing
between cities is more or less regular and that over long periods of time there is
a fairly consistent persistence of the hierarchical order cannot be inferred from
the nature and function of a single city. One has to search for processes that can
explain this emergent property at the level of the urban system according to the
rules of interactions occurring at the level of individual cities.

This has been interpreted in many models by simulating the competitive
growth process, which represents both the dynamics of each town and the resulting statistical aggregate in the form of the distribution of city sizes (Robson 1973;
Pumain 1982; Guérin-Pace 1993). However, the demonstration in these models is
statistical and non-spatial. We have formalised the complex dynamics of spatial
interactions in the growth process, including reversal of influence over time according to the definition of neighbouring places by a simple modular model (Page
et al. 2001). The relationships between cities' mass and spacing, and the evolution
of these relationships over time, have been studied by Anne Bretagnolle (1999),
who showed how urban interactions regulate the relative size of the elements in

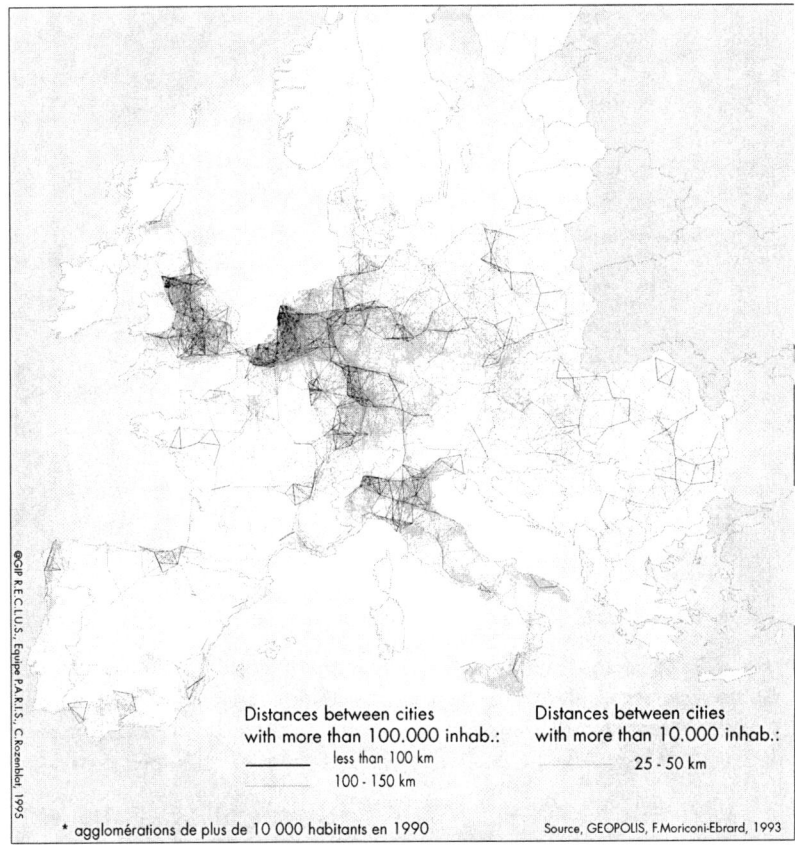

Distances between cities
with more than 100.000 inhab.:
_____ less than 100 km
_____ 100 - 150 km

Distances between cities
with more than 10.000 inhab.:
_____ 25 - 50 km

* agglomérations de plus de 10 000 habitants en 1990

Source, GEOPOLIS, F.Moriconi-Ebrard, 1993

©GIP R.E.C.LU.S., Equipe P.A.R.I.S, C.Rozenblat, 1995

Figure 8.10 Patterns of urban systems in Europe

an urban system, depending on the speed and intensity of spatial interactions. The hierarchisation is generated from the bottom by the short-circuiting of smaller intermediate centres linked to the process of space–time convergence, and from the top by the various processes of hierarchical diffusion of innovations in the urban system (Bretagnolle *et al.* 2002).

A link is then established with the second main structural property of urban systems, which is their qualitative socio-economic diversity, as expressed in typologies of the functional specialisation of towns and cities, which generally lasts without major changes for several decades. These slow dynamics of relative change in the inter-urban division of labour are produced through small deviations in a general process of diffusion of socio-economic changes, which is much more rapid (Paulus 2004). According to that extremely incremental process of interactive adjustments, all cities change in more or less the same direction and intensity in the space of activities phase, as illustrated by a correspondence analysis for French cities for the period 1962–90 in Figure 8.11. This result introduces a way for generalising central place theory in the broader framework of an evolutionary

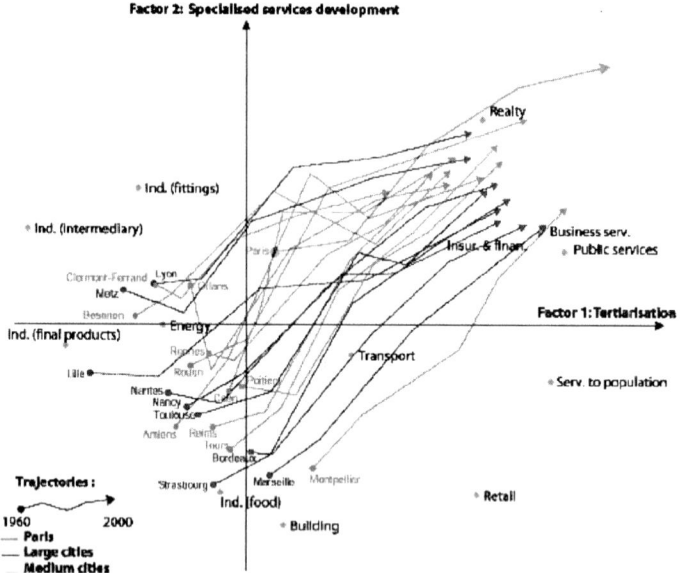

Figure 8.11 Trajectories of French cities in economic space (source: Paulus 2004)

theory of urban systems (Pumain 2000). Urban hierarchy and functional diversity are emergent properties stemming from a co-evolution process of towns and cities, which are cooperating and competing for access to socio-economic innovations, by attempting to secure better relative positions in spatial and social networks.

CONCLUSION

The observations we make about the location of multinational firms in Europe and their ownership linkages among cities, at two dates, 1990 and 1996, are a good example of urban co-evolution, even as these new connections contribute to the creation of a still emerging 'European system of cities'. The former hierarchical structure of the urban system has a strong regulation (or feedback) effect: the main capitals, London and Paris, are at the core of the process, whereas almost none of the smallest towns (population under 200,000) are involved in the diffusion process as yet. But two distinct processes are superimposed on the classical hierarchical diffusion process: one is a reinforcement of specialisation in multinational activities for cities already engaged in international business, especially financial business (as is the case for Brussels, Luxemburg and Zurich); the other, even more effective, process is linked to the existence of national borders, which guide the location of foreign firms towards national capitals, regardless of their relative size within the European urban system, and towards cities located close to international borders (which also gives an advantage to cities in the smallest countries, which are on the whole more 'open' to external exchanges than the largest countries). These two processes have the potential to reshape the European urban system, but to what extent? Two points in time are not enough to identify the possible consequences of these trends. A third survey, currently in progress, could bring clearer answers.

REFERENCES

Allen, J. (1997) 'Economies of power and space', in R. Lee and J. Wills (eds) *Geographies of Economies*, London: Arnold, pp. 59–70.

Allen, J., Massey, D. and Pryke, M. (1999) *Unsettling Cities*, London: Routledge.

Amin, S. (1973) *Le Développement inégal: essai sur les formations sociales du capitalisme périphérique*, Paris: Éditions de minuit.

Berry, B.J.L. (1964) 'Cities as systems within systems of cities', *Papers of the Regional Science Association*, 13: 147–63.

Bouinot, J. (2004) 'Des évolutions dans les comportements spatiaux des entreprises en 2003?', *Cybergeo*, http://www.cybergeo.presse.fr.

Bretagnolle, A. (1999) *Espace-temps et système de villes: effets de l'augmentation de la vitesse de circulation sur l'espacement et l'étalement des villes*, doctoral thesis, Université Paris I .

Bretagnolle, A., Paulus, F. and Pumain, D. (2002) 'Time and space scales for measuring urban growth', *Cybergeo*, 219.

Brunet, R. (1989) *Les Villes européennes*, Paris: Datar–La documentation Française.

Cattan, N., Pumain, D., Rozenblat, C. and Saint-Julien, T. (1999) *Le système des villes européennes*, 2nd edition, Paris: Anthropos.

Cohen, R.B. (1981) 'The new international division of labor, multinational corporations and urban hierarchy', in M. Dear and A.J. Scott (eds) *Urbanization and Urban Planning in Capitalist Society*, New York: Methuen, pp. 287–315.

Conti, S. and Spriano, G. (1990) *Effetto Città*, Turin: Fondazione Agnelli.

Crozet, M., Mayer, T. and Muchielli, J.L. (2004) 'How do firms agglomerate? A study of FDI in France', *Regional Science and Urban Economics*, 34: 27–54.

Crozier, M. and Friedberg, E. (1977) *L'Acteur et le système*, Paris: Editions Du Seuil.

Dicken, P. (1992) *Global Shift: The Internalization of Economic Activity*, 2nd edition, New York: Guilford Press.

Dicken, P. and Malmberg, A. (2001) 'Firms in territories: a relational perspective ', *Economic Geography*, 77: 345–63.

Dunning, J.H. (1977) 'Trade, location of economic activity and the MNE: in search of an eclectic approach ', in B. Ohlin, P.O. Hesselborn and P.M. Wijkman (eds) *International Allocation of Economic Activity*, London: Macmillan.

Dunning, J.H. (1992) *Multinational Enterprises and the Global Economy*, Wokingham: Addison Wesley.

European Communities (2001) *Unité de l'Europe, solidarité des peuples, diversité des territoires*, deuxième rapport sur la cohésion économique et sociale, Luxembourg: Office des publications officielles des communautés européennes.

ESPON (2003) 'The role, specific situation and potentials of urban areas as nodes in a polycentric development', *ESPON Project 1.1.1*, Third interim report, August, http://www.espon.lu/online/documentation/projects/thematic/index.html.

Francfort, I., Osty, F., Sainsaulieu, R. and Uhalde, M. (1995) *Les Mondes sociaux de l'entreprise*, Paris: Desclée de Brouwer.

Friedmann, J. (1978) 'The spatial organization of power in the development of urban systems', in L.S. Bourne and J.W. Simmons (eds) *Systems of Cities*, Oxford: Oxford University Press, pp. 328–40.

Friedmann, J. (1986) 'The world city hypothesis', *Development and Change*, 17: 69–84; also in P. Knox and P.J. Taylor (eds) (1995) *World Cities in a World System*, Cambridge: Cambridge University Press, pp. 317–31.

Graham, B. (1998) 'Liberalization, regional economic development and the geography of demand for air transport in the European Union', *Journal of Transport Geography*, 6: 87–104.

Guérin-Pace, F. (1993) *Deux siècles de croissance urbaine: la population des villes françaises de 1831 à 1990*, Paris: Anthropos.

Guimera, R., Mossa, S., Tutschi, A. and Amaral, A.N. (2005) ' The worldwide air transportation network: anomalous centrality, community structure, and cities' global roles', *Proceedings of the National Academy of the Sciences of the United States of America*, 102 (22): 7794–99.

Hall, P. (1995) 'Toward a general urban theory', in P. Hall, P. Newton, J. Brotchie, M. Batty and E. Blakely (eds) *Cities in Competition: Productive and Sustainable Cities for the 21st Century*, London: Longman, pp. 3–31.

Hirschmann, A. (1958) *The Strategy of Economic Development*, New Haven, CT: Yale University Press.

Jourdan, N. (2004) *Les Transferts interrégionaux d'établissements – Forte progression entre 1996 et 2001*. Insee Première 949 (February).

Juillard, E. and Nonn, H. (1976), *Espaces et régions en Europe occidentale*, Action Thématique Programmée 10, Paris: CNRS.

Krugman, P. (1988) *Strategic Trade Policy and the New International Economics*, Cambridge, MA: MIT Press.

Le Fillâtre, P. (1964) 'La puissance économique des grandes agglomérations françaises déduite de l'étude de la localisation des sièges et des succursales d'entreprises à établissements multiples', *Etudes et conjoncture*, 19 (1): 1–40.

Massey, D., Allen, J. and Pile S. (eds) (1999) *City Worlds*, London: Routledge.

Michalet, C.A. (1997) 'Strategies of multinationals and competition for foreign direct investment ', Foreign Investment Advisory Service, occasional paper 10.

Michalet, C.A. (1999) *La Séduction des nations ou comment attirer les investissements*, Paris: Economica.

Mignolet, M. and Pierre, I. (1998) 'Fiscalité et distribution des unités au sein de multinationales: l'exemple des groupes belges', *Revue d'Économie Régionale et Urbaine*, 2: 251–80.

Mintzberg, H. (1994) *The Rise and Fall of Strategic Planning*, New York: Prentice Hall.

Muchielli, J.-L. (1998) *Multinationales et mondialisation*, Paris: Seuil.

Muchielli, J.-L. and Puech, F. (2003) 'Internationalisation et localisation des firmes multinationales: l'exemple des entreprises françaises en Europe', *Économie et Statistiques*, 363–365: 129–44.

Myrdal, G. (1957) *Rich Lands and Poor*, New York: Harper and Row.

Page, M., Parisel, C., Pumain, D. and Sanders, L. (2001) 'Knowledge-based simulation of settlement systems', *Computers, Environment and Urban Systems*, 25: 167–93.

Pagetti, F. (1998) 'La rete bancaria nel sistema urbano europeo', in P. Bonavero and E. Dansero (eds) *L'Europa delle regioni e dele reti, i nuovi modelli di organizzazione territoriale nello spazio unificato europen*, Turin: UTET Libreria, pp. 361–71.

Paulus, F. (2004) *Coévolution dans les systèmes de villes : croissance et spécialisation des aires urbaines françaises de 1950 à 2000,* doctoral thesis, Université Paris I.

Porter, M. (1996) 'Competitive advantage, agglomeration economies and regional policy', *International Regional Science Review,* 19: 85–90.

Pred, A. (1973) 'Systems of cities and information flows', *Lund Studies in Geography,* Series B, 38.

Pred, A. (1977) *City-systems in Advanced Economies,* London: Hutchinson University Library.

Pumain, D. (1982) *La Dynamique des villes,* Paris: Economica.

Pumain, D. (1998) 'Les modèles d'auto-organisation et le changement urbain', *Cahiers de Géographie de Québec,* 117: 349–66.

Pumain, D. (2000) 'Settlement systems in the evolution', *Geografiska Annaler,* 82B (2): 73–87.

Pumain, D. (ed.) (2006) *Hierarchy in Natural and Social Sciences,* Dordrecht: Springer.

Pumain, D. and Saint-Julien, T. (1989) *Atlas des villes de France,* Paris: La Documentation Française.

Pumain, D. and Saint-Julien, T. (eds) (1996) *Urban Networks in Europe,* Paris: John Libbey–INED, Congresses and Colloquia, 15.

Reynaud, A. (1981) *Société, espace et justice,* Paris: PUF.

Robson, B. (1973) *Urban Growth, an Approach,* London: Methuen.

Rozenblat, C. (1992) *Les Réseaux des entreprises multinationales dans le réseau des villes européennes,* doctoral thesis, Université Paris I.

Rozenblat, C. (1995) 'Tissu d'un semis de villes européennes', *Mappemonde,* 4: 22–7.

Rozenblat, C. (1997) ' L'efficacité des réseaux de villes pour le développement et la diffusion des entreprises multinationales en Europe (1990–1996)', *Flux,* 27/28: 41–58.

Rozenblat, C. (1998) 'Commandement et dépendance', in T. Saint-Julien (ed.), *Atlas de France: l'industrie,* Reclus –La Documentation Française, 9: 74–8.

Rozenblat, C. and Cicille, P. (2003) *Les Villes européennes: analyse comparative,* Paris: Datar – La Documentation française.

Rozenblat, C. and Pumain, D. (1993) 'The location of multinational firms in the European urban system', *Urban Studies,* 10: 1691–1709.

Rozenblat, C. and Pumain, D. (2004) 'Articulated modes of integration: the structure of European urban system', in M. Pacione (ed.). *Changing Cities: International Perspectives,* Glasgow: Strathclyde University Publishing, pp. 91–105.

Sassen, S. (1991) *The Global City: New York, London, Tokyo,* Princeton, NJ: Princeton University Press.

Scott, A. (2001) *Global City-Regions: Trends, Theory, Policy,* Oxford: Oxford University Press.

Storper, M. (1997) *Regional World: Territorial Development in a Global Economy,* New York: Guilford Press.

Taylor, P.J. (2001) 'Specification of the world city network', *Geographical Analysis,* 33 (2): 181–94.

Taylor, P.J. and Walker D.R. (2001) 'World cities: a first multivariate analysis of their service complexes', *Urban Studies,* 38: 23–47

Taylor, P.J., Walker D.R.F, Catalano, G. and Hoyler, M. (2002) 'Diversity and power in the world city network', *Cities,* 19 (4): 231–41.

Veltz, P. (1998) 'Globalisation et territorialisation des groupes industriels: rapport de synthèse', Paris: Datar – La Documentation Française.

Wallerstein, I. (1980) *Le Système-monde du XVe siècle à nos jours*, Paris: Flammarion.

Zarifian, P. (1993) *Quels nouveaux modèles d'organisation pour l'industrie européenne? L'émergence de la firme coopératrice*, L'Harmattan.

Zimmermann, J.B. (1995) *L'Ancrage territorial des activités industrielles et technologiques*, Paris: Commissariat Général du Plan.

9 The metropolization of the European urban system in the era of globalization

Stefan Krätke

INTRODUCTION

The European Union forms a transnationally interlinked economic and institutional space that is being considered as an economic bloc in global competition with other large economic blocs in North America and Asia. With respect to its global competitiveness the EU has set up the ambitious aim of becoming a worldwide leading economic territory of the knowledge-intensive economy ('Lisbon process'). However, the development of the EU territory reveals considerable structural differences between the European regions. In particular, the regions' capacities in the field of innovation activity, research and technological development (see European Commission 1999, 2004; Cooke *et al.* 2000; Grotz and Schätzl 2001) are increasingly being considered as a key factor in regional economic success. The European urban and regional system today is subject to a double process which contains, on the one hand, a progressive economic integration within the EU territory ('Europeanization') and, on the other hand, the active involvement of the EU in a new development stage of the world economy ('globalization') which is marked by intensified worldwide economic interrelations. Thanks to their structural characteristics as centres of economic activity and prime nodes of transnational economic interaction, urban agglomerations and metropolitan regions are playing a prominent role in the globalization process. Furthermore, the economic integration of Europe happens quite substantially on the field of the transnational integration of formerly national urban systems. This leads to a revaluation of certain urban regions and to changes of their functional reach in the European and global context. The dominant economic developing centers of the EU territory are dynamic urban agglomerations and metropolitan regions in which in particular the knowledge-intensive services and research-intensive industrial activities concentrate. In these key sectors of an increasingly knowledge-intensive and innovation-driven economy the processes of selective regional concentration and cluster building are increasing the economic productivity and innovative capacity of the urban economic centers.

Against this background the economic development of the EU territory can be considered as a process of the *metropolization* of economic development potentials and innovation capacities. The urban agglomerations and metropolitan regions function as the 'motors' of economic development in the EU and at the same time as the prime nodes of Europe's integration in the global economy. However, the increasing concentration of economic developing potentials – in particular in the field of knowledge-intensive services and research-intensive industries – on dynamic urban agglomerations and metropolitan regions is also an essential driving force of increasing regional disparities in Europe. Not only underdeveloped rural regions but also many of the less dynamic urban regions of Europe threaten to stay behind in the process of metropolization if they are not connected with the urban agglomerations and metropolitan regions economically.

Not only the process of 'Europeanization' or the progressive economic integration of the EU countries has made the traditional thinking in national urban systems obsolete – even more the process of globalization has become a significant driving force of the *transnational* economic interlinkage of the European urban and regional system. Particularly in the field of technology- and knowledge-intensive economic activities local or regional clusters have won increasing importance for a successful performance on the world market (OECD 1999; Porter 2001). More and more the 'global players' anchor themselves with their own organizational units in the 'leading' regional enterprise clusters of certain activity branches in order to win direct access to the specific knowledge and innovation resources of the respective regions. This embedding of global enterprises in regional economic clusters carries at the same time the supra-regional and worldwide interlinking of the regional economies.

In this chapter metropolization and globalization are picked out as main trends of spatial development in Europe. The process of metropolization will be emphasized with respect to the selective concentration of the potentials of knowledge-intensive economic activities in the urban system of Europe. In the empiric investigation the different *sectoral profiles* of the European urban agglomerations in the field of the knowledge-intensive economy are worked out. With reference to the globalization process the different transnational connectivity of urban agglomerations and metropolitan regions of the EU territory will be highlighted with particular emphasis on the *specific geographic pattern of transnational interlinkages* of selected metropolitan regions within the EU territory. The global connectivity increases the economic power and development capacity of certain urban agglomerations and metropolitan regions, with the effect that the metropolization of the European urban system is further strengthened in the context of the globalization process.

METROPOLIZATION OF THE URBAN AND REGIONAL SYSTEM IN THE ERA OF THE 'KNOWLEDGE ECONOMY'

The large conurbations and metropolitan regions of Europe are being interconnected by transnationally extended networks of corporate organization which spread out not only over the European territory but also globally. In the pan-

European urban and regional system, particular metropolitan regions are taking on a high rank position if they qualify with respect to the economic potentials and functional reach of their economic relations as prominent *transnational* economic centers of Europe.

Many contributions to the economic-functional structures of the European and the global urban system are tending to characterize at least the high-ranking urban regions primarily with regard to their function as locational centers of advanced producer services (see Sassen 1996; Geppert 2005; Kujath 2005a; Paal 2005) and to neglect their role as industrial location centers. On the other hand, the debate on 'new industrial spaces' and technology districts as well as the debate on the formation of urban production clusters of the 'creative industries' has provided more and more evidence that the urban agglomerations and metropolitan regions are still functioning as significant locations of industrial activities (Scott 1988, 1998; Storper 1997): they are often location centers for new 'knowledge-based' production chains, for innovative production clusters in the field of information and communication technology, the pharmaceutical industry, medical engineering and biotechnology, the media industry, etc. (Berg *et al.* 2001; Scott 2001; Cooke 2002; Florida 2002, 2005; Krätke 2005). Also, the traditional technology-centred industrial branches like construction of vehicles and mechanical engineering are still a most important component of the economic potential of many urban agglomerations and metropolitan regions. Particularly in the field of 'innovative growth sectors,' industrial activities are frequently characterized not only by intensive intra-regional transaction and communication networks but also by strong linkages with the industrial innovation centers of other urban agglomerations and metropolitan regions within the national economic territory as well as on a transnational scale (Krätke 2002; Alvstam and Schamp 2005). Thus the linkages between innovative firms of the high technology clusters of, for example, Munich and San Francisco (Silicon Valley) contribute substantially to the worldwide interlinking of industrial innovation processes.

The prospects of a 'knowledge-based' regional development path are being emphasized as a central theme in economic geography and regional research with respect to the thesis that within an increasingly innovation-driven economy the development chances of urban regions today are determined in particular by their potentials and capacities in the field of *knowledge-intensive* economic activities (Keeble and Wilkinson 2000; Cooke 2002; Lo and Schamp 2003; Matthiesen 2004; Kujath 2005a). This debate refers to the importance of knowledge resources as well as research and educational infrastructures, to the significance of interactive knowledge generation within enterprise clusters for the competitiveness of the regions, and not least to the possible strengthening of the development prospects of cities and regions by the extension of *knowledge-intensive activity branches* of the regional economy ('profile shaping' of a region as a center of the knowledge economy). The knowledge-intensive economy comprises industrial activity branches with a high share of research and development activities for the generation of new technological knowledge, and those economic activities for which the generation and economic use of specific knowledge is a main focus, i.e. in particular the highly qualified enterprise services in the fields of business consultancy and organization management, financial services, high technology services, and not least the culture and media industry (which particularly depends on the generation and use of 'creative' knowledge).

The dominant economic development centers of the EU territory are dynamic urban agglomerations and metropolitan regions in which particularly the knowledge-intensive services as well as research-intensive industrial activities concentrate. In recent years the growth dynamics in many EU countries were mostly concentrated on the research-intensive and knowledge-intensive economic activity branches: An increase of jobs in the *industrial sector* (manufacturing activities) of the EU-15 has been noted for the most part in the research-intensive industries (whereas the less knowledge-intensive industrial activities revealed job losses on a scale that exceeds the gains in research-intensive industries); the increase of jobs in the *service sector* of the EU-15 has been concentrated for the most part on the knowledge-intensive services. In the research-intensive industrial branches and the knowledge-intensive service branches an ongoing process of selective locational concentration on urban agglomerations and metropolitan regions leads to the development of strong cluster potentials that raise the productivity and innovation capacity of these regional economic centers. In the course of this economic structural change the efficacy of agglomeration effects is rapidly increasing (owing to the social production and use of localization and urbanization economies), so that the spatial development of the European Union can be characterized as a process of *metropolization* of economic development potentials and innovation resources. 'Metropolization' is an expression for the increasing concentration of economic development potentials of the research-intensive industries and knowledge-intensive services on metropolitan regions and urban agglomerations.

However, the analysis of structural economic change in the European urban system should not be reduced to this overall development trend. The urban agglomerations and metropolitan regions of Europe distinguish themselves by *different profiles* (or a specific mix of sub-sectors) also in the knowledge-intensive economy. In order to analyze the contemporary European urban system with emphasis on the process of metropolization and the differentiation of economic profiles of urban agglomerations in the EU territory, the data on the knowledge-intensive economy provided by the Eurostat Regio database have been evaluated for the period 1997–2004. These data are available only for the NUTS 2 level, i.e. in a fairly rough spatial delimitation of the EU regions. Anyway, the use of NUTS 2 regions can be justified with regard to the investigation of the metropolization trend in the European urban system. A certain distortion in the analysis of urban regions might originate from the relatively ample delimitation of the NUTS 2 regions, which must be kept in mind for the proper interpretation of results – the data for urban regions are always to be understood as an aggregate of the respective central city's administrative territory and the administrative regions of its surrounding area. On the other hand, the NUTS 2 level is absolutely satisfactory for the purposes of this analysis if one takes into account that in the present time the economic functional spaces of the cities and in particular the metropolises of Europe have extended further and further.

The structural analysis of the potentials and development paths of European urban agglomerations and metropolitan regions in the field of knowledge-intensive economic activities includes a total of 60 urban agglomerations in the EU territory, of which approximately 20–25 can be categorized as 'metropolitan regions'. Because there is no uniform definition of metropolitan regions (see Kujath 2005b), this expression aims at a rough selection of the most 'outstanding' urban agglomerations in terms of their national functions and economic capacities. The 60 urban agglomerations were selected according to the criteria that the central city of the region has more than 450,000 inhabitants and that the population figure of the whole urban region amounts to more than 1 million inhabitants. In particular cases deviations were admitted to take up the capital regions of the Baltic countries, Slovakia and Slovenia in the analysis. Furthermore, for the purpose of this pan-European analysis several urban regions were put together in a widely delimitated urban agglomeration: These aggregated urban regions comprise Florence–Bologna, Manchester–Liverpool–Leeds–Sheffield, the 'Randstad Netherlands' (principally Amsterdam–Rotterdam–The Hague), and the Upper Silesian industrial district in Poland as well as the Rhine–Ruhr conurbation (from Dortmund through Essen, Duisburg, Düsseldorf to Cologne). As regards the interpretation of empirical findings it has to be taken into account that the aggregated analysis and representation conceal the *polycentric internal structure* of these urban agglomerations.

The European regional structure reveals above all a concentration of economic power in the urban agglomerations and particularly in the metropolitan regions among them. In 2002, 61 percent of the GDP and 56 percent of all employees of the EU-25 were concentrated on the 60 urban agglomerations. Contemporary economic development trends point to a further strengthening of this spatially selective concentration: In the period 1997–2002, 69 percent of the total increase in the EU-15's GDP was concentrated on the urban agglomerations of the EU-15. These preliminary findings concerning the metropolization thesis might lead back to the fact that the economic power of the prominent European urban agglomerations is based on a particularly strong and increasing concentration of knowledge-intensive economic activities.

Between 1997 and 2004, the 41 urban agglomerations of the EU-15 for which sectorally differentiated data are available in the Eurostat Regio database together reveal an *increasing share* of the EU-15's total employment in the knowledge-intensive economy, particularly in the subsectors of 'high-technology industry' and knowledge-intensive 'technology-related enterprise services', and beyond this also in the sub-sectors of knowledge-intensive market-related enterprise services and the media industry, and knowledge-intensive services in healthcare and education. With regard to the changing structure of employment we might expect a further accentuation of the metropolization tendency: Job increases in the *industrial sector* of the EU-15 (1997–2004) are concentrated on *research-intensive manufacturing activities,* in particular on the 'medium high-technology' industries. Some 67.6 percent of the total job increase in the *service sector* of the EU-15 (1997–2004) has to be ascribed to *the knowledge-intensive services*; the strongest relative (intra-sectoral) growth is to be recorded in the 'technology-related enterprise services' and the 'market-related enterprise services'.

According to these preliminary findings the urban agglomerations and in particular the metropolitan regions of the EU qualify *as the primary location centers* of knowledge-intensive industrial activity branches and advanced producer services. The metropolization of the urban and regional system might be understood as a spatial articulation of the increasing significance of knowledge-intensive activity branches in Europe's economic development.

In the following analysis the potentials of knowledge-intensive economic activities in the urban agglomerations and metropolitan regions of Europe are highlighted quite consciously in terms of *absolute* concentration and *absolute* changes of the number of workplaces. The normalization of regional data, e.g. with reference to numbers of inhabitants, is not very meaningful in the thematic framework of this study. This analysis is focusing on the selective concentration of particular economic activities in a system of competing locational centers – in this context absolute concentrations of activities are just as decisive as absolute changes, since they are of central importance for real concentration processes and their dynamics in terms of a strong 'pull effect'.

The overall representation of the knowledge-intensive economy's absolute concentrations in the urban agglomerations and metropolitan regions of the EU (see Figure 9.1) reveals that the so-called 'core area' of the EU economic territory (which is circumscribed by the pentagon London–Paris–Milan–Munich–Hamburg) encloses a large part of the location centers of the knowledge economy, wherein London and Paris take a leading position (and the Rhine–Ruhr conurbation shows a strong absolute concentration in *aggregated* representation likewise). Beyond it, however, the European economic territory contains a whole lot of other prominent centers of knowledge-intensive economic activities like, in particular, Barcelona, Florence–Bologna, Berlin, Copenhagen, Birmingham, and Manchester–Liverpool–Leeds–Sheffield. In the urban agglomerations of the middle and the north of the EU the 'knowledge economy' mostly reaches a share of 40–60 percent of the respective total regional employment.

The *development* of employment figures in the field of knowledge-intensive economic activities in the period 1997–2004 reveals an increase of the 'knowledge economy' in *all* urban agglomerations and metropolitan regions of the EU-15 (for the new EU member states, with few exceptions, no data are available). Strong absolute increases of employment figures in knowledge-intensive economic activities are to be noted in the urban agglomerations and metropolitan regions of the so-called 'core area' of the EU economic territory, and beyond this, nevertheless, also in urban regions like Nantes, Stockholm, Berlin, and Manchester–Liverpool–Leeds–Sheffield as well as in particular in the region of Dublin. Furthermore, there are clear signs of a process of catching up in the urban agglomerations and metropolitan regions in the south of the EU where among

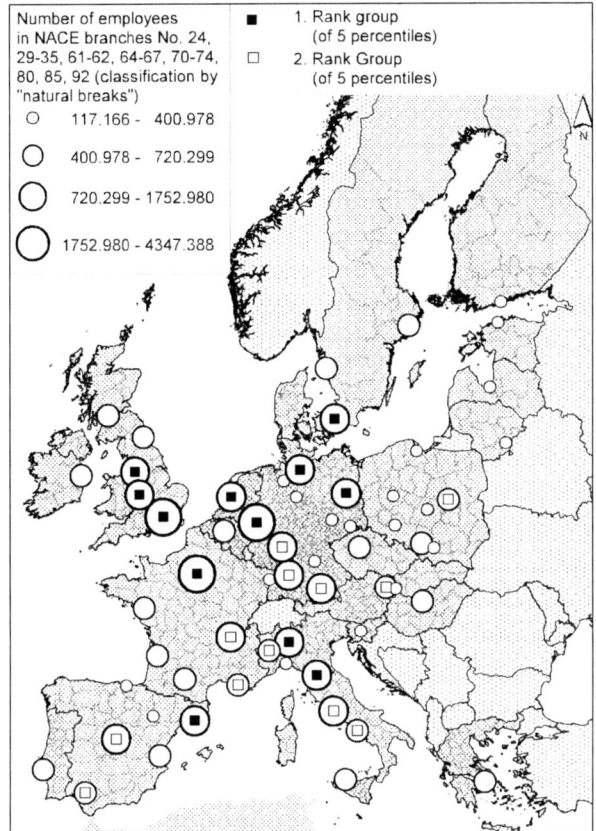

Figure 9.1 Absolute concentrations of the knowledge-intensive economy in the EU urban and regional system (2004)

others Madrid, Barcelona, Valencia, Florence–Bologna, and Rome belong to the regional 'winners' of the knowledge economy.

DIFFERENTIATION OF THE EUROPEAN URBAN AGGLOMERATIONS' AND METROPOLITAN REGIONS' ECONOMIC PROFILES

The preceding overall representation concealed important aspects of the knowledge-intensive economy's regional structure: The urban agglomerations and metropolitan regions of the EU have *different profiles* (or a specific 'mix' of branches) in the knowledge-intensive economy sector. The following sections focus on the European locational centers of selected sub-sectors of the knowledge-intensive economy with regard to employment figures.

In the sub-sector of *research-intensive high-technology industries* the metropolitan region of London turns out to qualify as the locational center with the strongest absolute concentration of employees (in 2004), followed by the Rhine–Ruhr conurbation on account of the aggregation effect of the cartographic representation. Other prominent locational centers of the high-tech industries are the regions of Paris, Lyons, Milan, Florence–Bologna, Munich, Stuttgart, and Frankfurt-Main as well as Dublin (see Figure 9.2). Moreover, the regions of Berlin, Hamburg, Manchester–Liverpool–Leeds–Sheffield, Barcelona, and Madrid belong to the strong 'second tier' locational centers of this activity branch. Altogether, it can be emphasized that a clear concentration of research-intensive *industrial branches* also appears in so-called 'service metropolises' like London, Paris, and Milan. This important fact is frequently passed over in superseded concepts of regional research that concentrate on the spatial articulations of the so-called 'service society' (see Geppert 2005; Paal 2005). In contrast, the representation of the locational centers of different sub-sectors of the knowledge-intensive

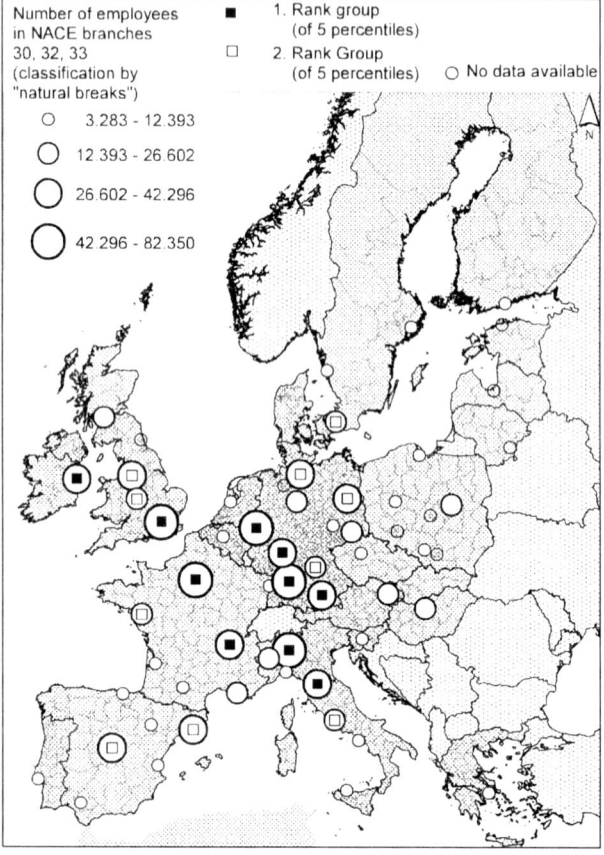

Figure 9.2 Absolute concentrations of research-intensive 'high-technology' industries in the EU urban and regional system (2004) (data for Brussels: 2002)

economy proves that the urban agglomerations and metropolitan regions of the EU are still prominent locational centers of research-intensive industrial activity branches and that their economic base is by no means reduced to services.

The development of regional employment figures in research-intensive high-tech industries is a partial component of the structural economic change toward the knowledge-intensive economy. The development of regional employment figures in the period 1997–2004 reveals that many metropolitan regions and urban agglomerations are characterized by a decline of the employment figure in this sub-sector (in particular the metropolitan regions of London and Paris, as well as Stockholm, Hamburg, and Vienna), while others note a clear increase: Among others, particularly Dublin, Copenhagen, Frankfurt-Main, Nantes, Milan, Florence–Bologna, Rome, Barcelona, Madrid, Seville, and Bilbao belong to this group of urban regions with job growth in high-technology industrial activity.

In the sub-sector of *medium high-technology industries* the metropolitan regions of Stuttgart and Milan as well as the Rhine–Ruhr conurbation appear as the locational centers with the strongest absolute concentration of employees (in 2004). Further prominent locational centers of the medium high-tech industries are the regions of Munich, Paris, Lyons, Barcelona, London, Birmingham, and Manchester–Liverpool–Leeds–Sheffield (see Figure 9.3). Again, a clear concentration of knowledge-intensive industrial activity branches also appears in so-called 'service metropolises' like London, Paris, and Milan.

The absolute change of the regional employment figures in this sub-sector in the period 1997–2004 points to a 'split' development pattern that contributes to the differentiation of development paths of the EU urban agglomerations and metropolitan regions in the course of the structural change toward a knowledge-intensive economy: Shrinking processes of the medium high-tech industries are to be registered above all in the urban regions of the northern parts of the EU with the exception of the urban regions of Dublin and Stockholm, while nearly all urban agglomerations and metropolitan regions in the south of the EU as well as in south Germany show an increase of employment figures in these industrial activity branches (see Figure 9.4).

In order to summarize briefly the findings on European centers of knowledge-intensive industrial activities, the sub-sectors of high- and medium high-tech industries and knowledge-intensive *technology-related services* might be aggregated: If we take into account that within the manufacturing sector there is a trend towards the relocation of technology-related service functions that previously have been performed within the manufacturing firm's organizational boundary to autonomous service firms (which in statistical terms are subsequently included in the 'service sector'), it might be reasonable to combine the sub-sectors of research-intensive manufacturing activity with the technology-related enterprise services into an aggregated group of 'knowledge-intensive industrial activities' closely related to manufacturing. This highlights the fact that research-intensive industries together with technology-related services make up a quite important economic base of the EU urban agglomerations and metropolitan regions. Technology-related services (like technical testing activities, lab services, etc.) are

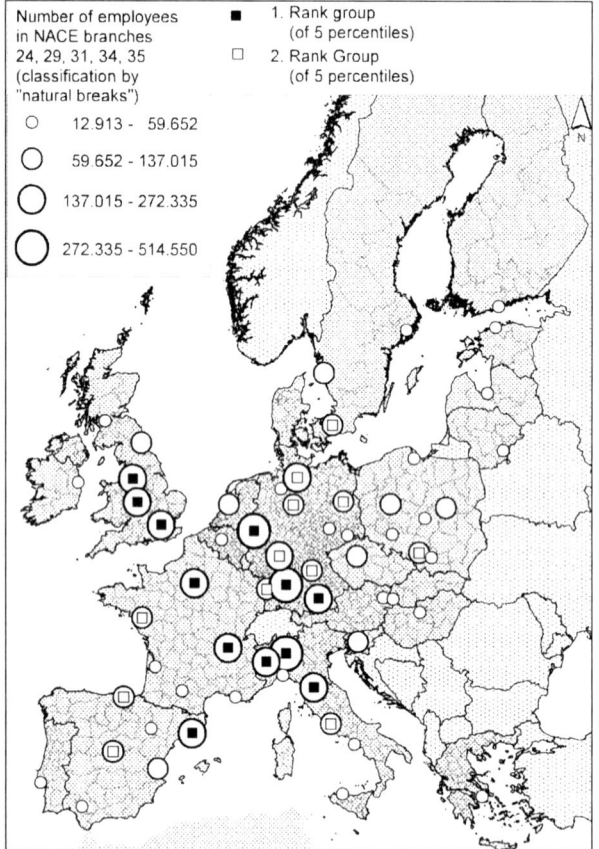

Figure 9.3 Absolute concentrations of the 'medium high-technology' industries in the EU
        urban and regional system (2004)

coupled functionally with the research-intensive industries and are expanding in
many urban agglomerations and metropolitan regions in connection with these
industrial activity branches. The following table contains a grouping of the EU urban regions according to five rank groups (based on percentiles) of total employment in this group of knowledge-intensive industrial activities (see Table 9.1).

As regards the *knowledge-intensive service sectors,* two selected sub-sectors
will be considered in the following section: the 'market-related enterprise services', which represent above all the so-called advanced producer services (economic consultancy, accountancy, legal advice, etc.), on the one hand, and the
'knowledge-intensive services in healthcare, education, and the media industry'
on the other. The financial services are not taken into closer consideration here,
because the European urban and regional system's major centers of financial
services already have been subject to detailed analysis (see Taylor and Walker
2001; Hoyler 2005). Likewise, the regional distribution of knowledge-intensive
technology-related services is not examined here (see above).

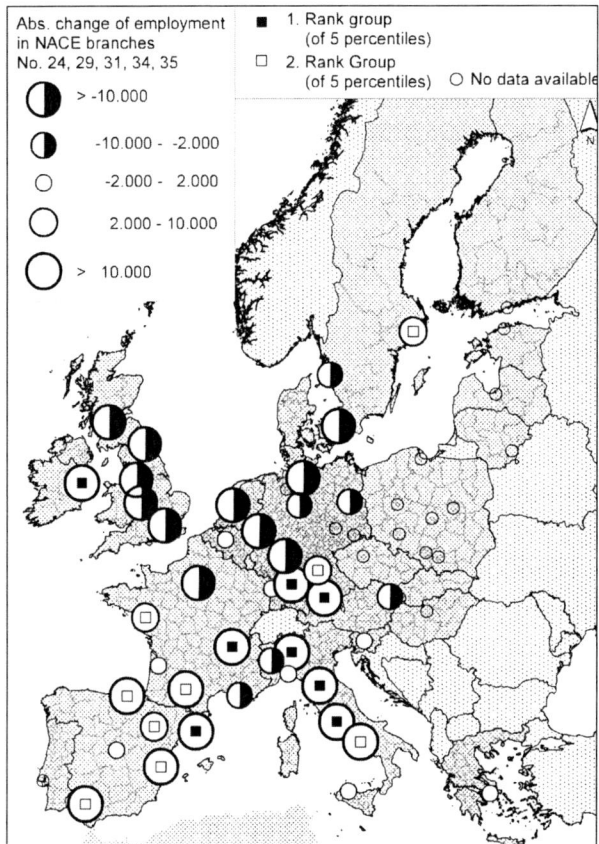

Figure 9.4 Increase/decrease of employment in 'medium high-technology' industries in the EU urban and regional system (1997–2004)

In the sub-sector of *market-related enterprise services* the metropolitan regions of London, Paris, Milan, Madrid, and Barcelona as well as the Randstad Netherlands are Europe's primary location centers (see Figure 9.5); the Rhine–Ruhr conurbation, the agglomerations of Manchester–Liverpool–Leeds–Sheffield, and Florence–Bologna reveal a considerable regional concentration of market-related enterprise services on account of the aggregation effect of summarizing several large urban locational centers. In the Federal Republic of Germany the primary locational centers of market-related enterprise services are spread out in a polycentric urban system of more than six metropolitan regions. Interestingly, with regard to employment figures, the metropolitan regions of Hamburg and Berlin today appear as stronger centers in this field of enterprise services than the metropolitan region of Frankfurt-Main, which functions as the leading center of financial services and global service firms in Germany. It does not need further comment at this point that the strongest concentration of market-related enterprise services has developed in prominent 'global cities' of the European territory like London,

Table 9.1 Grouping of European urban agglomerations and metropolitan regions according to aggregated employment in 'high-technology' industries, 'medium high-technology' industries and knowledge-intensive technology-related services in 2004: classification by five percentiles (the first percentile containing the highest values)

1	2	3	4	5
Barcelona	Amsterdam	Bilbao	Athens	Bratislava
Birmingham	Berlin	Gothenburg	Bordeaux	Bremen
Dusseldorf	Budapest	Hannover	Brussels	Gdansk
Florence	Copenhagen	Katowice	Dresden	Genua
Frankfurt-Main	Dublin	Marseille	Glasgow	Helsinki
London	Hamburg	Naples	Krakow	Lodz
Lyons	Madrid	Prague	Leipzig	Palermo
Manchester	Nantes	Seville	Lisbon	Riga
Milan	Nuremberg	Stockholm	Ljubljana	Talinn
Munich	Rome	Strasbourg	Newcastle	Vilnius
Paris	Turin	Toulouse	Poznan	Wroclaw
Stuttgart	Warsaw	Vienna	Valencia	Zaragoza

Source: Eurostat Regio Database; own calculations.

Paris, and Milan. The leading metropolitan regions of the EU are functioning not only as corporate service centers of the respective national economy but also as primary locational centers for the international or global firms in the sub-sector of market-related enterprise services (see below).

The sub-sector of *knowledge-intensive services in healthcare, education, and the media industry* reveals some deviations from the locational pattern of the primary centers of market-related enterprise services in the EU territory: The metropolitan region of London has an absolute concentration towering above all; concerning the Rhine–Ruhr conurbation the effect of the spatially aggregated representation comes through again. However, besides London, Paris, and Milan, the agglomeration of the Randstad Netherlands and the metropolitan regions of Berlin, Hamburg, and Copenhagen as well as the urban agglomerations of Birmingham, Manchester–Liverpool–Leeds–Sheffield, Lyons, and Florence–Bologna belong to the primary locational centers of services in the media industry, education, and healthcare (see Figure 9.6). Thus a strong concentration of knowledge-intensive services in this sub-sector is to be found also in metropolitan regions which do not belong to the prime European centers of market-related enterprise services. This leads to the conclusion that the so-called 'service metropolises' of the European urban system also have *different profiles* in their sectoral mix of knowledge-intensive service activities. In many urban agglomerations and metropolitan regions of the EU the media industry, education, and healthcare sector has a bigger weight in terms of employment figures than the market-related enterprise services.

The developmental dynamics of all knowledge-intensive services together in the period 1997–2004 is characterized by an increase of the employment figures *in all* urban agglomerations and metropolitan regions of the EU-15 (for the new EU

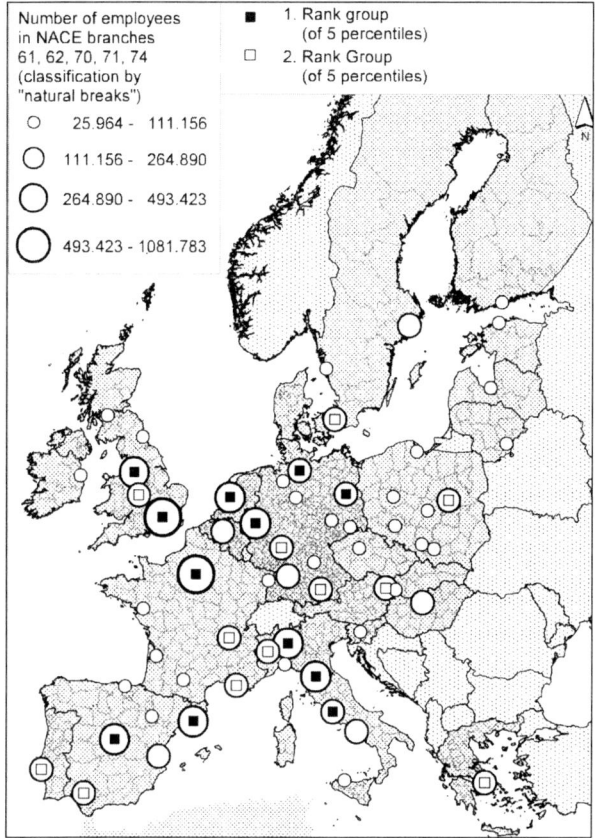

Number of employees
in NACE branches
61, 62, 70, 71, 74
(classification by
"natural breaks")

| ■ | 1. Rank group (of 5 percentiles) |
| □ | 2. Rank Group (of 5 percentiles) |

○   25.964 - 111.156

○   111.156 - 264.890

○   264.890 - 493.423

○   493.423 - 1081.783

Figure 9.5 Absolute concentrations of the knowledge intensive 'market-related enterprise services' in the EU urban and regional system (2004)

member states no data are available). In particular, strong absolute increases are to be noted in the urban agglomerations and metropolitan regions of the so-called 'core area' of the EU economic territory. Beyond it, nevertheless, a considerable increase of jobs in knowledge-intensive services also occurs in urban regions like Manchester–Liverpool–Leeds–Sheffield, Birmingham, Berlin, and Hamburg as well as in particular the region of Dublin. In the south of the EU territory, the urban regions of Madrid, Barcelona, Seville, Florence–Bologna, and Rome show a considerable increase of jobs in the knowledge-intensive service sector.

GLOBAL CONNECTIVITY AS A BOOSTING FORCE IN THE PROCESS OF METROPOLIZATION

As a starting thesis of this article the statement was made that the metropolization of the EU urban and regional system would be further strengthened in the context of globalization processes, which lead to a selective concentration of global

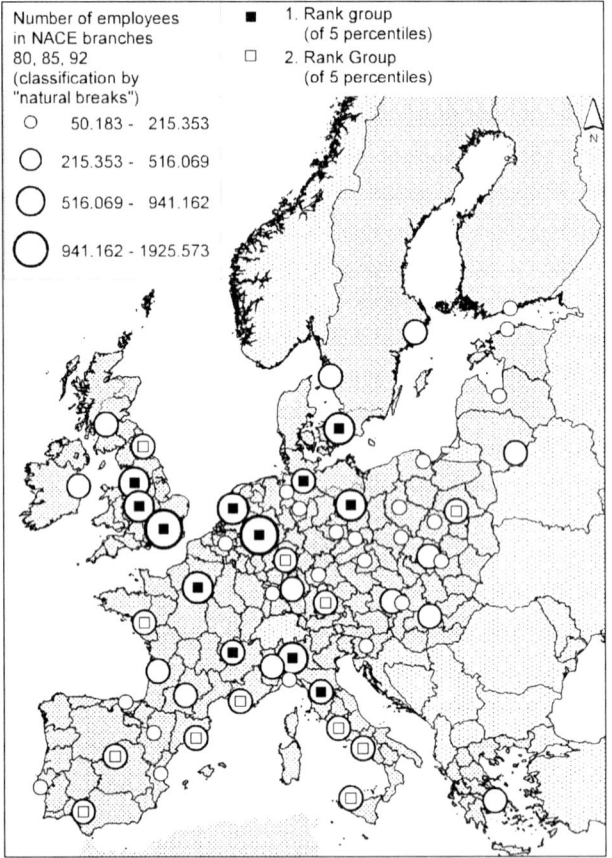

Figure 9.6 Absolute concentrations of the knowledge intensive services in education, healthcare, and the media industry in the EU urban and regional system (2004)

economic functions in the leading metropolitan regions of the EU territory. These urban regions are becoming the prime nodes of the organizational networks of worldwide operating enterprises and are functioning as major centers of Europe's integration into the world economy. Today there is widespread consensus among urban and regional researchers that large cities and urban regions would have to be analyzed in the context of present globalization processes as a part of a worldwide urban system, i.e. that the analysis of urban regions should not be restricted to their role within a single nation-state or national economic territory. In the framework of 'global city' research, London and Paris have always been classified as the indisputably outstanding global cities of the European economic territory (see Knox and Taylor 1995; Sassen 1991, 1996), highlighting their function as prime centers of the coordination and control of worldwide economic activities. However, in the meantime, research on the global urban system has extended its perspective: Today it is dealing less with world cities in terms of a classification category of rank orders of the global urban system, and more with processes of the *increasing global interlinking* of urban agglomerations and metropolitan regions (Taylor 2004) which include a whole lot of urban regions beyond the most prominent world cities (like London, Paris, New York, and Tokyo) and might be understood as a relevant factor of their regional economic development.

Today the EU urban agglomerations and metropolitan regions are the locational centers for new knowledge-based value chains and innovative production clusters in the field of research-intensive industrial activities and knowledge-intensive services (Scott 2001; Cooke 2002, 2003; Krätke 2005). Such urban or metropolitan 'knowledge clusters' are frequently characterized not only by dense inter-firm networks on the *regional* level but also by strong *supra-regional* connections with the 'knowledge clusters' of other regions, namely on a transnational and global scale (Rehfeld 2001; Krätke 2003; Zeller 2003; Alvstam and Schamp 2005). The global connections of urban or metropolitan 'knowledge clusters' are being established above all by the global players of the respective value chains, which in regional clusters often function as focal enterprises and anchor their amply extending locational network preferentially in the leading regional clusters of the respective activity branch (Ivarsson 2002; Krätke 2002; Lo and Schamp 2003). In this way specific regional knowledge and creativity resources are tied to the information and knowledge resources of other, geographically distant regions.

The transnational linkages between urban locational centers can be considered as an essential aspect of the phenomenon of 'globalizing cities' (i.e. the globalization of economic linkages of urban regions that do not belong to the small group of established prime global cities). Indeed, research on regional innovation networks in Europe have found out that the most successful regions in the field of 'innovative growth sectors' are those with strong regional (internal) *and at the same time* strong supra-regional and transnational interlinking of innovative enterprises (Arndt and Sternberg 2001). However, this research has not examined the role of the enterprise units of *global firms* located in the respective regions: The connection of a regional innovation network with global firms (as a particular resource of innovation and development) need not include a *direct supra-regional* connection, but can also contain *a regionally internal* interlinking with the enterprise units of global firms that are *located in the respective region.* This regional anchoring of global firms' organizational networks provides direct access to the respective regions' market potential, knowledge resources, and innovation impulses. By the global firms' strategy of setting up local anchoring points in a number of urban regions, the global interlinking of regional enterprise clusters (in terms of 'knowledge clusters') can be realized 'on site'.

A central (but nevertheless often too one-sidedly emphasized) aspect of urban development in the globalization process is the increasing concentration of global enterprise services in particular urban regions. Whereas the traditional notion of the 'service metropolis' emphasizes the importance of advanced producer services for the respective city's regional economy and the respective national economy's territory, global city research has been directed particularly to *the global reach* of an urban region's service capacities. For the long established global cities just as for the currently 'globalizing cities' it is typical that they are concentrating those corporate service firms upon themselves that have a 'global competence' (Beaverstock *et al.* 1999; Taylor and Walker 2001) in terms of their expert knowledge and transnationally extending organizational network.

The ongoing selective concentration of global service firms in the international urban system has been analyzed empirically by the Globalization and World Cities study group and network (for a comprehensive account of this research see Taylor 2004). Studies of the GaWC interpret the transnational locational network of global service firms as constituting a quite meaningful economic linkage between urban regions on which the extraction of *relational data* for the analysis of the global urban system can be based. The concentration of enterprise units of many global firms in a particular urban agglomeration or metropolitan region is at the center of the respective urban region's functional role as a more or less strong node of the organizational networks of global service firms. Further research on the global connectivity of urban regions might be directed to the analysis of the variety of globally interlinked economic activities in the urban system in order to discover the particular 'profiles' of the transnational connections of different urban agglomerations and metropolitan regions. This also applies in particular to the analysis of the global interlinkages of the European cities. Some steps in this direction have been made in the analysis of the transnational connectivity of the *global media cities* of the international urban system and the geographical patterns of their transnational linkages (see Krätke 2003; Krätke and Taylor 2004).

The thesis that globalization processes in the urban and regional system are boosting the tendency towards metropolization will be verified by demonstrating that the leading regional economic centers of the European territory – which have been identified in the preceding sections as the outstanding urban locational centers of the 'knowledge economy' – are at the same time characterized by the strongest degree of *global connectivity* in terms of their functioning as major nodes of the global firms' organizational networks. The most sophisticated empirical analysis of the transnational interlinkages of the global urban system has been made in the framework of the GaWC research (see above) with regard to global firms in the subsector of knowledge-intensive enterprise services. The following representation of the European urban regions' global connectivity is based on data provided by the GaWC (2005). These data were produced by P.J. Taylor and G. Catalano and constitute Data Set 11 of the GaWC Study Group and Network publication of inter-city data (http://www.lboro.ac.uk/gawc/). This data matrix includes 100 service firms with a globally extending organizational network and 315 cities from all parts of the world. The cartographic representation (see Figure 9.7) highlights the global connectivity measures of different urban regions particularly for *the European part* of the global urban system. This representation provides evidence for the finding that in the European economic territory global service firms have located their organizational units predominantly in the prominent metropolitan regions of the EU 'core area': The prime nodes of service firms' global organizational networks (as measured by the respective urban region's degree of global connectivity) in the EU territory are the metropolitan regions of London and Paris, followed by Milan, Frankfurt-Main, Brussels, and the Randstad Netherlands.

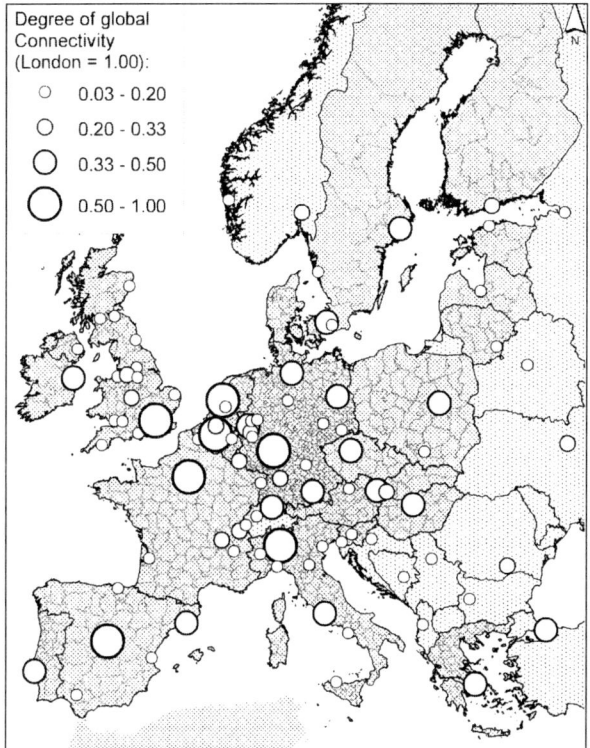

Figure 9.7 The global connectivity of metropolitan regions and urban agglomerations of the EU territory (2000): connectivity measures for each urban region in relation to the highest degree of connectivity of a European metropolis (London = 1.00)

Beyond these prime nodes, important secondary nodes have also formed outside the traditional EU core area, as for example in Madrid and Barcelona. Moreover, as an expression of the increasing integration of the urban regions of the EU accession countries of east central Europe in transnational economic networks, the metropolitan regions of Warsaw, Prague, and Budapest have taken on the role of important secondary nodes of global economic interlinking.

The representation of global connectivities is based on the geographic pattern of organizational networks of global service firms and combines numerous 'ego networks' of particular global firms with their specific anchoring points in order to extract an overall measure of connectivity for every urban region included in the analysis (concerning the methodical approach see Taylor 2004). However, in the framework of this chapter it might be useful to go beyond the aggregated measure of the urban regions' global connectivity and to concentrate on the *specific geographic patterns* of transnational interlinkages of *selected metropolitan regions* of Europe in a comparative perspective. This kind of analytic representation again can be based on the GaWC data set that was used in the previous step. However, for this purpose the GaWC data matrix has been extracted only with regard to the

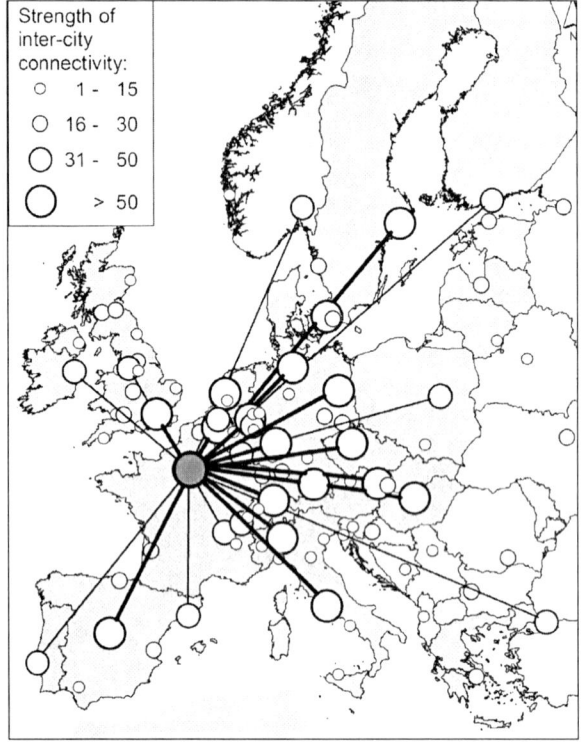

Figure 9.8 European inter-city linkages of the metropolitan region of Paris as mediated
by the organizational networks of global firms in the subsector of knowledge-
intensive market-related enterprise services (2000)

subsector of knowledge-intensive market-related enterprise services (furthermore,
the GaWC data matrix values for a global firm's presence in a particular city have
been recoded by the author to a range wherein the highest score combines the
valuation of a firm's global and regional headquarters). In the following cartographic representations (see Figures 9.8–9.13) the specific geographic pattern of
transnational interlinkages of selected European metropolitan regions *within the
European territory* – thus leaving out the cities' linkages to urban regions outside
the European territory – is filtered out by demonstrating the 'strength' of particular inter-urban economic linkages in Europe. The depicted *strength of singular
linkages* is based on each city's connectivity measure in relation to the selected
urban node, using a scale of classification that fits for inter-city comparisons. The
inter-city linkages of the two highest classification groups are made visible by
lines that emphasize the geographic pattern of the respective urban region's 'ego
network' within the European territory.

The selection of European urban regions for which a detailed cartographic
representation of their transnational inter-urban linkages within the European territory is to be worked out comprises (a) the metropolitan regions of Paris and

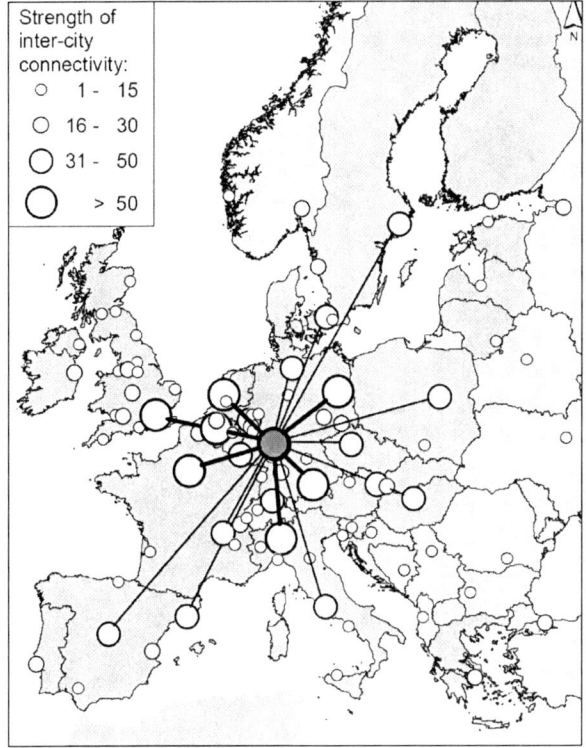

Figure 9.9 European inter-city linkages of the metropolitan region of Frankfurt-Main as mediated by the organizational networks of global firms in the subsector of knowledge-intensive market-related enterprise services (2000)

Frankfurt-Main in order to include two prime European centers of global connectivity in the particular subsector of knowledge-intensive market-related enterprise services, (b) the metropolitan regions of Berlin and Rome in order to get an impression of the transnational linkages of rather 'secondary' European centers of global connectivity, and (c) the metropolitan regions of Prague and Warsaw in order to emphasize the advanced integration of the east central European accession countries' metropolitan regions in the pan-European inter-regional economic linkages.

The metropolitan region of *Paris* has an overall degree of connectivity amounting to 65 percent of the connectivity measure of the European metropolis with the strongest global linkages (London = 100%). The cartographic representation of the European inter-city linkages originating from Paris (see Figure 9.8) reveals particularly strong linkages with nearly all of the prominent metropolitan regions and large urban agglomerations of Europe; these are particularly extended to the leading metropolitan regions of the EU accession countries in east central Europe. The metropolitan region of *Frankfurt-Main* is characterized by an overall degree of connectivity which amounts to 50 percent of London's connectivity measure

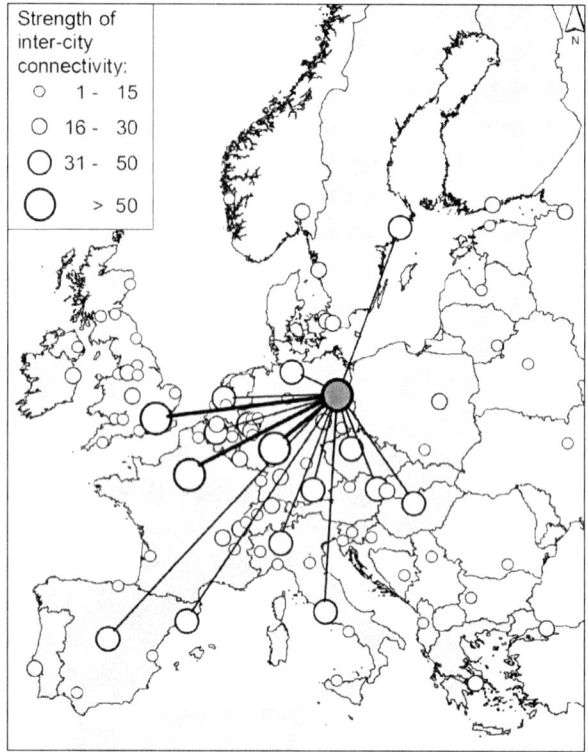

Figure 9.10 European inter-city linkages of the metropolitan region of Berlin as mediated
by the organizational networks of global firms in the subsector of knowledge-
intensive market-related enterprise services (2000)

and the geographic extension of Frankfurt's European inter-city linkages is comparable to those of the metropolitan region of Paris (see Figure 9.9). However,
compared to Paris, the metropolitan region of Frankfurt-Main reveals a smaller
number of inter-city links with the highest degree of connectivity.

As regards the metropolitan region of *Berlin,* which has an overall degree of
connectivity of 41 percent of the London measure, the geographic extension of
Berlin's most intensive inter-city links is oriented for the most part towards *western* Europe (despite the relatively strong links to Prague, Vienna, and Budapest),
with the strongest linkages being directed to the metropolitan regions of London,
Paris, and Frankfurt-Main (see Figure 9.10). In this respect Berlin is quite comparable to the metropolitan region of *Rome,* which has the same overall degree of
connectivity (41%) as Berlin and also the strongest single inter-city linkages with
London and Paris, followed by 'medium degree' linkages with most of the other
western European metropolitan regions (see Figure 9.11).

These findings might be taken as a basis for a view on selected metropolitan regions of the EU accession countries in a comparative perspective: At first,
the metropolitan region of *Prague* which already has the same overall degree
of connectivity (41%) as Berlin and Rome, reveals a geographic pattern of its

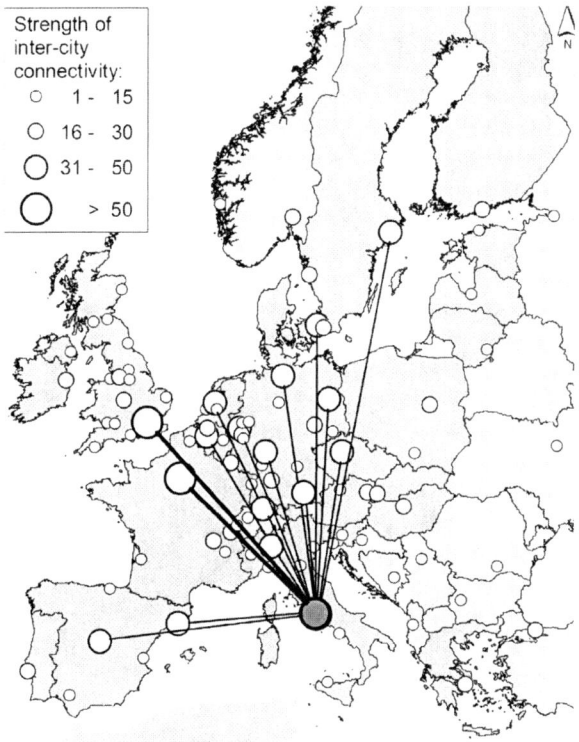

Strength of
inter-city
connectivity:
  ○   1 - 15
  ○   16 - 30
  ○   31 - 50
  ○   > 50

Figure 9.11 European inter-city linkages of the metropolitan region of Rome as mediated
by the organizational networks of global firms in the subsector of knowledge-
intensive market-related enterprise services (2000)

European inter-city linkages (see Figure 9.12) that is quite comparable to the extension and shape of Berlin's and Rome's pan-European connectivities. Again,
the strongest inter-city links of Prague are tied to London and Paris. Altogether,
the cartographic representation of Prague's European inter-city linkages might
be taken as a clear sign of the advanced state of integration of east central European metropolitan regions into the economic network of European cities. Just to
remember here: This representation has been restricted to the *European part* of a
globally extending urban network, so there are more transnational interlinkages of
the urban regions included that extend on the global scale of inter-city relations.
Recent research on globalization processes within the worldwide urban system
reveals that, since 1990, the capital city regions in particular of the EU accession countries in east central Europe have increasingly been integrated into the
transnational organization networks of global enterprises (Krätke 2002; Taylor
2004). Thus the conditions for a long-term 'rise' of the major urban regions of east
central Europe within the urban and regional system of the enlarged EU have been
developing in recent years.

As regards the metropolitan region of Warsaw, there is a lower overall degree
of connectivity (31% of the London measure) to be noted in comparison with the

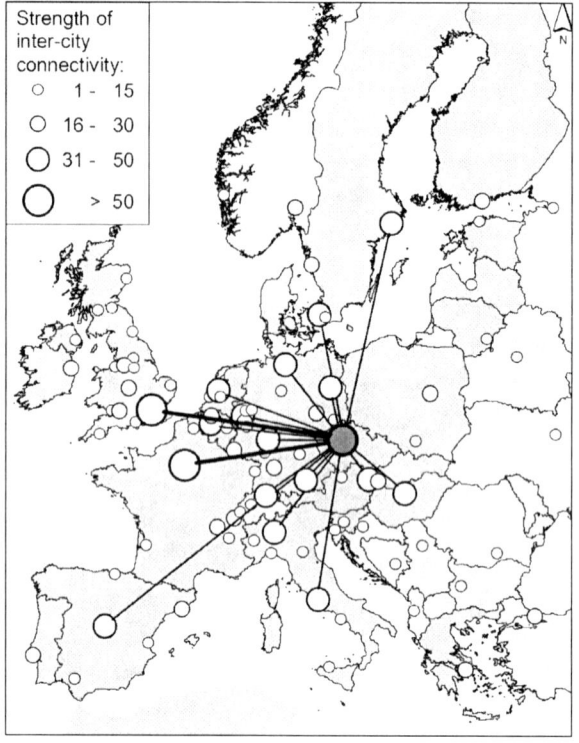

Figure 9.12 European inter-city linkages of the metropolitan region of Prague as mediated
by the organizational networks of global firms in the subsector of knowledge-
intensive market-related enterprise services (2000)

aforementioned urban regions. Warsaw's strongest European inter-city link is to
the metropolitan region of London (see Figure 9.13). Interestingly, the comparatively small number of 'high intensity' European inter-city links of Warsaw are
to prominent metropolitan centers of the market-related enterprise services in the
western part of Europe, whereas the links to metropolitan regions in geographical
proximity to Warsaw (like Berlin, Prague, Vienna, and Budapest) are characterized by a much lower grade of intensity.

As a short summary of the findings on the transnational connectivity of metropolitan regions and large urban agglomerations within the European part of
the global urban system, it should be emphasized that the economic inter-city
linkages that are mediated by the organizational networks of global firms are not
only being set up in the long established 'global cities' like London and Paris.
Today many other metropolitan regions and urban agglomerations of Europe are
being included in globalization processes in terms of becoming important nodes
of the organizational networks of global enterprises. Nevertheless, the strongest
concentration by far of global functions in the EU economic territory is still to be
found in London and Paris. The European urban regions are also characterized

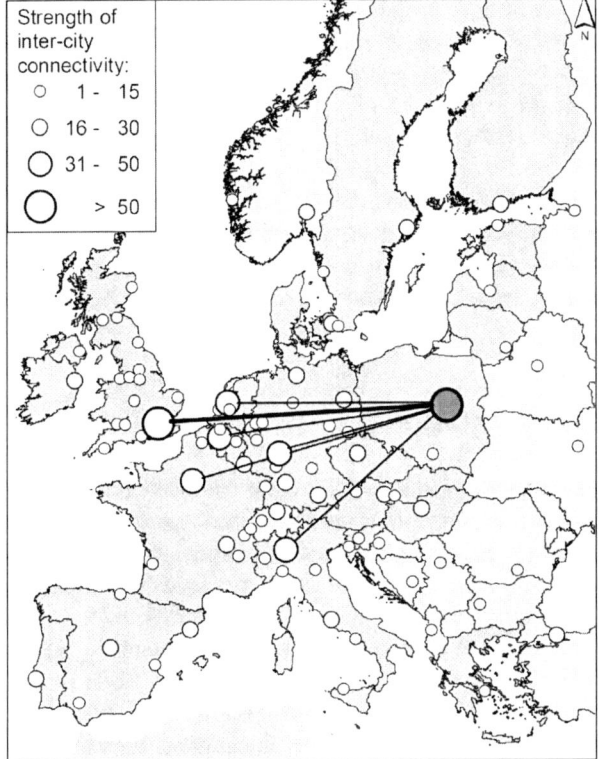

Figure 9.13 European inter-city linkages of the metropolitan region of Warsaw as mediated
by the organizational networks of global firms in the subsector of knowledge-
intensive market-related enterprise services (2000)

by *different geographical patterns* of their transnational inter-city linkages (in
terms of the extension and strength of the urban regions' particular ties). Moreover, the European urban regions are characterized by different sectoral profiles
of their global connectivity (Krätke and Taylor 2004), because the global firms
are tending to privilege different anchoring points in the global urban system with
regard to the business of market-related enterprise services, the media industry, or
research-intensive industrial activities.

Returning to the starting thesis of this section, i.e. the statement that the
metropolization of the EU urban system is being strengthened in the context of
globalization processes that lead to a selective concentration of global economic
functions in the leading metropolitan regions of the EU territory, we might prove
the empirical relation between the major European centers of knowledge-intensive
economic activities and the prime nodes of transnational inter-city linkages within
the European economic territory. Indeed, the correlation between total employment in knowledge-intensive *market-related services* and the overall connectic-
ity measure (extracted from GaWC Data Set No. 11 as mentioned above) for 49
urban agglomerations and metropolitan regions in Europe that could be included

in the analysis statistically reveals a highly significant and quite strong correlation coefficient of 0.77. Moreover, there is a highly significant and comparably strong correlation of 0.71 between the connectivity measure and the urban regions' total employment in *all sub-sectors* of the knowledge-intensive economy. Interestingly, the analysis also reveals a highly significant correlation coefficient of 0.60 with regard to the European urban regions' total employment in the sub-sector of knowledge-intensive high-tech industrial activities. This finding might give support to the thesis that global firms are anchoring their organizational networks in selected urban regions also in order to tap the respective regions' customer potential, specific knowledge resources, and innovation capacities in the sphere of knowledge-intensive industrial activity branches.

SUMMARY AND CONCLUSION

The preceding analysis has been focused on the thesis that the development of Europe's economic territory can be characterized as a process of metropolization of economic development potentials and innovation capacities. 'Metropolization' is used to denote the concentration of economic development potentials, particularly in the sub-sectors of research-intensive industries and knowledge-intensive services, on metropolitan regions and urban agglomerations. With regard to this development trend the metropolitan regions and urban agglomerations are functioning more and more as the 'motors' of the European economy as well as the 'prime nodes' of Europe's world-market integration. Moreover, with regard to the growing extent and impact of transnational inter-city linkages the economic integration process within the EU territory presents itself as a process of the cross-border integration of formerly national urban systems in terms of an intensively interwoven network of metropolitan economic activity centers. The urban regions' global functions are strengthening their economic power and thus have an amplifying effect on the process of metropolization.

Returning to the debate on the driving forces of regional economic development that today foster a tendency towards the *metropolization* of the urban and regional system (see the second section above), the conclusion might be drawn that, in theoretical terms, the particular 'strength' of the metropolitan regions and urban agglomerations in the context of the contemporary structural change towards an increasingly knowledge-intensive and innovation-driven economy lies in *the coupling of global and local interlinkages* within regional clusters of the knowledge-intensive activity branches (see Figure 9.14). This coupling allows a direct connection of processes of the 'hierarchical diffusion' of specific knowledge and innovation impulses between the centers of the European (as well as the global) urban system with processes of the 'neighbourly diffusion' of specific knowledge, learning and innovation impulses within the urban regions' clusters of the knowledge-intensive economy (see Hägerstrand 1967). Thus the present process of economic structural change carries altogether a pronounced amplification of the efficacies of spatial agglomeration.

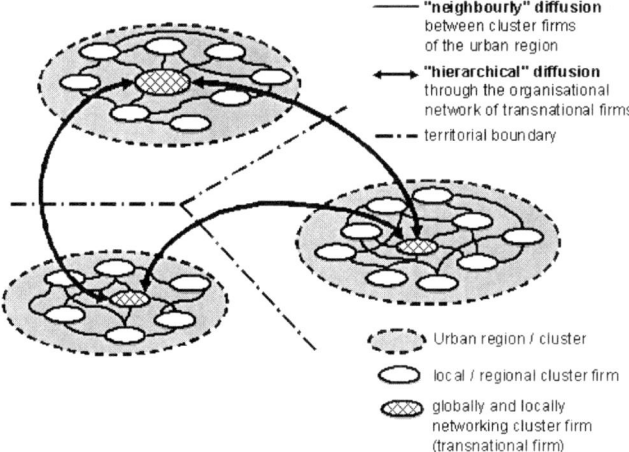

Figure 9.14 Coupling of the processes of hierarchical and neighborly diffusion of specific
knowledge and innovation impulses in metropolitan regions

With regard to the entire analysis the general conclusion might be drawn that
the traditional image of the EU economic space as representing a territorial mosaic of national economies just as the image of the EU nation-states as being
containers of 'national' urban systems is an increasingly questionable abstrac-
tion. As soon as the economy is understood as a spatially situated production
system, the EU economic territory presents itself primarily as an archipelago of
regional economic centers that constitute a transnationally interlinked 'network'
of dynamic urban agglomerations and metropolitan regions. This characteristic
feature of Europe's economic-spatial structure is being accentuated today by the
processes of metropolization and globalization.

REFERENCES

Alvstam, C. and Schamp, E.W. (eds) (2005) *Linking Industries across the World: Processes of Global Networking*, Aldershot: Ashgate.
Arndt, O. and Sternberg, R. (2001) 'Sind intraregional vernetzte Unternehmen erfolgreich-
er? Eine empirische Analyse auf der Basis von Industriebetrieben in zehn europäischen
Regionen', in R. Grotz and L. Schätzl (eds) *Regionale Innovationsnetzwerke im inter-
nationalen Vergleich*, Münster: LIT.
Beaverstock, J.V., Smith, R.G. and Taylor, P.J. (1999) 'A roster of world cities', *Cities*,
16(6): 445–58.
Berg, L., Braun, E. and Winden, W. (2001) 'Growth clusters in European cities: an integral
approach', *Urban Studies*, 38(1): 185–205.
Cooke, P. (2002) *Knowledge Economies: Clusters, Learning and Cooperative Advantage*,
London: Routledge.

Cooke, P. (2003) 'Biotechnology clusters, big pharma and the knowledge-driven economy', *International Journal of Technology Management*, 25: 65–81.

Cooke, P., Boekholt, P. and Tödtling, F. (2000) *The Governance of Innovation in Europe: Regional Perspectives on Global Competitiveness*, London: Pinter.

Florida, R. (2002) *The Rise of the Creative Class: And How It's Transforming Work, Leisure, Community and Everyday Life*, New York: Basic Books.

Florida, R. (2005) *Cities and the Creative Class*, New York: Routledge.

GaWC (2005) Data Set 11 of the GaWC Study Group and Network. Publication of inter-city data (http://www.lboro.ac.uk/gawc/), data produced by P.J. Taylor and G. Catalano.

Geppert, K. (2005) 'Die Position Berlins in der Hierarchie der europäischen Wirtschaftszentren', in H.-J. Kujath (ed.) *Knoten im Netz: Zur neuen Rolle der Metropolregionen in der Dienstleistungswirtschaft und Wissensökonomie*, Münster: LIT, pp. 203–24.

Grotz, R. and Schätzl, L. (eds) (2001) *Regionale Innovationsnetzwerke im internationalen Vergleich*, Münster: LIT.

Hägerstrand, T. (1967) *Innovation Diffusion as a Spatial Process*, Chicago: University of Chicago Press.

Hoyler, M. (2005) 'Finanzmetropolen im globalen Städtesystem: weltweite Vernetzungen und das Beispiel London/Frankfurt', in H.-J. Kujath (ed.) *Knoten im Netz: Zur neuen Rolle der Metropolregionen in der Dienstleistungswirtschaft und Wissensökonomie*, Münster: LIT, pp. 225–44.

Ivarsson, I. (2002) 'Collective technology learning between transnational corporations and local business partners: the case of West Sweden', *Environment and Planning A*, 34: 1877–97.

Keeble, D. and Wilkinson, F. (eds) (2000) *High-Technology Clusters, Networking and Collective Learning in Europe*, Aldershot: Ashgate.

Knox, P.L. and Taylor, P.J. (eds) (1995) *World Cities in a World-System*, Cambridge: Cambridge University Press.

Krätke, S. (2002) *Medienstadt: Urbane Cluster und globale Zentren der Kulturproduktion*, Opladen: Leske & Budrich.

Krätke, S. (2003) 'Global media cities in a worldwide urban network', *European Planning Studies*, 11(6): 605–29.

Krätke, S. (2005) 'Wissensintensive Wirtschaftsaktivitäten im Regionalsystem der Bundesrepublik Deutschland: Clusterpotenziale und Beitrag zur regionalen Wirtschaftsleistung', in H.-J. Kujath (ed.) *Knoten im Netz: Zur neuen Rolle der Metropolregionen in der Dienstleistungswirtschaft und Wissensökonomie*, Münster: LIT, pp. 159–202.

Krätke, S. and Taylor, P.J. (2004) 'A world geography of global media cities', *European Planning Studies*, 12(4): 459–79.

Kujath, H.-J. (ed.) (2005a) *Knoten im Netz: Zur neuen Rolle der Metropolregionen in der Dienstleistungswirtschaft und Wissensökonomie*, Münster: LIT.

Kujath, H.-J. (2005b) 'Deutsche Metropolregionen als Knoten in europäischen Netzwerken', *Geographische Rundschau*, 3: 20–8.

Lo, V. and Schamp, E.W. (eds) (2003) *Knowledge, Learning, and Regional Development*, Münster: LIT.

Matthiesen, U. (ed.) (2004) *Stadtregion und Wissen: Analysen und Plädoyers für eine wissensbasierte Stadtpolitik*, Opladen: VS-Verlag für Sozialwissenschaften.

OECD (ed.) (1999) *Boosting Innovation: The Cluster Approach*, Paris: OECD.

Paal, M. (2005) *Metropolen im Wettbewerb: Tertiärisierung und Dienstleistungsspezialisierung in europäischen Agglomerationen*, Münster: LIT.

Porter, M.E. (2001) 'Regions and the new economics of competition', in A.J. Scott (ed.) *Global City-Regions: Trends, Theory, Policy*, Oxford: Oxford University Press, pp. 139–57.

Rehfeld, D. (2001) 'Global strategies compared: firms, markets and regions', *European Planning Studies*, 9(1): 29–46.

Sassen, S. (1991) *The Global City: New York, London, Tokyo*, Princeton, NJ: Princeton University Press.

Sassen, S. (1996) *Metropolen des Weltmarkts: Die neue Rolle der Global Cities*, Frankfurt am Main: Campus.

Scott, A.J. (1988) *New Industrial Spaces: Flexible Production Organization and Regional Development in North America and Western Europe*, London: Pion.

Scott, A.J. (1998) *Regions and the World Economy: The Coming Shape of Global Production, Competition, and Political Order*, Oxford: Oxford University Press.

Scott, A.J. (ed.) (2001) *Global City-Regions. Trends, Theory, Policy*, Oxford: Oxford University Press.

Storper, M. (1997) *The Regional World: Territorial Development in a Global Economy*, New York: Guilford Press.

Taylor, P.J. (2004) *World City Network: A Global Urban Analysis*, London: Routledge.

Taylor, P.J. and Walker, D.R. (2001) 'World cities: a first multivariate analysis of their service complexes', *Urban Studies*, 38(1): 23–47.

Zeller, Ch. (2003) 'Restructuring knowledge acquisition and production in the pharmaceutical and biotech industries', in V. Lo and E.W. Schamp (eds) *Knowledge, Learning, and Regional Development*, Münster: LIT, pp. 131–67.

Part III Politics in inter-city relations

10 Positioning cities in the world

Towards a politics of flow

Phil Hubbard

> Cities and regions come with no automatic promise of territorial or systemic integrity, since they are made through the spatiality of flow.
>
> (Amin 2004: 34)

INTRODUCTION

The literature on globalisation is now quite bewildering in its breadth and complexity. Distilling the essence of this literature is thus nigh on impossible, yet one clear implication is that the contemporary world is characterised by new spatial formations in which network morphologies hold sway. From the communication infrastructures of global communications companies to the office structures of the advanced producer services, from the global web of airline routes to the movements of global migration, it seems we live in a world where everything flows. Given this, it has been argued that social scientists' traditional concern with the fixed and sedentary is being superseded by a preoccupation with the mobile and fluid. Bauman (2001) thus writes of *liquidity* as the defining characteristic of contemporary society: liquids may not bind or unite, but are extraordinarily mobile. These flows ooze, seep and flow around the world, often spilling over the 'dams' and 'defences' designed to impede their progress (such as immigration controls and border tariffs). The use of hydraulic metaphors implies a need for theories able to make sense of these new 'geographies of flow': a point underlined by Urry's (2000) recent proposal for a 'new mobilities' paradigm.

Accordingly, sedentary thinking in the social sciences is slowly being supplanted by mobile theories. Yet there is one area of critical enquiry which appears stubbornly recalcitrant to the onset of mobile thought. This is the realm of urban politics, where policy-makers and academics alike seem reluctant to accept that governance involves anything but the organisation and control of local assets. Dominant theories of urban politics – e.g. regime, coalition and regulation theories (see Hubbard *et al.*, 2002) – remain fixated on the partnership of 'locally embedded' actors and firms who develop a capacity to act through cooperation and networking that occurs within a prescribed territoriality. Likewise, the political mantra 'think globally, act locally' proliferated amongst urban governors in the 1990s, yet one major consequence was a rash of prestige projects designed to attract international investment (assumed to be global) to a particular city (assumed to be local). The result of this 'spatial atomism' (Doel *et al.*, 2002) is a world where cities imagine themselves to be islands of economic competitiveness, pitched in a global battle for jobs and dollars.

In this chapter, I take issue with this dominant conceptualisation of cities as bounded spaces in competition with other cities, working through some of the key contributions of the Globalization and World Cities group at Loughborough to tease out how cities might embrace this politics of flow. In doing so, I necessarily alight on some of the alternative strategies of globalisation being pursued by city governors as they struggle to adjust to a world where global flow is the norm and networking is the key to becoming more competitive in a global world. I begin, however, by reviewing the literature on world cities by way of exploring what it means to be a city in a global world.

CITIES IN THE WORLD

Summarising the links between globalisation and the city, Short and Kim (1998: 39) propose that 'globalization takes place in cities and cities embody and reflect globalization.' However, while the intensification of global processes seems to be providing a new context for urbanisation, the idea that cities have global reach and influence is not a new one. The term 'world city' was first coined by Patrick Geddes in 1915 to describe those cities in which a 'disproportionate' amount of the world's trade was carried out. Subsequent commentators on world cities – notably Hall (1966), Friedmann (1986) and Sassen (1991) – have reworked this definition in various ways, alerting us to the global economic and political power of a select cadre of cities (London, New York and Tokyo) which act as the command centres of the global economy. Beyond this, it is postulated there is a second tier of cities which serve as major regional centres in key globalisation arenas (e.g. Los Angeles in the Americas, Paris and Frankfurt in Europe). A third tier is deemed to comprise international cities of lesser importance, such as Seoul, Madrid, Sydney or Singapore. Beyond this, it is suggested there is a fourth level of nationally important cities with some strong international links (e.g. San Francisco, Osaka, Milan). Knox (1995: 239) adds a fifth tier of places like Rochester, NY, Columbus, OH, or the 'technopolises' of Japan, where 'an imaginative and aggressive leadership has sought to carve out distinctive niches in the global market place'.

The basis of such world city rosters has typically been the economic attributes of particular cities for which data are readily available, such as the number of corporate headquarters, stock exchange activity, presence of international banks, percentage of workforce employed by advanced producer services, air travel and so on (Friedmann and Wolff 1982). Aggregating such indicators produces a global urban hierarchy, with the most powerful and influential world cities at its apex. The fact that New York, London and Tokyo are commonly identified as the command centres of the global economy is not particularly surprising given their locational advantages (e.g. between them they offer 24-hour coverage of world financial markets).

For Sassen (1991), understanding the making of world cities demands serious consideration of the work of *advanced producer services*, defined as those corporate activities that enable the world economy to operate through the expert assistance they give to corporations. Covering such activities as banking, accountancy, insurance, advertising, public relations, law and management consultancy, these services went 'global' at the same time as their main clients, the transnational corporations (TNCs). The one thing these firms share is dependence on specialised knowledge. Their state-of-the-art commodities are produced by bringing together different forms of expertise to meet the specific needs of clients. In order to be able to put together such packages, it is understood that firms need to be embedded in knowledge-rich environments. Sassen (1991) suggests that world cities provide such environments, with face-to-face contacts between experts facilitated by the clustering together of knowledge-rich personnel in cities. Reflexivity and personal networking are also at the heart of Thrift's (1996) understanding of the global economy, with knowledge-rich workers, firms and institutions clustering to enable the transfer of embodied knowledge. Although new technologies have the potential to dematerialise these 'epistemic communities', making it possible for them to be located anywhere provided workers are electronically connected, Thrift emphasises a countervailing propensity for new forms of financial reflexivity to have gravitated together.

In sum, world cities are deemed to contain the economic, cultural and institutional infrastructure that channels national and provincial resources into the global economy and transmits the impulses of globalisation back to national and provincial centres. However, such conceptions of world city formation do not fundamentally challenge the idea that some cities are *places* crucial to the reproduction of global capitalism. A somewhat different perspective has been offered by Manuel Castells, whose *The Information Age* (1996) identifies a topology of networked flow as fundamental to informational capitalism. Arguing that global networks are becoming the new organising and managing principles (or 'social morphology') of capitalism, he identifies the emergence of a *space of flows*, an entangled skein of linkages, connections and relations across space. Castells (2000) describes this 'space of flows' as having three levels: the infrastructural, the organisational and the social. The first refers to the hard and soft technology that enables global communication, the second to the hubs and nodes that allow the network to operate, and the third to the networks of global elites that are of paramount importance in the information age. These three co-terminous networks provide the material support for time-sharing social practices, leading to an era in which *timeless time* exists in tension with chronological time – and a space of flows exists in tension with a bounded space of places.

Castells' rather simplistic opposition of global space and local place has left him open to criticism, particularly when his work is contrasted with the more nuanced takes on place identity offered by geographers (for an overview, see Cresswell

2004). Not withstanding these criticisms, Castells' structural take on globalisation has been widely influential in urban studies. Notably, Peter Taylor has sought to reconcile and clarify Castells' work in relation to his own interests on world cities, modernities and world systems. Taylor observes that Castells' conceptualisation of major world cities as 'command and control centres' is quite conventional in terms of the world cities literature, following, for example, Sassen's (1991) identification of London, New York and Tokyo at the top of a global urban hierarchy. According to Taylor (2003), the principal contribution of Castells to the world cities literature is instead to position the world city network in a richer theoretical context, making it one important network within a space of flows. Identifying world cities as the most direct illustration of hubs and nodes in the spatial logic of the informational age allows Castells to emphasise the unevenness of global flow, which he reads as critical for the distribution of wealth and power in the world.

Accordingly, Taylor (2003) insists that Castells' conception of global cities as 'embedded flow processes' rather than places usefully directs attention away from what cities contain (the case study approach) to their *connections* with other cities (the relational approach). It is this approach that informs the rosters of world cities generated by Taylor and members of the Globalization and World Cities (GaWC) network. Initially focusing on the role of world cities in supporting financial and business services (which Taylor regards as particularly central to contemporary patterns of economic change), GaWC has explored how leading service firms maintain their brand credibility by providing clients with a seamless service wherever the business takes them (so, for example, what begins as a London law firm becomes a successful global service firm by building a network of offices in cities spread across the globe). In this sense, London's service cluster may be considered as a node within a world city network in which the flows are the information, knowledge, instructions, personnel and ideas that move between the offices of particular firms.

Calculations of the service business links for individual cities and sectors therefore begin to identify the specific roles that individual cities play within an interlocking world city network. Whereas cities have traditionally been conceived as possessing bounded hinterlands in which they dominated, this shows that under conditions of contemporary globalisation servicing is worldwide (hence cities have 'global hinterlands'). Yet global hinterlands are not homogeneous or uniform: Frankfurt and London, for example, both have strong Pacific Asian links, yet London is more important in terms of servicing the US, while Frankfurt has a more European focus (Taylor 2001). This empirical specification highlights that London and Frankfurt have different roles as financial and business service centres within a world city network. Moreover, it begins to suggest that cities may connect to the world in different ways by virtue of the specific firms, workers, goods and services that pass through them.

Although this work is largely exploratory in nature, and has been criticised for illustrating strong empiricist tendencies (Thrift 1997), it does allow for the specification of a world city network based on the idea of relationality and connection. Further, it demonstrates that under conditions of contemporary globalisation *all* cities are global: they operate in a contemporary space of flows that enables them to have a global reach when circumstances demand such connections. Hence, while many urban geographers continue to use the term 'global city' for those leading metropolises that articulate global flow, work of this type suggests it is no longer possible to talk of 'non-global' cities.

THE SCALAR POLITICS OF ENTREPRENEURIALISM

The work of GaWC suggests that existing ideas of a world city hierarchy are flawed and need to be replaced by the idea of an interlocking city *network*. Nonetheless, GaWC rosters still imply hierarchical tendencies within network formation, suggesting that some cities are playing a more pivotal role than others in the articulation of globalism. The idea that there are winners and losers in the battle for global jobs and investment is one that has fuelled intense rivalry between different cities. According to Harvey (1989), this intensification of inter-city competition has led directly to the rise of *urban entrepreneurialism*. The motif of entrepreneurialism captures the increasingly 'businesslike' manner in which city governments operate, taking on characteristics once distinctive to the private sector – 'risk-taking, inventiveness, promotion and profit-motivation' (Hall and Hubbard 1996: 154). Fundamental is that city governments are paying increased attention to economic production, aggressively promoting investment within metropolitan areas.

While the entrepreneurial strategies designed to boost urban competitiveness are many and varied, we can identify at least two key ways in which entrepreneurial cities seek to improve their asset base so as to be more attractive to global investors. The first is to seek new 'location-specific' advantages for producing goods and services – for instance, through the installation of new physical, social and cybernetic infrastructures designed to promote agglomeration economies of scale and scope. The second is to find new sources of supply to enhance global competitive advantage. Examples here include re-skilling the extant workforce or, more commonly, seeking to attract highly skilled knowledge-rich workers (those whom Richard Florida, 2002, terms the 'creative class'). Both strategies thus embody a *neoliberal* ethos given they mobilise state power in the interests of encouraging private sector investment.

The alliances made between the state, police, businesses and developers suggest that entrepreneurial urban politics characteristically require collaboration between private and public sector actors. Deas and Giordano (2000) argue public–private partnerships are fundamental in the entrepreneurial promotion of the city because the public sector alone is unable to create, exploit, supplement and replenish city 'asset bases'. They contend that such actors include not just elected councillors and officials but also planners, local business people, police, Chambers of Commerce, universities, property consortia and media interests: in short, all those who have some sort of vested interest in promoting a particular city. Such public–private partnerships are generally termed *growth coalitions* as they represent elite groups of individuals who seek growth at almost any cost; elsewhere they are termed *urban regimes* (see Hubbard *et al.* 2002). Ultimately, what unites actors in these regimes is the perception that a project or policy would be good for them. Hence, it has been argued the key participants in urban regimes consist of locally dependent interests, whether these are the governments who want to increase their tax revenues, utility companies who want to boost spending on gas, water and electricity, retailers who want to increase local spending or rentiers who are seeking to enhance the value of their land.

Cox (1993) concludes that such local interests are forced to pursue entrepreneurial and outward-looking policies because they are largely *immobile* in an era where jobs and workers appear infinitely mobile. Crucial here is the idea of local dependence, which Cox (1993: 434) defines as 'some non-substitutability of local social relations'. The term describes the unique nature of particular local conditions and the consequent immobility of particular forms of capital and labour between geographical locations. In the theory of local dependency, capital production is embedded in a locality through its dependence on natural resources, physical infrastructures, labour markets and relations with suppliers and buyers found within a given territory. In turn, the labour force is dependent on a particular locality for amenities such as housing and for social relations, including a network of family and friendship. This type of territorial constraint can also be recognised in the spatial structure of the state (for example, in the location of services such as policing and libraries, which are necessarily locally embedded).

Harvey (1989) provides a similarly spatialised account of capitalist processes through an exploration of spatial *flows* (of money, capital, goods, credit and migrant labour) and the spatial *fixities* which embed capitalist production into particular locations. These fixities include the natural resources of an area, its built environment (physical infrastructure) and the spatial limits attached to the social reproduction of the workforce (i.e. the travel-to-work area of a city). Consequently, he contends that capitalist accumulation exhibits a tendency to concentrate in particular localities that become characterised by particular patterns of production and economic specialisms. Harvey (1989) argues that such 'production complexes' operate primarily on the scale of the 'urban-region', with stability within an urban region being an advantage for both workers and capitalists who have a stake in the locality. Consequently, both have an interest in maintaining the 'structured coherence' of production systems, leading to the development of a regional class alliance (including homeowners, the local state, real estate developers) that seeks to stabilise processes of production and consumption.

Such ideas about scale dependence are intuitively attractive, and allow us to make connections between contemporary urban politics and the increasingly globalised organisation of capitalism (Cox 1993). Consequently, they exercise a considerable influence on academics and policy-makers, informing a multitude of local economic development strategies designed to embed 'footloose' capital. Yet many of the assumptions on which such policies rest remain unproven. For instance, the claim that capital can flow 'at the twinkle of an eye from one fragmented place to another as the Starship Enterprise moves from one end of the Galaxy to another' (Swyngedouw 1989: 39) needs to be tempered with the observation that much capital remains embedded in property. Nor are all workers dependent on a given locality for their housing and social needs; transnationalism is a reality for many. MacLeod (1999) consequently accuses theories of local dependency of buying into a naive notion of scale whereby dependency is mapped unproblematically onto the local. As he details, dependencies may exist at other scales, and one notable trend is for class alliances to extend across space and time to include actors and institutions whose 'stake in place' takes different forms. In particular, the idea that urban governance is being *rescaled* suggests that a reassessment of local dependency theory is overdue.

RESCALING THE CITY

Though most entrepreneurial policies seek to promote urban growth by manipulating the assets that are contained *within* the city, both theory and practice are converging around the idea that most effective scale of territorial competition is often larger than established units of city government because these units seldom correspond to economically functional areas (Cheshire 1999). Consequently, one increasingly important strategy for enhancing urban competitiveness is *upscaling* the city. This refers to the process by which cities seek to redefine their spatiality, normally by working collectively with other localities towards common ends. Through such mutual cooperation, city governors may seek to transcend the boundaries of urban space, possibly 'jumping scales' (so that, for example, a city with a regional profile becomes part of a more significant national or international conurbation, and more competitive and attractive in global terms). Such rescaling provides evidence that urban governance needs to be understood not only in terms of networks of association *within* cities, but also in terms of urban policy-makers' relations with other actors, spaces and, indeed, geographical scales.

In broad terms, it is possible to identify three distinct types of upscaling: the multi-tiering of government, the polycentric linking of urban functions and the creation of complex urban networks (see Savitch and Kantor 2002). The first of these implies a process by which localities can band together to establish an umbrella institution with metropolitan-wide functions. Under this arrangement, extant localities may retain a number of existing functions, but another level of government make take over many of its policy-making functions. The umbrella tier can be independently elected or separately appointed: either way it provides an institutional layer that typically supplants or supplements the bureaucratic responsibilities of lower-tier urban governments. For example, in the British context it is significant that Labour governments since 1997 have embarked on a programme of constitutional change that has involved the establishment of the Greater London Authority with an elected mayor who has responsibility for the London Plan (also known as the Spatial Development Strategy for London). This covers topics ranging from strategic housing targets and locations to green belts, economic development and policies for improving movement and accessibility.

As such, the GLA provides a strategic framework to guide the work of the 32 London Boroughs, effectively supplanting the strategic planning functions of the boroughs. Beyond the capital, Regional Development Agencies and Regional Chambers (as well as the creation of Assemblies in Wales and Northern Ireland and the establishment of the Scottish Parliament) suggest that this process works in two directions, with the nation-state 'scaling down' as cities 'scale up'.

In part, this 'new regionalism' is based on an understanding that the city-region can operate as a competitive spatial unit. Theoretically, it is fuelled by some 'quite convincing claims' that the region has re-emerged to challenge the taken-for-granted position of the nation-state as the 'pre-eminent site and scale for territorial economic organisation in contemporary capitalism' (Ward and Jonas 2004: 2119). This contrasts with the rhetoric of earlier times when city-regions were seen as cumbersome and uncompetitive units, which capital and investment bypassed as they sought to locate in either the nation-state or the individual firm. 'Scaling up' by becoming part of a larger city-region is a way that municipalities seek to enhance their competitiveness, benefiting from the agglomeration economies that exist at a regional scale:

> For certain cities, those located broadly in the same economic region (if not the same country), it can be economically advantageous to develop collaborative arrangements, networks, alliances, resources, etc with each other. . . . This enables the more fortunate city-regions to specialise around clusters of economic activity, exploit comparative advantages, and out-compete cities located in other economic regions.
>
> (Ward and Jonas 2004: 2122)

Of course, being subsumed within a larger city-region might involve some loss of local identity, yet many urban governors see these losses as compensated by the economic benefits of being part of a larger – and more competitive – unit. Significantly, many of those who make important investment decisions for large transnational firms start by looking at cities with at least 1 million population, as they consider this the minimum population needed to support the type of services and infrastructure they require. This type of logic encouraged French planners to create eight 'regional capitals' – Lyons–St-Étienne, Marseilles, Bordeaux, Lille–Roubaix–Tourcoing, Toulouse, Strasbourg, Nantes and Metz–Nancy – each of which typically captured two or three urban centres within a wider metropolitan boundary. For example, the boundary of the RUL (Région Urbaine de Lyon), otherwise known as l'Aire Urbaine de Lyon, has created a 30-mile metropolitan area around Lyons containing all the facilities and infrastructures of a large urban region of European dimensions. This scale bestows on Lyons a heightened visibility, its population of over 1 million symbolically qualifying it, according to Lyons's political leaders, as a 'European mêtropole'.

However, it is important to stress that countervailing tendencies also exist, with some cities having 'downscaled'. For example, Greater Manchester Council (GMC) was formed in 1974 through the amalgamation of central Manchester

and nine towns in the north-west (Bolton, Oldham, Salford, Wigan, Ashton under Lyne, Trafford, Stockport and Rochdale). However, in the absence of clearly
defined and demarcated responsibilities between GMC and the district councils
within its jurisdiction, local politicians and officers were left to negotiate these for
themselves. *Streamlining the Cities*, a White Paper published in 1983, suggested
that abolition would remove such conflicts, insisting that the metropolitan coun-
ties were a wasteful and unnecessary tier of government in any case. Hence, while
London regional government, after being abolished in 1986, was re-established
in the 1990s to produce an expanded 'unit of territorial competition', the towns
and cities of Greater Manchester have moved in the other direction. As Randels
and Dicken (2004) imply, this is because central Manchester already considers
itself as a competitive urban region, and feels it does not need to collaborate with
Bolton or Rochdale to compete with continental European cities such as Lyons
and Milan. In this context, it is interesting to note that Manchester is involved
in a number of domestic (e.g. UK Core Cities) and international networks (e.g.
Eurocities) which do not include any of the other towns in the Greater Manchester
area.

This brings us to a second form of rescaling, which involves the linking of
specific functions across a number of towns and cities within a given region.
These linked functions usually depend upon an agreement being made about
the specialisms that might be best provided in specific locales. For instance, the
French government encourages this form of functional differentiation through
a system of contracts with individual cities (*contrats de villes*). These contracts
may advance a particular industry (*technopole*) or rejuvenate neighbourhoods or
adopt strategies for combating social exclusion. In the aforementioned example
of Lyons, for example, the discourse of Le Reseau de Villes is that each city
will promote its particular sector strengths: Lyons in health and the environment,
Grenoble in electronics, St-Etienne in architecture and the built environment,
and so on. Similarly, in the English East Midlands, the Three Cities Town Net
project is seeking to build on Leicester, Derby and Nottingham's specific specialisms (such as Leicester's leadership in space research, Nottingham's excellence
in biotechnology and science and Derby's tradition of aeronautical engineering).
This initiative is jointly funded by Derby City Council, Leicestershire Promo-
tions, Nottingham City Council, the European Commission, the Department of
Communities and Local Government, Nottingham East Midlands Airport and the
East Midlands Development Agency, in keeping with the Development Agency's
aim for the region to become one of Europe's top 20 performers by 2010. Again,
it is significant that each of the cities involved has a population of around 300,000:
by collaborating, they can claim to have over 1 million population within their
travel-to-work area, meaning they are able to compete with better-known cities
across the world.

This form of upscaling ties in with the identification of polycentric urban regions (PUR) as globally competitive spatial units *par excellence*. These are defined as regions with two or more historically and politically separate cities which
are nonetheless functionally interconnected and spatially proximate. A notable

example here might be the urban network formed by the 'Flemish Diamond', whose three largest cities—Brussels, Antwerp, Ghent—are not merely physically proximate but also functionally integrated, in the sense that they perform different functions (Albrechts and Lievois 2004). Despite the question marks remaining about the definition of polycentric urban regions (see especially Bailey and Turok 2001), considerable stress has been laid on the supposed economic advantages of the PUR, particularly in terms of its capacity to foster cooperation and to permit the efficient exchange of goods, services and information. Nonetheless, Parr (2002) argues it is very difficult to accept that these advantages are unique to the polycentric urban regions and are therefore not present in economic systems based on alternative spatial structures, particularly in an age of continually improving transportation and telecommunications systems. Related to this is the focus on the internal network structure of the polycentric urban regions to the neglect of external links. Given the current concern about urban competitiveness, Parr concludes more attention should be paid to the *international* links of the polycentric urban region and the firms within it, particularly with respect to the sources of inputs and the location of markets.

Hence, although many planners regard polycentric urban regions as offering a model for sustainable urban development and sustained capitalist growth, the creation of local inter-city alliances has limitations similar to those of the urban entrepreneurialism that has persisted since the 1980s. Indeed, the idea that international competitiveness may be best enhanced through bringing together cities that are already well connected is something of an oxymoron. Nonetheless, this idea is widespread in the policy literature, where urban competitiveness is frequently understood to be a function of the assets contained *within* a specific territory. Developing the work of management guru Michael Porter, Kresl (1995), for example, is adamant that external aspects must be excluded from any analysis of the determinants of a city's competitiveness. Likewise, he insists that a city's international competitiveness is quite different from the concept of an international city, arguing that the latter concerns connectivity at the international scale, while the former concerns only the city in question. In his view, a city can be extremely competitive without being connected to a network of other cities, just as a city can be fully plugged into a network without being at all competitive. It is on this basis that Kresl (1995: x) claims 'a city may dramatically increase its competitiveness, even its international competitiveness, without being or increasing the degree to which it is an international city.'

For Kresl, city competitiveness and success unquestionably are a function of the city's asset base. Likewise, in the influential work of Krugman (1998), which dismisses the notion of competition and holds to the idea of strategic economic complementarity, urban success is deemed to derive from agglomeration economies at the local scale. Against this, Phelps (2004) insists that *agglomeration economies* now appear to be available over much less localised geographical scales than the city-regions classically identified as competitive economic territories. His argument is that the growth of diffuse urban forms suggests a need to move away from classical agglomeration theory and the tendency to simply up-scale the territory over which external economies are perceived to be available. Parr concurs, detailing the sources of competitive advantage associated with particular forms of external economy:

> External economies of *scale* ... are based on the existence of specialised servicing activities, the possibilities for cooperative research and development activity, and the advantage of industry-wide marketing. By contrast, external economies of *scope* refer to cost savings to the individual firm which are dependent on the existence of firms in other industries. ... A third externality is concerned with external economies of *complexity*. These result from the fact that a firm is linked in input/output terms to firms in other industries to form an identifiable production entity. ... Each of these external economies may exist in a variety of spatial settings, *including ones in which the relevant activities have a dispersed pattern of location.*
>
> (Parr 2002: 271, emphasis added)

Developing this argument, Parr (2002) details the type of economic advantage that may accrue to a firm locating in a given city-region. Yet he also notes that there are spatial diseconomies associated with locating in cities themselves, and that many of the benefits of being located in a metropolitan region can actually be enjoyed at a distance. This is particularly the case in the service sector, where economies of scale may be less important than in the case of manufacturing. As Coe and Townsend's (1998) examination of new firm formation and clustering in the UK makes clear, localised agglomeration economies make little sense in relation to the service sector, where many firms enjoy the advantages of access to London's global markets even though they may be highly decentralised (located in cities throughout the 'greater' south-east, including Peterborough, Hemel Hempstead and Milton Keynes).

COMPLEX URBAN NETWORKS: TOWARDS A POLITICS OF FLOW?

Debunking the myth that agglomeration economies are only created through physical proximity, work in the 'new economic geography' suggests more attention needs to be devoted to the networks of interdependence and synergy created between firms that may be located at a distance. Phelps (2004) thus argues that the use of existing theory by economists and geographers, which emphasises external economies of scale, provides an increasingly inadequate means of understanding contemporary forms of urban economic agglomeration. His suggestion is that more attention is paid to *network externalities*, noting that there are many instances where firms agglomerate for reasons that are nothing to do with the presence of other firms. The upshot here is that extra-regional linkages may be more important than intra-regional ones.

In spite of these observations, many examples of urban upscaling appear to be based on a rather straightforward conception of international competitiveness, whereby a city region can outcompete its rivals by offering a distinctive and significant *spatial agglomeration* of know-how and institutional capacity within a given territory. In contrast, a third option for rescaling lies in the 'complex network' approach. Here, no tier of government is added, and there is no attempt to redraw the boundaries of the city. Rather, officials of existing independent localities make decisions to cooperate through multiple, overlapping agreements. This idea emphasises the importance of synergy, allowing localities to trade on each other's strengths. In effect, networking of this type emphasises that 'cities cannot make it on their own in an era of globalisation and increased interurban competition, but need to cooperate' (Leitner and Sheppard 2002).

What is perhaps most significant here is that these complex municipal networks may be spatially diffuse, involving cities that are not physically proximate. For example, the Lyons–Marseilles 'Cooperation Charter' was agreed in 1997, bringing these distant cities together to 'reinforce . . . the strengths of each of these metropoles, cornerstones in the new territory of prosperity of the Greater South-East of Europe' (Aderly 1997). Underpinning this were Marseilles's wish to regenerate its port activities and Lyons's wish to renovate and extend its inland port terminal, taking freight off congested roads. Other networks may be cross-national, such as the 'Transmanche Metropole', a 'cooperative initiative involving strategy between the city authorities of Caen, Le Havre and Rouen in France and the English local authorities of Southampton, Portsmouth, Bournemouth and Poole' (Church and Reid 1996: 1307). This network aims to share knowledge between city governments on approaches to economic development. Church and Reid explain that these areas face a unique type of regional economic problem based on restructuring ports, depressed resort towns and a rural hinterland with marginal activities. Likewise, the Town-Net project is one of the transnational projects in the EU's Interreg IIIB programme for the North Sea area. In this project seven participating city-regions, from the Netherlands, the United Kingdom, Norway, Sweden, Denmark and Germany, have teamed up to carry out a wide range of activities falling within four themes: networks and linkages; spatial quality; economic cooperation; and regional/town-network identity. The project's main objective is to strengthen the participating regions as a whole by stimulating development of each town's specialism, encouraging greater cooperation between towns, cities and regions without competitive overlap and resulting in differentiation and complementarity.

These examples suggest that policy-makers are beginning to embrace a politics of cooperation where the intention is to enhance connectivity between cities. In many ways, these strategies acknowledge the economic realities of a world where global networks provide the dominant social morphology. However, it is surely dangerous to argue that work *beyond* the confines of the city should to be prioritised over work within. More prosaically, perhaps, there seems to be a case for academics and policy-makers alike to question inherited understandings of competition and agglomeration, inside and outside, local and global, near and far. Entrepreneurial politics, polycentric urban networking, regional alliances and transnational networks may all play a role in enhancing city competitiveness, but what seems to be badly lacking at the moment is any monitoring of how these strategies succeed in establishing new flows – which is crucial if cities are to be more competitive in these globalised times.

CONCLUSION

This chapter has explored the literature on globalisation and the city to stress that cities exist in a space of flows. Accordingly, geographers are slowly learning that it is important not merely to map a city's fixed position within an increasingly interconnected global network but to explore the way it stretches through time and space. Even so, the idea that cities have an 'inside' and 'outside' remains predominant in urban political thinking, as does the idea that cities are fixed locales in time and space. As Amin (2004) notes, 'it will take a lot to displace the A–Z or concentric-circle image of London by a *relational* map that incorporates the network of sites around the world that pump fresh food into a distribution centre called Covent Garden, that draws neighbourhood boundaries around settlements in post-colonial countries with which social and kinship ties remain strong, that makes us see sites such as Heathrow airport or Kings Cross station as radiations of trails shooting out across the land and far beyond'. Yet perhaps this relationality is what is truly interesting about cities. The implications of this form of thinking for the conceptualisation of cities are profound: working through these issues thus represents a major challenge for contemporary urban geography – and urban politics.

REFERENCES

Aderly (1997) *Lyon et sa region: faits et chiffres*. Lyons: Aderly.

Albrechts, G. and Leivois, G. (2004) 'The Flemish Diamond: urban network in the making?', *European Planning Studies*, 12(3): 350–72.

Amin, A. (2004) 'Regions unbound: towards a new politics and place', *Geografiska Annaler B*, 86(1): 31–42.

Bailey, N. and Turok, I. (2001) 'Central Scotland as a polycentric urban region: useful planning concept or chimera?', *Urban Studies*, 38(7): 697–715.

Bauman, Z. (2001) 'Consuming life', *Journal of Consumer Studies*, 1(1): 9–29.

Castells, M. (1996) *The Rise of the Network Society: The Information Age: Economy, Society, and Culture: Volume I*. Oxford: Blackwell.

Castells. M. (2000) 'Materials for an exploratory theory of the Network Society', *British Journal of Sociology*, 51(1): 1–24.

Cheshire, P. (1999) 'Cities in competition: articulating the gains from integration', *Urban Studies*, 36(5/6): 843–64.

Church, A. and Reid, P. (1996) 'Urban power, international networks and competition: the example of cross-border co-operation', *Urban Studies*, 33(8): 1297–1318.

Coe, N. and Townsend, A. (1998) 'Debunking the myth of localised agglomerations', *Transactions, Institute of British Geographers*, 23(3): 385–404.

Cox, K. (1993) 'The Local and the Global in the New Urban Politics: a Critical View', *Environment and Planning D: Society and Space*, 11: 433–448.

Cresswell, T. (2004) *Place – A Short Introduction*, Oxford: Blackwell.

Deas, I. and Giordano, B. (2000) 'From the "new localism" to the "new regionalism"? The implications of regional development agencies for city-regional relations', *Political Geography*, 19(3): 273–92.

Doel, M., Beaverstock, J., Taylor, P. and Hubbard, P. (2002) 'Attending to the world: co-efficiency and collaboration in the world city network', *Global Networks*, 2(2): 96–116.

Florida, R. (2002) *The Rise of the Creative Class*. New York: Basic Books.

Friedmann, J. (1986) 'The world city hypothesis', *Development and Change*, 17(1): 69–83.

Friedmann, J. and Wolff, G. (1982) 'World city formation: an agenda for research and action', *International Journal of Urban and Regional Research*, 3(2): 309–44.

Hall, T. and Hubbard, P. (1996) 'The entrepreneurial city: new urban politics, new urban geographies?', *Progress in Human Geography*, 20(2): 153–74.

Hall, P. (1966) *The World Cities*, London: Weidenfeld and Nicolson.

Harvey, D. (1989) 'From managerialism to entrepreneurialism: the transformation of governance in late capitalism', *Geografiska Annaler B*, 71: 3–17.

Hubbard, P., Kitchin, R., Bartley, B. and Fuller, D. (2002) *Thinking Geographically: Space, Theory and Contemporary Human Geography*, London: Continuum.

Knox, P. (1995) 'World cities and the organisation of global space', in R.J. Johnston, P.J. Taylor and M.J. Watts (eds) *Geographies of Global Change*, Oxford: Blackwell Publishers.

Kresl, P.K. (1995) 'The determinants of urban competitiveness: a survey', in P.K. Kresl and G. Gaert (eds) *North American Cities and the Global Economy*, London: Sage.

Krugman, P. (1998) 'What's new about the new economic geography?' *Oxford Review of Economic Policy*, 14(2) (Summer): 7–17.

Leitner, H. and Sheppard, E. (2002) 'The city is dead, long live the network', *Antipode*, 34(3): 495–518.

MacLeod, G. (1999) 'Place, politics and "scale dependence": exploring the structuration of Euro-regionalism', *European Urban and Regional Studies*, 6(3): 231–54.

Parr, J.B. (2002) 'Agglomeration economies: ambiguities and confusions', *Environment and Planning A*, 34(6): 717–31.

Phelps, N.A. (2004) 'Clusters, dispersion and the spaces in between: for an economic geography of the banal', *Urban Studies*, 41(5/6): 971–89.

Randels, S. and Dicken, P. (2004) '"Scale" and the instituted construction of the urban: contrasting the cases of Manchester and Lyon', *Environment and Planning A*, 36(11): 2011–32.

Sassen, S. (1991) *The Global City: New York, London, Tokyo*, Princeton, NJ: Princeton University Press.

Savitch, H.V. and Kantor, P. (2002) *Cities in the International Marketplace: The Political Economy of Urban Development in North America and Western Europe*, Princeton, NJ: Princeton University Press.

Short, J.R. and Kim, Y.H. (1998) 'Urban crises/urban representations: selling the city in difficult times', in T. Hall and P.J. Hubbard (eds) *The Entrepreneurial City*, Chichester: John Wiley and Sons.

Swyngedouw, E. (1989) 'The heart of the place: the resurrection of locality in an age of hyperspace', *Geografiska Annaler B*, 71: 31–42.

Taylor, P.J. (2001) 'Urban hinterworlds: geographies of corporate service provision under conditions of contemporary globalization', *Geography*, 86(1): 51–60.

Taylor, P. (2003) *World City Network: A Global Urban Analysis*, London: Routledge.

Thrift, N.J. (1996) 'New urban eras and old technological fears: reconfiguring the goodwill of electronic things', *Urban Studies*, 33: 1463–94.

Thrift, N.J. (1997) 'Cities without modernity, cities with magic', *Scottish Geographical Magazine*, 113(2): 138–49.

Urry, J. (2000) *Sociology beyond Societies: Mobilities for the Twenty First Century*, London: Routledge.

Ward, K. and Jonas, A. (2004) 'Competitive city-regionalism as a politics of space: a critical reinterpretation of the new regionalism', *Environment and Planning A*, 36: 2119–39.

11 Political world cities

Where flows through entwined multi-state and transnational networks meet places

Herman van der Wusten

POLITICAL CENTERS AND WORLD CITIES

Central places come in very different guises. They range from bustling metropoles to sleepy crossroads. They may be concentrations of the headquarters of important companies and aggregates of providers of advanced business services; or places with many political offices; or palaces of major religious dignitaries and their support staff; or all of them together. With the compression of space–time there are perhaps also virtual central places where some of these functions are very directly linked despite physical distances, but the question remains unresolved how indispensable the opportunity for unmediated, direct personal contact still is. Central places in some way dominate their geographical environment or are well connected parts of networks linking divergent places. Many have both features.

In this chapter I concentrate on real places that are central thanks to a political function. Emphasis will be on political central places as units in a widely spread, global network. Central places have most often been studied from an economic perspective emphasizing transport cost and travel time. But the symbolic force of place, the social value of congregation, and the functional attraction of an in-formation-rich milieu have to be added particularly in the case of political and cultural central places. While economic central places have long been recognized as an appropriate topic of study, centrality in politics and religious institutions has often been taken for granted in social science (but even Christaller (1933) recognized other than purely economic considerations in his central place model by adding the 'administrative principle'). The relations of economic, political, and cultural institutions in central places are obviously also an important question to be considered in this context (for a contemporary example outside the state order involving large-scale networks in south Morocco, see Pascon *et al.* 1984).

Classical functions and institutions within a polity like heads of state, delibera-tive assemblies, legal courts, treasuries, agencies of law enforcement and defense of the realm, corps of diplomats, and at a later stage the management of public services tend to aggregate in capital cities. These capitals are the foremost com-mand and control centers of the state system. The aggregation is not complete. In some countries these functions are somewhat distributed (e.g. the Netherlands,

South Africa, Germany) (see van der Wusten 2003 for additional examples). But the aggregation is also not perfect because not all political functions are completely embodied in states.

The state system now dominates alternative forms of political order, but the state as its basic building block is not the only form of polity we know. Cities, city leagues, and empires are historical alternatives with contemporary potential and the European Union has emerged as an unprecedented new form that still has to find its enduring shape. In addition, the state system could never function as a mere set of mutually isolated state containers. There always had to be some cohesive force. This came in two ways: by official contacts from state to state apparatus through shared conventions in diplomacy (thus producing an 'international society of states', Watson 1992) and by 'transnational' cooperation between citizens and their associations from different countries (originally called 'internationalism', a word that was later largely captured by state forces). Contact and cooperation can be on a bilateral or a multilateral basis (from the perspective of the number of states involved). Bilateral links will most easily begin and end in the same capital cities; there is more latitude – but perhaps not more preference – for multilateral institutions to be established in another place.

Politics has its own set of rank-ordered and interrelated central places. They can be command and control centers of delimited territories or nodes in widely spaced networks or both. Many regional capitals (Kooij and Pellenberg 1994) are home to management offices for a range of services within delimited territories below state level. A few cities host a multitude of offices of international organizations. They are mutually linked and also related to the externally oriented parts of state apparatuses. In these networks power, influence, and authority are produced and distributed. Despite the nominal equality of all states in terms of 'club membership' of the system – the family of nations as the UN endearingly but thoroughly falsely has it – there have always been clear status differences between states and accordingly there has always been a ranking of their capitals on the basis of national power: the stock of power assets in the territory plus the assets that can be mobilized thanks to a country's international situation. This translates to a variation of positions in the networks that politically link states, their capitals and other political centers. International organizations also differ in importance, have different inputs in the networks in which they are active, and contribute differently to the centrality of the places where they are hosted.

From the bottom up people are politically and administratively linked to lower-tier political central places (district and regional capitals etc.), then often in a more encompassing and intense fashion to the institutions in their national capital as a symbol of national attachment (but for inhabitants of Catalonia, Scotland, and Bavaria there is competition between Barcelona and Madrid, Edinburgh and London, Munich and Berlin), and finally – in most cases less intimately – to over-arching political central places like Brussels, Strasbourg, Geneva, New York, and Washington as the residences of even more widely important political institutions. These places also act as political wholesale depôts or ultimate headquarters for state functionaries and national politicians. From the top down state leaders are, according to the basic rules of the state system, the upper tier of the hierarchy and they are consequently by definition located in the highest-ranked political centers. But they are involved in all kinds of interdependencies with their colleagues. In fact most are to various degrees dependent on others asymmetrically. Therefore, some centers are much more central than others. In addition functionaries in international organizations wield certain powers irrespective of their state origin but at least partially derived from their institutional position. Leaders of sub-national and transnational units (e.g. mayors of major cities and chairpersons of vocal nongovernmental associations) may also have considerable authority in their respective spheres of interest and expertise. The places that host those institutions are also political central places of some consequence.

Political centers are thus nodes in a political network that stretches globally. The network is differentiated as to policy sector and also geographically, but parts of it are seamless webs with worldwide range. There is no doubt that some of the cities in which these centers are located are political global cities in the full sense while others are more tangentially involved. All capital cities are global to the extent that they are linked in a bilateral diplomatic network that eventually covers the globe and that the attention of their ministry of external affairs is somehow geared to the UN institutions in which they take part. It should however be added that the bilateral diplomatic network has a very strong regional bias while at the same time a few globally operating states, particularly the United States, stretch their bilateral relations across all regional divides. This makes Washington into a truly global diplomatic center, whereas most capital cities are in their bilateral relations primarily involved in their own region.

Most of the analyses on hierarchies and networks in the political field worldwide have been done on the basis of states as the appropriate unit of analysis. It would be worthwhile in my view to additionally concentrate on cities with state-belongingness as an obviously important contextual and explanatory factor. There are four reasons for this. Preferences for political sites are not only in terms of states, but also of cities for their perceived atmosphere and their amenities (Henrikson 2005). City governments have a certain agency in attracting these institutions although state governments are generally more important in the case of state-related international organizations (van der Wusten 2006). The hosting of political functions has sizeable local impacts economically and culturally that may hardly be visible at the state level (Elmhorn 2001). The aggregation of all political functions at state level neglects significant divisions on the basis of geographical separation (think of Washington and New York; Paris and Strasbourg; Berlin, Frankfurt, and Karlsruhe etc.).

In the remainder of this chapter I will first provide an overview of the attention on politics generally and political institutions in particular in recent research on world cities. I will then concentrate on one particular instance of political central place formation in a network context: central places specialized in the hosting of organizations oriented to multilateral interstate cooperation. This is followed by the cases of four European cities (Geneva, Brussels, Vienna, The Hague) that have emerged as important 'political world cities' of this kind during the last century. These European cities are of global significance but Europe is not the only interesting continent in this respect. We should not rashly generalize conclusions drawn from this limited set of cases but look at them primarily as the European contribution to the emergence of an extra category of world political cities over and above the nodes in the interstate network of capitals. I close with a few notes concerning unilateral and multilateral policy-making in light of the apparently emerging geography of world political cities.

FINDINGS OF RECENT RESEARCH

The ambition of research into world/global cities has always been to understand their emergence from a political economy perspective with culture included. The main driving force of global city formation has been ascribed to changes in the organization of production and distribution of goods and services. From the start, there have additionally been the political economy questions of how this economic restructuring reshapes politics and how political activity affects the local outcomes of global economic restructuring. Related questions have been how a global culture is particularly articulated in global cities, how it is conditioned by the alteration in the political economy, and how it provides feedback to the further enrichment of city life.

Such a wide set of concerns also spurs the 1995 volume edited by Knox and Taylor, which takes stock of current research practice at the time. In the introductory chapters Knox tries to integrate an economic with a political and a cultural perspective on world cities summarized in a three-dimensional figure, which shows the divergent positions on these three dimensions of London, New York, and Tokyo, as the three perceived frontrunners for the global city championships of the period, and other cities. Taylor has an initial statement of the strains between the (political) state order and (economic) world city networks that are much further elaborated in his more recent work (Knox and Taylor 1995: 48–62; Taylor 2004a). In his latest papers he has moved away from a complete opposition of (political) spaces of places and (economic) spaces of flows as residing in the mosaic of states versus the networks of cities (Taylor 2004b). The remainder of the book comes in two parts: one deals with the network ('cities in systems') of global cities, the other with 'politics and policy'. This last part is mostly concerned with the politics of world cities: the management of their services but also their efforts to connect politically with other cities in transnational frameworks. It also deals with the cultural perspective on globalization that emphasizes the perception and recognition of the globe as a meaningful context of social life and how this particularly takes shape in world cities and underlines a globally functioning economy.

In fact, much of the world cities literature has been written from an economic perspective with politics often added and culture not entirely forgotten. This follows the emphasis in the work of Friedmann (1986), Sassen (1991), and Castells (1996), the framers of this work, and is reflected in the list of 165 papers so far shown on the GaWC website (www.lboro.ac.uk/gawc). Slightly more than

50 percent of these papers have been produced by the Loughborough group or with their substantial cooperation; the remainder have come from elsewhere. More than half of Loughborough's own output deals with advanced business services; in the papers originating elsewhere this is about 40 percent (with often less well demarcated categories). About 10 percent of both Loughborough's own output and the papers from elsewhere are concerned with the manifestation of global culture in world cities mainly through research on international migration of skilled workers and expat communities. The remainder of the papers deal with individual case studies (often of single cities), non-economic institutions, or theoretical issues.

In this last category quite a few papers deal in different ways with politics and political institutions. They reiterate the concerns mentioned earlier and deal with, in effect, three topics:

a. city government (including place marketing and branding): how globalization is locally managed;
b. transnational city networks as a possible political alternative to the territorial state order: how transnational networks of intercity cooperation may undermine the state order, prone to violence and dysfunctional for economic development as it is purported to be;
c. centrality produced by political institutions: how interstate organizations plus supra-national bodies and transnational associations produce city-based political nodes and networks.

Papers on city government raise several research questions. The first one is the appropriate way of governing urban regions that are deeply involved in globalization. Functional requirements of scope, style and structure of government are important in that connection. A subcategory is aimed at the way cities should be externally represented in order to function as the privileged nodes in the global space: the branding of cities (secure? tolerant? creative? knowledge-based? entrepreneurial? social, perhaps, after all?) is deemed essential (e.g. bulletins 37 (reprinted as Hubbard 2001), 64 (reprinted as Doel and Hubbard 2002a), 81 (reprinted as Doel and Hubbard 2002b)).

There is another stream of papers mainly propagated by Peter Taylor that squarely opposes the city networks to the territorial states as part of the preceding historical political order and as a potentially attractive alternative for current political practice. He tends in that connection to be particularly dismissive of the role of capital cities as a kind of 'unnatural' traitor, which either are economically unsuccessful or suck the lifeblood of the preferred city networks (see also Taylor 2004a: 186–8). In a similar vein there is an interest in cities playing a positive role in conflict resolution and in more pacific international relations than territorial states have apparently been able to conduct (e.g. bulletins 107 (reprinted as Taylor 2004c), 123 (Stanley 2003)).

Finally a few papers deal with political institutions as center producers in their own right. A few describe single cases: Washington (Gerhard 2003) as a dense tapestry of national and international political communication and The Hague as a global judicial center in search of a proper branding (van der Wusten 2006). One deals with the distribution of political centrality in the European state system over time (van der Wusten 2004) and a few others draw the contemporaneous structure of political centrality on the basis of interstate bilateral links, multilateral organizations with state membership, and global civic associations (in particular Taylor 2004b)

There are some exploratory, follow-up attempts to show the overall interrelations of globalization in the centrality of cities based on economic, political and cultural institutionalization. One by Taylor (2004d), particularly suggestive and based on 16 datasets, portrays the upper level of the world city hierarchy as a small array of global cities from an economic as well as a political, cultural, and social point of view. Again, as in earlier work on advanced producer services, London and New York reign supreme, now as multidimensional global cities. On a somewhat lower rung Brussels and Geneva stand out as global cities particularly specialized in the political sphere (together with Washington and in contrast to Tokyo, Singapore, and Hong Kong, which are at the same level but economically specialized). Vienna, again a bit lower, is also a specialized global city in politics as an articulator of one globally operating sub-network.

In the general literature on political globalization the interest in cities has so far been minute (e.g. the seminal work by Held *et al.* (1999) has no time for contemporary cities at all). There is one major monograph that deals with a contemporary city's political institutions from the perspective of the world and global cities literature, particularly with respect to Sassen's work and the debates it has provoked. This is Elmhorn's book on Brussels (2001). It is a very interesting effort to make sense of Brussels' growing international position from the perspective of the concentration of transnational companies' offices, advanced producer services, and the EU institutions and their interrelations. The study may be somewhat overplaying the importance of international business and be too little interested in the size of the different Belgian governments themselves in Brussels, yet at the same time it is very well connected to the world cities literature. In addition, by means of some detailed case studies, the book deals with the local consequences of these globalizing tendencies. It maintains the inevitability of clashes between the space of places (of regional government and local inhabitants, some of them international civil servants residing in Brussels) and the space of flows (of international business and the production of political regulations) as they have to intersect somewhere.

In conclusion, there is a recognizable though small segment of the world cities literature dealing with political institutionalization. Some of it is interwoven with economics (e.g. the income generated by the hosting of those institutions), another part not (e.g. the flows of authority and loyalty generated by political 'command and control centers'). There is definitely room for further work in this area. An obvious question is how and where politics, other than state system politics, is articulated. This is important because it brings into play globalizing forces additional to those emanating from the dominant political order (the state system) itself. In the next two sections I concentrate on places where multilateral politics is conducted. These concentrations of political institutionalization are certainly not outside the state system but they combine bilateral, multilateral, and transnational elements in such a way that the classic rules of the state system (the state is endowed with integral sovereignty, external relations are only meant to maintain this state of affairs) do no longer strictly apply.

MULTILATERAL POLITICAL WORLD CITIES

Politics has always built, dressed up, and used places for consultation, deliberation, decision-making, and expression of authority. For most of these functions 'no place' is an option (even then places are used, but participants do not congregate) but it is rarely taken. Politics still needs direct face-to-face contact in many instances. Such places may be provisional, in the case of one-off meetings, conferences, and spectacles, or permanent when they accommodate headquarters of international organizations and offices for the external services of the state apparatus. Levels of centrality among those places are bound to arise. This also applies to politics transgressing state levels. A major part of such politics is still macro-regionally oriented, often with inputs from globally active actors, but these macro-regions have fuzzy boundaries and their outputs tend to resonate in other parts of the world as well. In addition a global level, as a politics with high relevance everywhere and/or participation from all different parts of the world, has emerged.

Multilateral diplomatic activity at least occasionally congregating in one place is as old as the state system. In fact, the Peace of Westphalia in 1648 resulted from conferences in the German cities of Münster and Osnabrück attended by delegates of the aspiring members of the new club of state sovereigns. During the following period major interstate conflicts were often finished by peace congresses that again brought together many members of the European state system in a setting of multilateral negotiation. At the Vienna Congress of 1815 some rules for the treatment of diplomats generally and also for the conduct of such multilateral negotiations were agreed, but permanent congress machinery was still lacking.

In the course of the nineteenth century a few permanent interstate organizations were established, particularly to cope with the growing intensity of transnational contacts. They provided some very small, fragmented, permanent organizations allowing regular multilateral contact and meetings at short notice if need be. The League of Nations, still a small organization in terms of staff by today's standards and to a large extent still European, extended the functions of permanent multilateral machinery in various directions, also toward preventive measures against interstate war. The United Nations continued this process on a global scale. Currently some additional 240 interstate organizations function in all fields of state activity. They vary widely in membership and geographical scope, but they are mostly regional in nature. The UN and other interstate organizations attract their own diplomatic representations from individual states. This gives cities like New

York, Geneva, and Brussels in particular an extra boost as political world cities but this applies also to others. Receiving cities and national governments consider the hosting of international organizations as one way to enhance their diplomatic status because extra diplomats are posted. An important difference between most UN organizations and most other interstate organizations has to do with their office structures. The others are often only established in one place. They maintain links with capital cities in member states and with other states through diplomatic representations. But the nature of those links differs from those within multi-office organizations. These offices are as a rule vertically ordered and the resulting networks consequently show a steep hierarchy (Taylor 2004b).

In networks of bilateral relations hierarchy is produced by differences in numbers of links drawn and sent by each node. I have already mentioned Washington's predominant position in that network worldwide. In multilateral politics hierarchy is produced in three different ways.

First of all, a multilateral agreement of cooperation can be read as a series of bilateral links between all participants. For a number of those agreements all those links add up producing network attributes like centrality of all different nodes. These nodes are as a rule considered to be states. But they can also be cities, mostly capital cities. Nierop's (1994) analysis follows this format. Compared to the level of hierarchy in the bilateral diplomatic network, this one turns out to be pretty flat (Nierop 1994: 119) based on a number of fuzzy macro-regions.

Second, as multilateral agreements become organized in a permanent fashion they get headquarters and/or secretariats and possibly other offices. These result cumulatively in more or less centrality depending on the number and importance of offices in places. We lack a comprehensive study of centrality on the basis of the concentration of those offices. Taylor's (2004d) analysis is limited to UN organizations, but misses some (e.g. the judicial institutions in The Hague, see below). Their cumulation in cities shows a clear hierarchy headed by New York and Geneva. Van der Wusten (2004) describes the distribution of intergovernmental organization secretariats in Europe and the US (this covers less than half of all IGOs and only one office per organization). The four top cities (Paris, Brussels, London, Geneva) cumulatively host at least 40 percent of all secretariats in Europe, but their mutual differences are not very large and the remainder is widely distributed. Centrality is thus clearly present, but not so clear cut as in the case of the UN.

Third, each individual multilateral organization with a number of offices also produces its own network between cities. The nodes in these interlocking networks have connectivities, and suggest levels of centrality for each individual place. Taylor has done such an analysis again for a series of UN organizations modeled after his earlier work on global producer services (Taylor 2004b). Geneva turns out to be preponderant, with New York lagging behind, and Brussels plus some developing countries' capitals in a much more prominent position according to this measure. The hierarchy at the top is less steep in this instance. For many multilateral organizations such analyses can not be performed as they do not have a network of offices.

In sum, various hierarchies of centers of multilateral political cooperation can be drawn. Chances of a clear-cut conclusion are not bright but much work remains to be done. Geneva is the only incontrovertible political global city at the top of the hierarchy in this field.

Multilateral interstate organizations (bringing together the voices of larger or smaller numbers of individual states) were not the only cohesive force that brought or held the state system together. As mentioned earlier, the traditionally more normal form of diplomatic contact, in line with the basic ideology of the system, was bilateral (dividing up all contacts between states in pairwise interactions). A third way to keep the system of state containers intact through cohesion was, paradoxically, the organization of the mixing of their contents, the flow of transnational contacts (of individuals and collective non-state actors across borders). Multilateral diplomacy enabled the birth of the state system and transnational contacts actually precede it. In fact those contacts only became transnational as the state system was imposed. Multilateral diplomacy and transnational contacts add importantly to the basic rules of the state system but equally contribute to a less one-dimensional structure. In this way they may also potentially undermine the territorial order of sovereign states. Multilateralism brings to life additional relevant players, the international organizations, that are concerned with issues transgressing single state borders, introduce new regulatory devices, and therefore change the ways international politics is conducted. The most formidable current example is obviously the European Union. Transnational cooperation does the same. It comes in very different guises from very loose associations to highly formal organizations to movements campaigning on burning questions (Sommier 2003).

The networks of multilateralism, interstate bilateral diplomacy, and transnational activity are in fact interwoven and they sometimes blend. They produce to some extent shared nodes. This was the case long before the twentieth century when special envoys who also acted as diplomats in bilateral relations were convened to multilateral peace congresses simultaneously meeting with interested and concerned cosmopolitan citizens. The last occasions in such a setting were perhaps the Hague Conferences in 1899 and 1907 but new forms were found: the world of ornamented town halls in spas, urban palaces of the nobility and royal quarters, inns and coffeehouses transformed in luxury hotels, multimedia conference rooms and photo ops. Washington, Brussels, and Geneva are the best current examples of specialized global 'political company towns' that we have, with incessant political consultation, demonstration, deliberation, and decision-making on matters of global significance in their international quarters and elsewhere. Washington's position has grown from its importance as a diplomatic center initially based on bilateral relations. It has subsequently also hosted multilateral organizations and finally transnational associations as well. Brussels has become the unrecognized but real capital of Europe's so far most successful venture in multilateral cooperation. This has additionally made it into an important diplomatic center for bilateral contacts and has drawn a multitude of transnational associations. Geneva grew from a place with a specific cosmopolitan tradition that then was articulated by its hosting of an important transnational movement (now the International Red Cross and Red Crescent Movement emanating from a still existing Swiss body upholding humanitarian law: the International Committee of the Red Cross) into a center of multilateral interstate cooperation. The presence of diplomatic representations of very many countries finally opened avenues for a multitude of bilateral contacts as well.

EUROPE'S PREMIER POLITICAL WORLD CITIES: GENEVA, BRUSSELS, VIENNA, THE HAGUE

To further discuss Europe's contribution to the set of specialized political world cities particularly geared towards multilateralism I add two more cases to Geneva and Brussels: Vienna and The Hague. Vienna proudly carries the title of third UN city and has a high-profile center for the UN organizations that it has hosted since 1979. Additionally it is home to organizations like the Organization of Petroleum-Exporting Countries (OPEC) and the Organization for Security and Co-operation in Europe (OSCE). The Hague has been an important center of multilateral diplomatic activity in the modern era since the Hague Conferences of 1899 and 1907 that resulted in the construction of the Peace Palace and the hosting of judicial institutions practicing international law (prompting a slogan with two versions: 'judicial/legal capital of the world'; van der Wusten 2006). In addition the city is home to some other UN institutions and to EU and NATO offices. Two questions will frame my discussion:

• Why have these four cities been particularly successful as seats of multilateral politics?
• What are the major differences among these cities in terms of hierarchy and profiles of multilateral organizations hosted in these cities and what accounts for those differences?

Without much doubt these cities are now the most important world cities of multilateral politics within Europe. The European part of the current state system is old and multilateral politics has a long tradition, also evident in terms of interstate organizations. For this and other reasons the conditions for significant multilateral politics may be different from those in other parts of the world. For example the prominent position of cities like Addis Ababa and Bangkok as centers of multilateral politics in other continents (Taylor 2004b) is apparently related to the continuous independence of the states of which they are the capitals. This translates into the experience of their diplomats and their early availability when locations were selected in those parts of the world in the era of decolonization.

The main initial impetus for the establishment of multilateral political organizations in Geneva and The Hague was in the first quarter of the twentieth century. For Brussels and Vienna this development started around 1960. Factors at state and city level apparently predisposed these cities to this role. In addition political actors from the local level but particularly from the national level or from outside the state pushed the initial decisions that set the following process of center formation in motion.

At state level Switzerland, Belgium, Austria, and the Netherlands shared at those different moments a position as small members in the core area of the European state system, formally neutral or maximally at the fringes of the alliances of major powers. In this way they can be considered as reasonably neutral ground for the conduct of multilateral politics. It is interesting that precisely these four countries have been the main examples of Lijphart's (1968) thesis on 'consociational democracies', a type of polity that results not only from internal cleavage structures but also from a sensitive international position that needs to be commonly defended (Katzenstein 1985). This results in a largely tamed elite-oriented domestic politics and shows definite structural resemblances with multilateral politics at the international level. This may increase the appetite for hosting such politics inside the country.

The four cities in these four small states have for a long time known a sustained international atmosphere. This seems even more important than seating the national government, as the example of Geneva shows. Geneva has since Calvin been a cosmopolitan center with many famous inhabitants and visitors. The beauty of the lake view has often been sung as an attractive force. The city also had a university of repute and attracted many foreign students. Around 1900 a large proportion of the student body was in fact Russian. Russian revolutionaries like Lenin and Plekhanov found here a welcome home (Kudryavtsev *et al.* 1969). The Hague had for centuries been the seat of the Dutch government, originally selected as a neutral spot among 18 competing cities. From the late nineteenth century it functioned as the residence of different well-to-do groups, e.g. those who returned rich from the colonies. Around 1900 it transformed into a well furnished small city with a lot of parks and modern hotels, theatres and clubs close to Scheveningen, newly built up as a seaside resort (Eyffinger 1999: 102–9). Brussels was also remade into a beautiful, thriving city around this time but it lost a few competitions as a candidate for the hosting of multilateral institutions. In the nineteenth century it had frequently been a refuge for banned socialists and a home to committed internationalists. Since 1894 the city boasted its own (liberal oriented) university. Brussels got a second chance after World War II and a successful world exposition in which the city attracted a lot of international attention and was also physically prepared to receive international institutions on its territory (Dumoulin 2001). Vienna had long been a capital of a major power and an important diplomatic center symbolized by the Congress of 1815. The extension of the inner city by the 'grand project' of the Ringstrasse had made it into one of the most admired cities of Europe and around 1900 it was a cultural center of the first rank (Schorske 1980; Olsen 1986), not least by the presence of a great university and a magnet for flows of all sorts of immigrants. After the demise of the empire Austria became a small state, neutralized in the aftermath of World War II with an oversized capital city that still retained its physical treasures and its renown.

Brussels and Vienna were certainly the most internationalized cities in their respective countries at the time. How important this was became clear in the early 1950s when the Belgian government tried to locate the European Coal and Steel Community in Liège to the dismay of its negotiating partners. In the case of Geneva and The Hague their national predominance as cities with an international atmosphere is less assured. Within Switzerland Zürich (particularly as an international financial center) and the capital Bern (which was after all the diplomatic center and had already received the seats of a few of the earliest interstate organizations) could have provided the internationalized background deemed necessary once Switzerland was an option. A similar argument can be made for the Netherlands, where The Hague could be seen as having Amsterdam as a competitor. The Hague even lacked a university, which Amsterdam did have. It is perhaps significant that in this case the internationalized atmosphere in the end induced transnational associations to have their seats most frequently in Amsterdam while The Hague has remained far more important as a host of multilateral international organizations. In sum, to emerge as a multilateral political world city within a country, one does not need to have the most international atmosphere in the country, but a certain cosmopolitan air is indispensable.

Politicians, experts, and their preferences and personal influence in the small bodies that make the choices have also been of importance in initiating and then further extending a city's role as a multilateral political center. Let me concentrate on the decisions that put these cities on the road to becoming political world cities.

In the earlier cases of The Hague and Geneva outside actors apparently dominated decision-making, particularly in the case of The Hague. In the more recent cases the hand of powerful national politicians particularly at state level is visible, although they obviously always needed the approval of international meetings to have it their way. The Hague conference of 1899 was initiated by the Russian Czar and agreed among foreign ministers with hardly any input from the Dutch government. It merely acceded to requests to act as host. The Peace Palace then came about as a result of representations made by some non-Dutch internationalists to Andrew Carnegie. The physical fact of the Peace Palace plus the pressure from the international professional community of international legal scholars provided the impetus to make The Hague the seat of the Permanent International Court of Justice in 1922, hardly the lobbying of the Dutch government, which did not have much of a position during the negotiations on the League of Nations (Northedge 1986).

Competing with Geneva, Brussels was for a brief period a candidate for the seat of the League. The Belgian minister of foreign affairs, present at the negotiations, was supported by the French, but the Americans were very much in favour of Geneva from the outset. They saw it as the city with the right reputation for this organization. In League matters they usually prevailed. The Swiss government had also discreetly lobbied for Geneva and the Swiss president, Gustave Ador, immediately sent a telegram to the city government when the decision was taken in Paris (Fatio 1924: 97).

The key decisions on European cooperation that put Brussels on the path to slowly becoming the 'capital of Europe' were the location in Brussels of the main negotiations leading to the Treaty of Rome, the 1958 decision to make Brussels one of the two seats of the councils and commissions of the three European bodies, and the establishment of the merged councils and commissions in Brussels in 1965, realized in 1967. It was all provisional but tended very much to last. While the Commission and Council of Ministers landed in Brussels, the Parliament was hosted in Strasbourg. The Court and some services came to Luxemburg (Dumoulin 2001: 55–66). It was around the Commission and the Council that growth concentrated and additional organs were located. It is not difficult to presume the hand of Paul Henri Spaak, one of the major architects of European cooperation and the Belgian minister of foreign affairs with the stature of a grand statesman as a former prime minister, in this arrangement.

In the case of Vienna, following the State Treaty of 1955, which provided new breathing space for Austrian autonomy and consequently a new development of Austrian diplomacy, there was an impetus to use its official neutral status to attract the seats of international organizations (also conferences and urban tourism). From the early 1950s there was already a tiny UNHCR office in the city, which then gradually expanded. In 1956 it was agreed to put up the newly established International Atomic Energy Association (IAEA) headquarters in Vienna. Subsequently OPEC, launched in Baghdad 1960 but at first seated in Geneva, was induced to move to Vienna in 1965. Then Austria attracted some headquarters of UN organizations and offered space in a new International Conference Centre, part of a major extension plan for Vienna across the Danube that was opened in 1979 (Lichtenberger 1993). The complex was appealingly called UNO City. A major political figure in this connection was Bruno Kreisky, foreign secretary in the early part of the 1960s and Austrian chancellor 1970–83. He was most probably assisted in the attraction of UN organizations by the Austrian Kurt Waldheim, secretary-general of the UN 1972–81. Waldheim's later loss of reputation as Austrian state president made the further extension of Vienna's international position problematic for a while, but after Waldheim's departure Vienna knew how to acquire the seat of OSCE by offering the Hofburg, one of the most prestigious locations in the city. Austria has a reputation for largesse in providing special advantages and exemptions for diplomats who are posted in Vienna.

Beyond the initial decision-making that puts certain cities on the map as centers of multilateral politics, they have had to maintain and extend that role to be world political cities. This, again, results from favorable conditions and the outcome of international decision-making. In particular the further evolution of the organizations that have initially established their headquarters is now a major factor in this equation. They are important for their own growth, for the example they set, and for the services they attract that in their turn improve the conditions for other international organizations to follow. There are also negative effects that tend to slow down further growth, particularly price increases in various local markets, but they are secondary. Local popular protests against newly arriving organizations, their housing needs, and their further intrusions in existing neighborhoods have also occurred.

Geneva and Brussels have come out as by far the most important multilateral political cities because of the development of the League and its UN successor and the most important institutions of the extremely successful European cooperation. Belgian and Swiss local and national politics have always been forthcoming in accommodating the needs of these institutions after they have taken up residence. This has always been extremely difficult in the case of Belgium because of the inability of European partners to agree on a definitive decision to make Brussels the actual capital of European cooperation. The European institutions grew informally inside a certain part of the city that was incrementally restructured, destroying an existing neighborhood with a vocal population. 'Bruxellisation' became for a while a rallying cry to oppose grand-scale, unplanned urbanism that destroys the fabric of existing urban life (Thiry 1982; Elmhorn 2001: 207–68). In Geneva a prominent location was put at the disposal of the League on the outskirts of the city, where a huge office and congress facility was ready in 1937. The *Palais des Nations* was the anchor that kept a major part of the League apparatus, concerned with functional and technical international cooperation, largely in Geneva after 1945 while the political center of gravity of the new UN moved to New York. But when the local UN operations further extended, there was citizen protest in Geneva in the 1960s about the perceived negative consequences (Favez and Raffestin 1974: 373). In the current period the ever more extended security measures put a large extra charge on the possibilities of producing fruitful combinations of public city life and the presence of international institutions.

After its initial successes in attracting international organization in the period 1965–80 Vienna has made one additional gain in attracting OSCE, but the future of that body is uncertain. Vienna has not been able to become the global center for technical arms limitation and disarmament institutions following the arrival of IAEA. The chemical weapons control organization (OPCW) has finally landed in The Hague. It has also not become the most important center for development organizations after the arrival of the United Nations International Development Organization. Other very important organizations in that field are still based in Geneva, as well as in New York and Washington and in Nairobi.

The Hague's judicial institutions somewhat faded after their failures in the 1930s (see for example Shaw's (1939) scathing attack in a play called *Geneva*, largely set in the Peace Palace) and their inability to play a significant role in most of the Cold War period. International sympathy for the court and the palace was maintained by a strong global professional community of international law practitioners in academia and the foreign ministries. In the 1990s the interest in the arbitration facility of the Permanent Court of Arbitration and in the International Court of Justice as one possible means of dispensing justice picked up again, the Yugoslav Tribunal came to The Hague and then so did the International Criminal Court as the product of UN and global decision-making after Dutch lobbying. In the realm of European cooperation Europol and Eurojust as manifestations of an increasing level of institutional cooperation in the justice field in the EU also arrived. I have already mentioned the OPCW. In brief The Hague's reputation as 'judicial capital of the world', first established in the early part of the twentieth century and then waning for a long period, finally came to some fruition towards the end of the century. It remains to be seen if this complex will now grow any further. The Yugoslav Tribunal will in any case wind down in the coming years (van der Wusten 2006).

UNILATERALISM, MULTILATERALISM, AND THE EMERGING GEOGRAPHY OF GLOBAL POLITICS

By way of concluding note let me just wonder how the high-profile turn towards unilateralism in the foreign policy of the American administration since 2000 (Skidmore 2005) affects the structures of multilateralist global politics as expressed in the world political cities that I have especially stressed in the preceding section. Actual unilateralism can only be conducted under a condition of hegemony in a state system. Unilateralism can also be executed on the basis of enforcement pure and simple. In that case the state order is definitively undermined as state sovereignty is no longer respected and one moves toward an imperial relationship. In the case of hegemony enforcement is combined with attraction by the subjected actors, which implies a willingness to accede to pressure on a positive basis.

In hegemonial circumstances a largely unilateralist policy can be successfully conducted through multilateral channels by the sheer attractiveness and persuasive capacity of the unilateralist leadership. In imperial circumstances multilateral structures are less attractive from the perspective of the unilaterally operating actor. It looks as if the hegemonial period in the external relations of the US that started after World War II is largely gone and a stark choice approaches: unilateral policy in an imperial structure that sheds multilateral centers or a return to a multilateralist foreign policy using the existing frameworks manifested in the multilateral world political cities. This depends on the strength of those structures, their regional or global reach and the willingness of American politics to tread that path. If that path is not taken (as suggested for example by the American refusal to ratify the treaty establishing the International Criminal Court in The Hague), a perceptible change in the global distribution of political world cities is to be expected. But if the past is anything to go by this time the nature of the change will be more incremental than shift-like. Reputation will long remain but count for somewhat less. Geneva, Brussels, Vienna, and The Hague will lose some of their luster like their equivalents on other continents, Brussels less than the others as its dynamics are more tied to a European macro-region. Functions oriented toward Europe might become relatively more prominent if a truly imperial stance were to be opposed. Washington will become even more important than it already is, but the nature of its political clout would have to be based on military links even more than it already is (Priest 2003). What this would mean for New York as a global political center is a big question mark.

REFERENCES

Castells, M. (1996) *The Rise of Network Society*, Oxford: Blackwell.

Christaller, W. (1933) *Die zentralen Orte in Süddeutschland: eine ökonomisch-geographische Untersuchung über die Gesetzmässigkeit der Verbreitung und Entwicklung der Siedlungen mit städtischen Funktionen*, Jena: Gustav Fischer.

Doel, M.A. and Hubbard, P.J. (2002a) 'Marketing world cities in a global space of flows: collaboration or competition?' in A. Mayr, M. Meurer & J. Voigt (eds) *Stadt und Region: Dynamik von Lebenswelten*, Leipzig: Deutsche Gesellschaft für Geographie, pp. 87–97.

Doel, M.A. and Hubbard, P.J. (2002b) 'Taking world cities literally: marketing the city in a global space of flows', *City*, 6 (3): 351–68.

Dumoulin, M. (ed.) (2001) *Bruxelles, l'européenne. Regards croisés sur une région capitale/Brussel, hart van Europa. Een verkennende blik op een kapitale regio*, Brussels: Tempora.

Elmhorn, C. (2001) *Brussels: A Reflexive World City*, Stockholm: Almqvist & Wiksell International.

Eyffinger, A. (1999) *The 1899 Hague Peace Conference: 'The Parliament of Man, the Federation of the World'*, The Hague: Kluwer.

Fatio, G. (1924) *Genève, siege de la Société des Nations*, Leiden: Sijthoff.

Favez, J.-C. and Raffestin, C. (1974) 'De la Genève radicale à la cité internationale', in P. Guichonnet (ed.) *Histoire de Genève*, Toulouse: Privat, pp. 299–386

Friedmann, J. (1986) 'The world city hypothesis', *Development and Change*, 17 (1): 69–84.

Gerhard, U. (2003) *Local Activity Patterns in a Global City: Analysing the Political Sector in Washington D.C.* GaWC bulletin 99

Held, D., McGrew, A., Goldblatt, D. and Perraton, J. (1999) *Global Transformations: Politics, Economics and Culture*, Cambridge: Polity Press.

Henrikson, A.K. (2005) 'The geography of diplomacy', in C. Flint (ed.) *The Geography of War and Peace: From Death Camps to Diplomats*, Oxford: Oxford University Press, pp. 369–394.

Hubbard, P.J. (2001) 'The politics of flow: on Birmingham, globalization and competitiveness', *Soundings*, 17: 167–71.

Katzenstein, P.J. (1985) *Small States in World Markets*, Ithaca, NY: Cornell University Press.

Kudryavtsev, A., Muravyova, L. and Sivolap-Kaftanova, I. (1969) *Lenin's Geneva Addresses*, Moscow: Progress Publishers.

Knox, P.L. and Taylor, P.J. (eds.) (1995) *World Cities in a World-System*, Cambridge: Cambridge University Press.

Kooij, P. and Pellenberg, P.H. (eds.) (1994) *Regional Capitals: Past, Present, Prospects: Ghent, Groningen, Munster, Norwich, Odense, Rennes*, Assen: Van Gorcum.

Lichtenberger, E. (1993) *Vienna: Bridge between Cultures*, London: Belhaven Press.

Lijphart, A. (1968) *The Politics of Accommodation: Pluralism and Democracy in the Netherlands*, Berkeley, CA: University of California Press.

Nierop, T. (1994) *Systems and Regions in Global Politics: An Empirical Study of Diplomacy, International Organization and Trade 1950–1991*, Chichester: Wiley.

Northedge, F.S. (1986) *The League of Nations: Its Life and Times*, Leicester: Leicester University Press.

Olsen, D. (1986) *The City as a Work of Art: London, Paris, Vienna.* New Haven, CT: Yale University Press.

Pascon, P., Arrif, A., Schroeter, D., Tozy, M. and van der Wusten, H. (1984) *La Maison d'Iligh et l'histoire sociale du Tazerwalt*, Rabat: SMER.

Priest, D. (2003) *The Mission: Waging War and Keeping Peace with America's Military*, New York: Norton.

Sassen, S. (1991) *The Global City*, Princeton, NJ: Princeton University Press.

Schorske, C.E. (1980) *Fin de Siècle Vienna: Politics and Culture*, New York: Knopf.

Shaw, G.B. (1939) *Geneva: A Fancied Page of History in Three Acts*, London: Constable.

Skidmore, D. (2005) 'Understanding the unilateralist turn in U.S. foreign policy', *Foreign Policy Analysis*, 2: 207–28.

Sommier, I. (2003) *Le Renouveau des mouvements contestataires a l'heure de la mondialisation*, Paris: Flammarion.

Stanley, B. (2003) *City Wars or Cities of Peace. (Re-)integrating the Urban into Conflict Resolution.* GaWC bulletin 123

Taylor, P.J. (2004a) *World City Network: A Global Urban Analysis*, London: Routledge.

Taylor, P.J. (2004b) *New Political Geographies: Global Civil Society and Global Governance through World City Networks.* GaWC bulletin 149.

Taylor, P.J. (2004c) 'Homo geographicus: a geohistorical manifesto for cities', *Review* (Fernand Braudel Center), 27 (1): 37–60.

Taylor, P.J. (2004d) *Leading World Cities: Empirical Evaluations of Urban Nodes in Multiple Networks.* GaWC bulletin 146.

Thiry, J.-P. (1982) *L'Europe à Bruxelles. Etude des ASPECTS Économiques, sociaux, culturels et urbanistiques de la presence des institutions des communautés européennes à Bruxelles*, Brussels: ERU.

Watson, A. (1992) *The Evolution of International Society*, London: Routledge.

van der Wusten, H. (2003) 'De adressen van het nationaal bestuur. Keuze en effect van vestigingsplaatsen ten behoeve van de politieke functie', in C. Cortie, J. Droogleever Fortuijn and M. Wagenaar (eds.) *Stad en land. Over bewoners en woonmilieus*, Amsterdam: Aksant, pp. 160–73.

van der Wusten, H. (2004) 'The distribution of political centrality in the European state system', *Political Geography*, 23, 677- 700.

van der Wusten, H. (2006) ' "Legal capital of the world": political centre-formation in The Hague, *TESG Journal of Economic and Social Geography*, 97 (3): 253–66.

12 Inter-city relations and the 'war on terror'

Stephen Graham

INTRODUCTION

Programmes of organized, political violence have always been legitimized and sustained through complex imaginative geographies. This term – following Foucault (1970), Said (1978) and Gregory (1995) – denotes the ways in which imperialist societies are constructed through normalizing, binary judgements about both 'foreign' and colonized territories and the 'home' spaces which sit at the 'heart of empire'.

Such imaginative geographies tend to be characterized by stark binaries of place attachment. These tend to be particularly powerful at times of war. As Ken Hewitt (1983: 258) has argued, 'war . . . mobilizes the highly charged and dangerous dialectic of place attachment: the perceived antithesis of "our" places or homeland and "theirs" '. Very often, such polarizations are manufactured discursively through racist and imperial discourses and propaganda which emanate from both formal state and other media sources. To Hewitt, these work to produce 'an unbridled sentimentalizing of one's own while dehumanizing the enemy's people and land.' Such binaried constructions 'seem an essential step in cultivating readiness to destroy the latter' (ibid.).

The purpose of this chapter is to demonstrate that the Bush Administration's 'war on terror' rests fundamentally on such two-sided constructions of (particularly urban) place. The chapter argues that the discursive construction of the 'war on terror' since 11 September 2001 has been deeply marked by attempts to rework imaginative geographies separating the urban places of the US 'homeland' and those Arab cities purported to be the sources of 'terrorist' threats against US national interests. Such reworkings of popular and political imaginative geographies have worked by projecting places, and particularly cities, into two mutually exclusive, mutually constitutive, classifications: those, in Bush's famous phrase, who are either 'with us' or 'against us' (*sic*) (see Graham 2004).

In a world of intensifying transnational migration, transport and media flows, and economic globalization, however, such attempts at constructing a mutually exclusive binary – a securitized 'inside' enclosing the urban places of the US Empire's 'homeland', and an urbanizing 'outside' where US military power can preemptively attack places deemed sources of 'terrorist' threats – are inevitably ambivalent. Reconstructing imaginative geographies separating a putative urban 'homeland' from the 'terrorist' spaces requiring US military intervention is thus inseparably bound up with efforts to reshape material and political economic connections between 'homeland' and cities and the rest of the world.

Whilst dramatic, these discursive constructions are far from original. In fact they revivify long-established colonial and Orientalist tropes to represent Middle Eastern culture as intrinsically barbaric, infantile, backward or threatening from the point of view of Western colonial powers (Gregory 2004a). Arab cities, moreover, have long been represented by Western powers as dark, exotic, labyrinthine and structureless places that need to be 'unveiled' for the production of 'order' through the superior scientific and military technologies of the occupying, colonizing West. By burying 'disturbing similarities between "us" and "them" in a discourse that systematically produces the Third World as Other', such Orientalism deploys considerable 'symbolic violence' (Gusterson 1999). This is done, crucially, in order to produce both 'the Third World' and 'the West'.

The Bush administration's language of moral absolutism is, in particular, deeply Orientalist. It works by separating 'the civilized world' – the 'homeland' cities which must be 'defended' – from the 'dark forces', the 'axis of evil' and the 'terrorist nests' of Islamic cities, which are alleged to sustain the 'evildoers' that threaten the health, prosperity and democracy of the whole of the 'free' world (Tuastad 2003). The result of such imaginative geographies is an ahistorical, essentialized and deeply Orientalist projection of Arab civilization that is very easily worked to 'recycle the same unverifiable fictions and vast generalizations to stir up "America" against the foreign devil' (Said 2003: vi). The Orientalist notions of racial worth that helped to shape the real and imagined geographies of Western colonialism are thus being reworked as fundamental foundations for the 'war on terror' (Gregory 2004a). As Paul Gilroy suggests, these

old, modern notions of racial difference appear once again to be active within the calculus [of the 'war on terror'] that tacitly assigns differential value to lives lost according to their locations and supposed racial origins or considers that some human bodies are more easily and appropriately humiliated, imprisoned, shackled, starved and destroyed than others

(Gilroy 2003: 263)

Discourses of 'terrorism' are crucially important in sustaining such differential values and binaried notions of urban places (Collins and Glover 2002). Central here is the principle of the absolute eternality of the 'terrorist' and of 'terrorist' places – the inviolable inhumanity and shadowy, monster-like status of those deemed to be actual or dormant 'terrorists', those sympathetic to them or the places essentialized as being the home, source or refuge of 'terrorists' (Puar and Rai 2002). The unbound diffusion of terrorist labelling within the rhetoric of the 'war on terror', moreover, works to allow virtually any political opposition to the sovereign power of the US and its allies to be condemned as 'terrorist' or addressed through emergency 'anti-terrorist' legislation. Protagonists of such opposition are thus dehumanized, demonized and, above all, delegitimized. 'Without defined shape, or determinate roots,' Derek Gregory writes, the mantle of 'terrorism' can now be 'be cast over *any* form of resistance to sovereign power' (2004a: 219, original emphasis).

In such a context, this chapter seeks to trace in detail the ways in which the deep-rooted dialectics of urban place attachment, and the imaginative geographies that fuel them, are at the very heart of the 'war on terror'. To achieve this, two brief case studies are developed. These address especially important 'sites' of place construction in the 'war on terror': the reworking of imaginative geographies of US cities as 'homeland' spaces which must be re-engineered to address supposed imperatives of 'national security', and the intensified imaginative construction of Islamic cities as little more than 'terrorist nest' targets to soak up US military firepower.

REIMAGINING 'HOMELAND' CITIES AS NATIONAL SECURITY SPACES

> Everything and everywhere is perceived as a border from which a potentially threatening Other can leap.
>
> (Hage 2003: 86)

The first key element in the imaginative geographies of the 'war on terror' is an appeal by the Bush administration to securitize the everyday urban spaces and technics of a newly 'rebordered' US 'homeland' (Lutz 2002). Here, discourses of 'security,' emphasizing endless threats from an almost infinite range of people, places and technologies, are being used to justify a massive process of state-building. Vast efforts are being made by US political, military and media elites in order to 'spread . . . generalized promiscuous anxiety through the American populace, a sense of imminent but inexact catastrophe' lurking just beneath the surface of normal, technologized, (sub)urbanized, everyday life in the US (Raban 2004: 7). Despite the unavoidable and continuing interconnections between US cities and more or less distant elsewheres, 'the rhetoric of "insides" needing protection from external threats in the form of international organizations is pervasive' (Dalby 2000: 5). This reimagining of 'homeland' cities involves four simultaneous processes.

The 'domestic front' in the 'war on terror'

First, the homeland security drive is being organized as a purported attempt to protect those 'insides' – the bodies and everyday spaces of valued, non-threatening, full US citizens – from demonized Others apparently lurking, with a wide range of threatening technologies and pathogens, both within and outside US national space. Fuelled by the larger mobilization of 'terrorist' discourses discussed above, and the blurring of the boundary separating law enforcement from state military activity (Kraska 2001), this process has 'activate[d] a policing of points of vulnerability against an enemy who inheres within the space of the US' (Passavant and Dean 2002, cited in Gregory 2003). The 'enemy' here are constructed as dormant 'terrorists' and their sympathizers, a rhetoric that easily translates – in the context of the wider portrayals of the 'homeland at war' against secretive and unknowable Others – into an overall crackdown on criticism and dissent, or those simply deemed to be insufficiently 'patriotic'. As a result, to put it mildly, 'cosmopolitan estrangement and democracy-enriching dissent are not being prized as civic assets' in the US (or UK) in the early twenty-first century (Gilroy 2003: 266).

A 'domestic front' has thus been drawn in Bush's 'war on terror'. Sally Howell and Andrew Shryock (2003) call this a 'cracking down on diaspora'. This process involves deepening state surveillance and violence against those seen to harbor 'terrorist threats,' combined with radically increased efforts to ensure the effective filtering power of national and infrastructural borders. After decades in which the business press and politicians endlessly celebrated the supposed collapse of boundaries through neoliberal globalization, 'in both political debates and policy practice, borders are very much back in style' (Andreas 2003: 1). Once again, Western nations – and the securitized cities now seen once again to sit hierarchically within their dominant territorial patronage – are being normatively imagined as bounded, organized spaces with closely controlled, and filtered, relationships with the supposed terrors of the outside world. In the US, for example, national immigration, border control, transportation and social policy strategies have been remodelled since 9/11 in an

> attempt to reconstitute the [United States] as a bounded area that can be fortified against outsiders and other global influences. In this imagining of nation, the US ceases to be a constellation of local, national, international, and global relations, experiences, and meanings that coalesce in places like New York City and Washington DC; rather, it is increasingly defined by a 'security perimeter' and the strict surveillance of borders.
>
> (Hyndman 2003: 2)

Securitizing everyday spaces and systems

Second, as well as further militarizing national territorial borders, the US homeland security drive is also attempting to re-engineer the basic everyday systems and spaces of US urban life – even if this is sometimes a stealthy and invisible process. As a result, urban public life is being saturated by 'intelligent' surveillance systems, checkpoints, 'defensive' urban design and planning, and intensifying security (Johnson 2002; Williams 2003). In the wake of 9/11, and the Homeland Security drive, the design of buildings and streets, the management of traffic, the physical planning of cities, migration and refugee policy, transportation policing, the design of social policies for ethnically diverse cities and neighborhoods, even the lending policies of neighborhood libraries, are being brought within the widening umbrella of US 'homeland security'.

In cities like Washington DC, completely new (and tellingly titled) 'urban design and security plans' have been brought in. These emphasize that one of the most important objectives of public urban planning in such strategic centres is now the 'hardening' of all possible terrorist targets. Once again, it seems, geopolitical and strategic concerns are directly shaping the day-to-day practices of US urban professionals. Jonathan Raban, writing of everyday life in post-9/11 Seattle, captures the palpable effects of this militarization on urban everyday life and landscape:

> To live in America now, at least to live in a port city like Seattle, is to be surrounded by the machinery and rhetoric of covert war, in which everyone must be treated as a potential enemy until they can prove themselves a friend. Surveillance and security devices are everywhere: the spreading epidemic of razor wire, the warnings in public libraries that the FBI can demand to know what books you're borrowing, the Humvee laden with troops in combat fatigues, the Coast Guard gun boats patrolling the bay, the pat-down searches and X-ray machines, the nondescript grey boxes equipped with radar antennae, that are meant to sniff pathogens in the air.
>
> (Raban 2004: 6)

US cities within anti-cosmopolitan constructions of 'homeland'

Third, this attempted reconstruction of national boundaries, as well as being sustained by material and technological investments in and around strategic urban spaces, relies on massive linguistic work (Kaplan 2003: 85). Tom Ridge, the Homeland Security Secretary, for example, has widely argued that, after 9/11, 'the only turf is the turf we stand on' (cited ibid.). This 'rebordered' discourse constructs an imaginary, domesticated, singular and spatially fixed imagined community of US nationhood (Andreas and Biersteker 2003). Such an imagined community – tied intrinsically to some purported, familial, 'turf' – centres on valorizing an exclusive, separated and privileged population. It therefore contrasts starkly with previous US state rhetoric which centred on notions of boundless mobility and assimilation (Kaplan 2003: 86).

Such discourses are central to reimagining the actual and normative geographies of what contemporary US urban life actually consists of or what it might become. Amy Kaplan, in analysing the languages of 'homeland security', detects a 'decidedly antiurban and anticosmopolitan ring' to this upsurge of nationalism after 9/11 (2003: 88). Paul Gilroy goes further and suggests that the widespread invocation by the Bush administration, following Huntingdon (1993), of the idea of a 'clash of civilizations,' necessarily '*requires* that cosmopolitan consciousness is ridiculed' in the pronouncements of the US state and the mainstream media (2003: 266, emphasis added). Post 9/11, he diagnoses a pervasive 'inability to conceptualize multicultural and postcolonial relations as anything other than on-tological risk and ethnic jeopardy' (261).

The very term 'homeland security', in fact, serves to rework the imaginative geographies of contemporary US urbanism in important ways. It shifts the empha-sis away from complex and mobile diasporic social formations, sustaining large metropolitan areas, towards a much clearer mapping which demarcates clear, es-sentialized geographies of entitlement and threat. On many scales – from neigh-borhoods through cities and nations to the international – this separation works to define those citizens who are deemed to warrant value and the full protection of citizenship, and those deemed threatening as real or potential sources of 'terror-ism': the targets for the blossoming national security state.

As Amy Kaplan suggests (2003: 84), even the very word 'homeland' suggests some 'inexorable connection to a place deeply rooted in the past'. It necessar-ily problematizes the inherently diverse and mobile fabric of the diasporas that actually constitutes the social fabric of US urbanism. Such language offers a 'folksy rural quality, which combines a German romantic notion of the folk with the heartland of America to resurrect the rural myth of American identity' (88). At the same time, it precludes 'an urban vision of America as multiple turfs with contested points of view and conflicting grounds upon which to stand' (ibid.).

Such a discourse is particularly problematic in global metropolitan cities like New York, constituted as they are by massive constellations of diasporic social groups. 'In what sense,' asks Kaplan (2003: 84), 'would New Yorkers refer to their city as the homeland? Home, yes, but homeland. Not likely.' Ironically, even the grim casualty lists of 9/11 – in which nationals from 41 countries died (see http://www.september11victims.com/september11victims/COUNTRY_CITI-ZENSHIP.htm, accessed 4 July 2005) – revealed at least temporarily the impossi-bility of separating some purportedly pure, 'inside,' homeland city from the wider international flows that now constitute cities like New York – even with massive state surveillance and violence. 'If it existed, any comfortable distinction between domestic and international, here and there, us and them, ceased to have meaning after that day' (Hyndman 2003: 1). For, as Tim Watson writes:

> global labor migration patterns have . . . brought the world to lower Manhat-tan to service the corporate office blocks: the dishwashers, messengers, cof-fee-cart vendors, and office cleaners were Mexican, Bangladeshi, Jamaican and Palestinian. One of the tragedies of September 11th 2001 was that it took such an extraordinary event to reveal the everyday reality of life at the heart of the global city.
>
> (Watson 2003: 109)

Since 9/11, however, representations of the dead of 9/11 have widely tried to reinscribe simplistic notions of homogeneity and patriotism upon the silent vic-tims. Perversely, the non-US dead have been routinely accorded US citizenship – a posthumous erasure of transnational difference that emerges under the endless claims that the attack led to '3,200 American dead'. And so the 'rebordering' of transnational space associated with the homeland security drive works to sanctify the purity of national 'homeland' communities by erasing the signature of the diverse transnational diasporic networks that together constitute global cities like New York. As we shall shortly discuss, such forced amnesia, of course, makes the calls to violence of some purportedly unified community against an essentialized enemy much more effective.

'Homeland security' policies have been associated with a considerable growth in state and non-state violence against immigrant and Arab-American groups. Indeed, 'the notion of the homeland itself contributes to making the life of immigrants terribly insecure' (Kaplan 2003: 87). Systematic state repression and mass incarceration have been brought to bear on Arab-American neighbourhoods like Dearbon in Detroit – the first place to have its own, local, office of Homeland Security (Howell and Shryock 2003). Such Arab-American neighborhoods are now overwhelmingly portrayed in the US national media as 'zones of threat'. Arab-Americans are widely represented as 'clearly being in' their local cities and 'with us' (*sic*), but the point is almost always stressed, as Howell and Shryock (2003: 444) put it, that 'their hearts might still be over there, "with them"' (*sic*). Thousands of US citizens have also effectively been stripped of any notion of value, to be thrown into extra- or intra-territorial camps as suspect 'terrorists for potentially indefinite periods of time without trial. More than ever, the discourses and practices of the "war on terror" therefore work to make '"Arab" and "American" all but antithetical adjectives' (Watson 2003). As we shall see shortly, this situation is immutably bound up with the widespread demonization of Middle Eastern and Arab cities more generally within 'war on terror' discourses.

€verydoy sites ond spoces os sources of feor

The final element of the homeland security drive is the production of permanent anxiety around everyday urban spaces, systems and events that were previously banalized, taken for granted or ignored in US urban everyday life (Luke 2004). With streams of vague warnings, omnipresent colour-coded alerts and endless media coverage of purported threats to US urban life, everyday events, malfunctions or acts of violence in the city – which would previously have been seen as the results of local social problems, individual pathologies, bureaucratic failings or simple accidents – are now widely assumed be the results of 'terrorist' action. The 'homeland' is thus cast in terms of a constant 'state of emergency' (Armitage 2002). In this the only things that can be guaranteed are new sources of fear and oscillations on Tom Ridge's colour-coded threat monitor. In the process, parked vans, delayed trains, envelopes with white powder, people with packages, 'Arab'-looking people, colds and 'flu, low-flying aircraft, electricity outages, subway derailments and Internet viruses are now sources of mass, geopolitically charged, anxiety. Homeland security thus depends, ironically, on the indefinite promulgation of a pervasive and radical *in*security. This fuels acceptance that the everyday sites and spaces of daily life within the continental US must now be viewed as battlegrounds – the key sites within a new, permanent and boundless war.

Cindi Katz (2004) calls this the 'routinization of terror talk and the increasing ordinariness of its physical markers'. She argues that such processes generate a radical ontological insecurity because they create pervasive feelings of vulnerability and threat in everyday urban life. In the process, 'terror talk' helps to define reimagined communities of nationhood as well as normative imaginative geographies of 'homeland' and 'target' cities. As Giroux (2003: ix) suggests, 'notions of community [in the US] are now organized not only around flag-waving displays of patriotism, but also around collective fears and ongoing militarization of visual culture and public space'.

'TERROR CITIES': ORIENTALIST CONSTRUCTIONS OF ARAB URBAN PLACES AS MILITARY TARGETS

Which leads us to our second case study: an analysis of the way in which (selected) Arab cities are being overwhelmingly constructed within 'war on terror' discourses as targets for US military firepower. Far from being isolated from the securitization of US cities, this process is inseparable from it. As Edward Said (2003: xxiii) stressed just before his death, from the point of view of the discursive foundations of both US foreign policy and dominant portrayals of Arabs in the US media, the devaluation and dehumanization of people in the 'target' cities of the Arab world cannot be separated from the securitization of the (re)imagined communities in 'homeland' ones. As the Iraq invasion was prepared, Said wrote that 'without a well-organized sense that these people over there were not like "us" and didn't appreciate "our" values – the very core of the Orientalist dogma – there would have been no war' in Iraq. Thus, crucially, a powerful relation exists

> between securing the homeland against encroachment of foreign terrorists and enforcing [US] national power abroad. The homeland may contract borders around a fixed space of the nation and nativity, but it simultaneously also expands the capacity of the United States to move unilaterally across the borders of other nations.
>
> (Kaplan 2003: 87)

Examples of such fabricated imaginative geographies connecting perceptions of the vulnerabilities of US urban life with the United State's aggressive and pre-emptive war against Arab cities deemed to be somehow *intrinsically terroristic* are not hard to find. In early 2004, for example, General Sanchez, US commander in Iraq, urged that 'every American needs to believe this: that if we fail here in this [Iraqi environment] the next battlefield will be the streets of America.' Paul Bremer, head of US civilian command in Iraq, stressed a few months earlier that 'I would rather be fighting [the terrorists] here [in Iraq] than in New York' (both quoted in Pieterse 2004: 122).

The discursive construction of selected Arab cities as targets for US military firepower occurs in at least four interrelated ways.

Vertical representations of Arab cities as collections of military targets

First, the voyeuristic consumption by Western publics of the US urban bombing campaigns that have been such a dominant feature of the 'war on terror' is itself based on mediated representations where cities are actually constructed as little more than receiving points for the dropping of murderous ordnance. Verticalized web and newspaper maps, for example, have routinely displayed cities like Baghdad as little more than impact points, where GPS-targeted bombs and missiles either are envisaged to land or have landed, grouped along flat cartographic surfaces (Gregory 2004a). Meanwhile, the weapons' actual impacts on the everyday life for the ordinary Iraqis or Afghanis who are caught up in the bombing, as 'collateral damage', have been both marginalized and violently repressed by the US military. (Most famously this has involved the bombing of Al-Jazeera transmission facilities because they transmitted images of the dead civilians that resulted from the bombing.)

In this verticalized imaginative geography, which is strongly linked to the wider history of colonial bombing and repression by Western powers, Arab 'cities' are thus reduced to the

> places and people you are about to bomb, to targets, to letters on a map or co-ordinates on a visual display. Then, missiles rain down on K-A-B-U-L, on 34.51861N, 69.15222E, but not on the eviscerated city of Kabul, its buildings already devastated and its population already terrorized by years of grinding war.
>
> (Gregory 2004b)

Strikingly, the failure to even *count* the 100,000 or so dead Iraqi civilians that had resulted, by the end of 2004, from the war's bombing campaigns and urban battles (Roberts *et al.* 2004) reveals that the civilians of targeted cities are 'cast out' so that they warrant no legal status or discursive or visual presence (Gregory 2004b). Their sacrifice can go largely unremarked; their bloody deaths can be blindly unrepresented.

Constructing Arab cities as 'terrorist nests'

Second, such casting out of the lives and suffering of ordinary civilians is legitimized and obscured in the 'war on terror' by a wider discourse in which the entire cities of such victims are essentialized as little more than 'factories' or 'nests' sustaining 'terrorists' and 'extremists' (Graham 2005). To achieve this, huge discursive and material work is being done by both the US military and the mainstream US media to construct Islamic cities as dehumanized 'terror cities' – nest-like environments whose very geography undermines the high-tech, orbital mastery of US forces. For example, as a major battle raged there in April 2004 in which over 600 Iraqi civilians died, General Richard Myers, Chair of the US Joint Chiefs of Staff, labelled the whole of Fallujah a dehumanized 'rat's nest' or 'hornet's nest' of 'terrorist resistance' against US occupation that needed to be 'dealt with' (quoted in News24.com 2004, see Graham 2005).

Such disclosures have been backed up by widespread popular geopolitical representations of Iraqi cities. Derek Gregory (2004b: 202), for example, analyses how, in their pre-invasion discussions about the threat of 'urban warfare' to US forces in Iraq, mainstream news media like *Time Magazine* repeatedly depicted Orientalized streets where 'nothing was what it seemed, where deceit and danger threatened at every turn' and where the US forces' high-tech weapons and surveillance gear were the key to 'reveal the traps' and 'lift' the Orientalized veil obscuring Iraqi urban places (ibid.).

In the bloody urban battles of 2004 for Saddam City, Fallujah and Najaf, the promulgations of the US military forces fighting in Iraq – and their leaders back in the US – have also routinely blended Islamophobic racism and crude Orientalism. Again, this worked to continually reinforce the perception that these cities are little but 'nests' of terrorist violence that necessitate targeting by superior US surveillance technologies and military firepower, which will somehow act to 'cleanse' or redeem the intrinsically terroristic urban places of Iraq. 'The Iraqis are sick people and we are the chemotherapy,' boasted one US Marine to the *New Statesman* in April 2003. Leavened in here have been widespread invocations of some essentialized 'Arab mind' (see Patai 1983). 'You have to understand the Arab mind,' suggested Captain Todd Brown, a company commander with the US Fourth Infantry Division in Baghdad in early December 2003. 'The only thing they understand is force – force, pride and saving face' (quoted in Wilkins 2003).

Widespread pronouncements of the fighting US soldiers themselves illustrate these imaginative geographies all too clearly. US Marine snipers after the battle of Fallujah, for example, talked exultantly about their 'kills' of 'rag-heads' and 'sand niggers' in Fallujah (Davis 2004). Shocked senior British officers in Iraq – whose forces are far from blameless in terms of brutality against Iraqi civilians – even alleged, anonymously, that American forces often viewed Iraqi civilians 'as *untermenschen* [the Nazis' word for subhuman]' (quoted in Rayment 2004). This view, of course, has been reinforced by the extending list of prison torture scandals that have erupted since the end of the 2003.

Added to these street-level discourses, a large group of professional 'urban warfare' commentators, writing regular columns in US newspapers, have routinely projected deeply racist notions implying that the inhabitants of targeted Iraqi cities are merely subhuman pests requiring extermination. An important example comes from the highly influential 'urban warfare' commentator, Ralph Peters, writing in the neoconservative *New York Post*. To Peters and many like him cities like Fallujah and Najaf are little more than killing zones which challenge the US military to harness its techno-scientific might to sustain hegemony. This must be done, he argues, by killing 'terrorists' as rapidly and efficiently – and with as few US casualties – as possible. During the battle of Fallujah, Peters (2004a) labeled the entire city a 'terror-city' in his column. Praising the US Marines 'for hammering the terrorists into the dirt' in the battle, he nevertheless castigated the ceasefire negotiations that, he argued, had allowed those 'terrorists' left alive to melt back into the civilian population (ibid.).

In a later article Peters (2004b) concluded that a military, technological solution was available to US forces that would enable them to 'win' such battles more conclusively in the future: killing faster, before any international media coverage is possible. 'This is the new reality of combat,' he wrote. 'Not only in Iraq. But in every broken country, plague pit and terrorist refuge to which our troops have to go in the future'. Arguing that the presence of 'global media' meant that 'a bonanza of terrorists and insurgents' were allowed to 'escape' US forces in Fallujah, US forces, he argued 'have to speed the kill'. By 'accelerating urban combat' to 'fight within the "media cycle" before journalists sympathetic to terrorists and murderers can twist the facts and portray us as the villains,' new technologies were needed, Peters suggested. This was so that 'our enemies are overwhelmed and destroyed before hostile cameras can defeat us. If we do not learn to kill very, very swiftly, we will continue to lose slowly' (ibid.).

Othering by simulation I: 'urban warfare' video games

> In a world being torn apart by international conflict, one thing is on everyone's mind as they finish watching the nightly news: 'Man, this would make a great game!'
>
> (Jenkins 2003)

Third, the construction of Arab cities as targets for US military firepower now sustains a large industry of computer gaming and simulation. Such simulations – which are devised especially to create positive images for the US military amongst younger computer game users – 'propel the player into the world of the gaming industry's latest fetish: modern urban warfare' (DelPiano 2004). They work to further reinforce imaginary geographies equating Islamic cities with 'terrorism' and US military intervention.

Such games serve to blur the boundaries separating war from entertainment. Worse still, they demonstrate that 'the entertainment industry has assumed a posture of co-operation towards a culture of permanent war' (Deck 2004). Within such games, Arab cities are represented merely as 'collections of objects not congeries of people' (Gregory 2004b: 201). When people *are* represented they are the shadowy, subhuman, racialized figures of absolutely external 'terrorists' to be annihilated repeatedly in sanitized 'action' as entertainment or military training (or both). Andrew Deck (2004) argues that the proliferation of urban warfare games based on actual, ongoing US military interventions in Arab cities works to 'call forth a cult of ultra-patriotic xenophobes whose greatest joy is to destroy, regardless of how racist, imperialistic, and flimsy the rationale' for the simulated battle.

The US Army – which now brands itself as 'the world's premier land force' – itself works hard and at many levels to demonize Arab urbanism per se through the medium of video games. In fact it is now one of the world's biggest developers of video games, which it deliberately deploys as aids to training and recreation amongst US soldiers and the generation of both recruits and revenue (Gaudiosi 2004).

The products of this work are dominated by scenarios which depict US soldiers fighting dark, animalistic figures in highly unrealistic portrayals of supposedly Arab cities. Here, once again, the only role for the everyday sites and spaces of the city is as environments for military engagement. 'Cars are used as bombs, bystanders become victims [although they die without spilling blood], houses become headquarters, apartments become lookout points, and anything to be strewn in the street becomes suitable cover' (DelPiano 2004). Indeed, the actual physical geographies of Arab cities are being digitized to provide the three-dimensional 'battlespace' for each game. One games developer boasts that 'we've built a portion of the downtown area of a large Middle Eastern capital city where we [*sic*] have a significant presence today' (cited in Deck 2004).

These representations, of course, resonate strongly with the pronouncements of military urban warfare specialists in the wider media like those of Ralph Peters discussed above. They also blur with increasing seamlessness into news reports about the actual Iraq war. Kuma Reality Games, for example, which has actually sponsored Fox News's coverage of the 'war on terror' in the US, uses this sponsorship to promote an urban combat game. In their words, this centres on US Marines fighting 'militant followers of radical Shiite cleric Muqtaqa al-Sadr in the filthy urban slum that is Sadr city' (quoted in Deck 2004).

Not to be outdone, the US Army itself now gives games such as *America's Army* – with its simulations of 'counter terror' warfare in densely packed Islamic cities in a fictional country of 'Zekistan' – free to millions over the Internet as an aid to recruitment. 'The mission' of America's Army, writes Steve O'Hagan (2004)

> is to slaughter evildoers, with something about 'liberty' . . . going on in the background . . . . These games may be ultra-realistic down to the caliber of the weapons, but when bullets hit flesh people just crumple serenely into a heap. No blood. No exit wounds. No screams.

America's Army has been followed up by the even more elaborate game, *Full Spectrum Warrior*, another ex-military training video game in which US forces again wage urban warfare in simulations of Middle Eastern cities whilst this time dispensing racist and Islamophobic expletives. Even some video game reviewers have commented that 'this game would have been fine without the tawdry 4 letter words and negative racist remarks' from the simulated US soldiers (Peterson 2004). Such racist remarks have done little to inhibit the game's popularity, however. Writing in a chat room on the neoconservative FreeRepublic.Com, one reviewer of the game gushes that, 'given the current state of the world, it's amazingly relevant, not to mention fun to fire on raghead terrorist wanna-bes'.

Othering by simulation II: 'urban warfare' training sites

Finally, to parallel such virtual, voyeuristic, Othering, US and Western military forces have constructed their own simulations of Islamic cities as targets – this time in physical space. A chain of 80 mock 'Islamic' urban districts have been built across the world since 9/11 designed purely to hone the skills of US forces in fighting and killing in 'urbanized terrain'. Taking 18 months to construct, these simulated 'cities' are then endlessly destroyed and remade in practice assaults that hone the US forces for the 'real thing' in sieges such as those in Fallujah.

Replete with minarets, pyrotechnic systems, loop-tapes with calls to prayer, donkeys, hired 'civilians' in Islamic dress wandering through narrow streets, and olfactory machines to create the smell of rotting corpses, this shadow urban system simulates not the complex cultural, social or physical realities of real Middle Eastern urbanism, but the imaginative geographies of the military and theme park designers that are brought in to design and construct it.

CONCLUSIONS

This essay has demonstrated some of the ways in which the political, discursive, material and geographical dimensions of the Bush administration's 'war on terror' rest fundamentally on dialectical constructions of urban place. Such constructions, essentially, invoke both political and public reworkings of imaginative geographies. These are shaped and legitimized to do geopolitical work. And they tend to be ignored in literatures which overwhelmingly separate discussions about the imaginative geographies constructed for 'homeland' cities from those of cities targeted for US military attack in the Middle East or Central Asia. Moreover, it has been shown that the dialectical constructions of urban place which underlie the 'war on terror' can only really be understood if analysis stretches to cover the *mutually constitutive* representation of both 'homeland' and 'target' cities.

In fact, it is crucial to stress that the imaginative geographies promulgated in the 'war on terror' are noticeably antiurban *in general.* They serve to demonize Arab cities and essentialize them as little but receiving points for (real or virtual) military ordnance. At the same time, they problematize and deny the cosmopolitan mixing that is the very essence of urban life in metropolitan America and provide the basis for a deeply reactionary and entrenching national security state backed up by tax cuts for the super-rich and a dramatic dismantling of urban welfare, social and fiscal programmes aimed at the urban poor (Giroux 2003).

In achieving this unusually broad approach, the current chapter allows us to make three further conclusions. First, and crucially, there are extremely strong resonances between the dialectical constructions of urban places in official US 'war on terror' pronouncements and those in the 'popular geopolitical' domains of news media, novels, Internet chat rooms, films and, most notable of all, video games. This points to the increasing integration of the prosecution, representation and imagination of 'asymmetric' urban warfare in the early 21st century. The growth of the 'military–industrial–media–entertainment network' (Der Derian 2001) that sustains this blurring is occurring as reporters become 'embedded' in urban combat (with the language of 'they're moving out' becoming a language of 'we're moving out'); theme park designers construct 'mock' Islamic cities for US urban combat training; voyeuristic media both ratchet up fear about attacks in the 'urban homeland' and legitimize pre-emptive war in 'target cities'; private military corporations soak up huge contracts for both 'homeland security' and overseas military aggression; and the military themselves construct Orientalist and racist video games where virtualized Arab cities are experienced as mere environments for the killing of 'terrorists' as entertainment for US suburbanites in the 'homeland'. Importantly, this complex of discourses and representations – themselves the product of deeply militarized popular and political cultures – work, on the one hand, to problematize urban cosmopolitanism in 'homeland cities' and, on the other, to essentialize and reify the social ecologies of 'target' cities in profoundly racist ways.

Second, this essay has demonstrated that the production of this highly charged dialectic – the forging of exclusionary, nationalist, imagined communities and the Othering of whole swathes of our urbanizing planet – has been a fundamental prerequisite for the legitimization of the entire 'war on terror'. Worryingly, such fundamentalist and racist constructions of urban place have their almost exact shadow in the charged dialectics of urban place routinely disseminated by al-Qa'eda itself. Here, however, the 'targets' are the 'infidel,' 'Christian' or 'Zionist' cities of the West or Israel. The sentimentalized spaces of the Islamic 'homeland', meanwhile, are to be violently 'purified' of Western presence in order to create a transnational Islamic space or *umma* which systematically excludes all diversity and Otherness through continuous, murderous force.

The real tragedy of the 'war on terror', then, is that it has closely parallelled al-Qa'eda in invoking homogeneous, anticosmopolitan and profoundly exclusionary notions of 'community' as a way of legitimizing massive violence based on this charged place dialectic fuelled by hypermasculine notions of assymetric war (Gilroy 2003). In so doing, the 'war on terror' has worked to construct a self-reinforcing cycle of terrorist atrocity and counter-terrorist atrocity, as the daily carnage on the streets of Iraq's cities makes plain. Once set in train, as we see in both Iraq and in Israel–Palestine, such cycles are extraordinarily difficult to reverse. As Zulaika (2003: 198) suggests, the problem here is that:

> such a categorically ill-defined, perpetually deferred, simple minded Good-versus-Evil war ['against terror'] echoes and re-creates the very absolutist mentality and exceptionalist tactics of the insurgent terrorists. By formally adopting the terrorists' own game – one that by definition lacks rules of engagement, definite endings, clear alignments between enemies and fiends,

or formal arrangements of any sort, military, political, legal, or ethical – the inevitable danger lies in reproducing it endlessly.

Finally, the imaginative geographies that underpin 'war on terror' discourses, which stress separateness and the total difference between 'homeland' and 'target' cities, are, of course, being overlaid by much more complex geographies of connection and disconnection. Thus, a revivified Orientalism is used to invoke apparently 'common-sense' binary imaginative geographies of 'inside' and 'outside' just as a wide range of processes – inter-city economic relations, flows of reconstruction capital, flows of natural resources, diasporic, migration and labour networks, electronic mediascapes – intensify to demonstrate that insides and outsides cannot be separated, and that only the relational understandings, grounded in the inter-city imaginaries of flow that are the focus of this book can ever help capture the realities of contemporary global change.

Thus, not surprisingly, the racist imaginary geographies at the centre of Bush's 'war on terror' become ridden with deep contradictions whenever they are compared to even the most cursory evidence about the realities of contemporary change. Constructions of 'homeland cities' as endlessly vulnerable spaces open without warning to an almost infinite range of distanciated threats actually work to *underline* the integration of such spaces into the manifold flows and processes of globalization. The techno-scientific systems that allow US military forces to undertake trans-global military operations as part of the 'war on terror' increasingly treat 'home' and 'target' domains as a single, transnational and increasingly urban 'battlespace.' And the increasingly aggressive attempts to stake claim to the world's dwindling fossil fuel reserves, to bolster the profligate consumption practices in North America, only deepen the sense that urban lives right across the world are intrinsically enmeshed together within integrated political ecologies of resource extraction, resource use and resulting global pollution and climate change.

In exposing the dangerous absurdities of the imaginative geographies at the root of Bush's war on terror, and the huge state violence predicated on them, perhaps critical social science and urban research can play some part in undermining their terrifying levels of success in mobilizing and radicalizing domestic political support. To achieve this, however, research on inter-city networks will need to move substantially beyond its original foci on economic, corporate and technological networks. In addition, it will need to expose how global imaginative geographies of cities are constructed and attached to vast realms of popular geopolitical texts to legitimize the geopolitical projects and programmes of violence of both state and non-state actors alike.

NOTE

Some of this text draws from the article Cities and the 'War on Terror' in *International Journal of Urban and Regional Research* 30: 255–276.

REFERENCES

Andreas, P. (2003) 'A tale of two borders: the US–Canada and US–Mexico lines after 9/11', in P. Andreas and T. Biersteker (eds) *The Rebordering of North America*, New York: Routledge.

Andreas, P. and Biersteker, T. (eds) (2003) *The Rebordering of North America*, New York: Routledge.

Armitage, J. (2002) 'State of emergency', *Theory, Culture and Society*, 19: 27–38.

Collins, J. and Glover, R. (2002) *Collateral Language: A User's Guide to America's New War*, New York: New York University Press.

Dalby, S. (2000) 'A critical geopolitics of global governance', paper available at www.ciaonet.org/isa/das01/.

Davis, M. (2004) 'The Pentagon as global slum lord', http://www.tomdispatch.com/.

Deck, A. (2004) 'Demilitarizing the playground', *No Quarter*, http://artcontext.net/crit/essays/noQuarter/.

DelPiano, S. (2004) 'Review of *Full Spectrum Warrior*', *Games First*, http://www.gamesfirst.com/reviews/07.10.04/FullSpectrumRev/fullspectrumreview.htm.

Der Derian, J. (2001) *Virtuous War: Mapping the Military–Industrial–Media–Entertainment Complex*, Boulder, CO: Westview.

Foucault, M. (1970) *The Order of Things*, London: Tavistock.

Gaudiosi, J. (2004) 'Army sets up video-game studio', *Wired*, June: 23.

Gilroy, P. (2003) '"Where ignorant armies clash by night": homogeneous community and the planetary aspect', *International Journal of Cultural Studies*, 6: 261–76.

Giroux, H. (2003) *Public Spaces, Private Lives: Democracy beyond 9/11*, Oxford: Rowman and Littlefield.

Graham, S. (ed.) (2004) *Cities, War and Terrorism: Towards an Urban Geopolitics*, Oxford: Blackwell.

Graham, S. (2005) 'Remember Fallujah: demonising place, constructing atrocity', *Environment and Planning D: Society and Space*, 23: 1–10.

Gregory, D. (1995) 'Imaginative geographies', *Progress in Human Geography*, 19: 447–85.

Gregory, D. (2003) 'Defiled cities', *Singapore Journal of Tropical Geography*, 24: 307–26.

Gregory, D. (2004a) *The Colonial Present*, Blackwell: Oxford.

Gregory, D. (2004b) 'Who's responsible? Dangerous geography', *ZNet*, www.znet.org, 3 May.

Gusterson, H. (1999) 'Nuclear weapons and the Other in the Western imagination', *Cultural Anthropology*, 14: 111–43.

Hage, G. (2003) '"Comes a time we are enthusiasm": understanding Palestinian suicide bombers in times of Exigophobia', *Public Culture*, 15 (1): 65–89.

Hewitt, K. (1983) 'Place annihilation: area bombing and the fate of urban places', *Annals of the Association of American Geographers*, 73: 257–84.

Howell, S. and Shryock, A. (2003) 'Cracking down on dispaora: Arab Detroit and America's "war on terror"', *Anthropological Quarterly*, 76: 443–62.

Huntingdon, S. (1993) *The Clash of Civilizations and the Remaking of World Order*, New York: Simon and Schuster.

Hyndman, J. (2003) 'Beyond either/or: a feminist analysis of September 11th', *ACME: An International E-Journal for Critical Geographies*, www.acme-journal.org.

Jenkins, H. (2003) 'A war of words over Iraqi video games', *Guardian*, 13 November: 18.

Johnson, J. (2002) 'Immigration reform, homeland defense and metropolitan economies in the post 9/11 environment', *Urban Geography*, 23: 201–12.

Kaplan, A. (2003) 'Homeland insecurities: reflections on language and space', *Radical History Review*, 85: 82–93.

Katz. C. (2004) 'Banal terrorism', paper available from the author, ckatz@gc.cuny.edu.

Kraska, P. (2001) *Militarizing the American Criminal Justice System: The Changing Roles of the Armed Forces and the Police*, Boston: Northeastern University Press.

Luke, T. (2004) 'Everyday techniques as extraordinary threats: urban technostructures and nonplaces in terrorist actions', in S. Graham (ed.) *Cities, War and Terrorism: Towards an Urban Geopolitics*, Oxford: Blackwell.

Lutz, C. (2002) 'Making war at home in the United States: militarization and the current crisis', *American Anthropologist*, 104: 733–5.

News24.com. (2004) 'Fallujah a "rat's nest" ', www.news24.com, 21 April.

O'Hagan, S. (2004) 'Recruitment hard drive', *Guardian Guide*, 19–25 June: 12–13.

Passavant, P. and Dean, J. (2002) 'Representation and the event', *Theory and Event*, 5 (4), http://muse.jhu.edu/journals/theory_and_event/voo5/5.4passavant.html.

Patai, R. (1983) *The Arab Mind*, New York: Macmillan.

Peters, R. (2004a) 'He who hesitates', *New York Post*, 27 April, www.nypost.com.

Peters, R. (2004b) 'Kill faster!', *New York Post*, 20 May, www.nypost.com.

Peterson, B. (2004) '*Full Spectrum Warrior* review for X-box', http://www.gaming-age.com/cgi-bin/reviews/review.pl?sys=xbox&game=fullspectrumwarrior.

Pieterse, J. (2004) 'Neoliberal empire', *Theory, Culture and Society,* 21: 119–40.

Puar, J. and Rai, A. (2002) 'Monster, terrorist, fag: the war on terrorism and the production of docile patriots', *Social Text*, 20: 117–48.

Raban, J. (2004) 'Running scared', *Guardian*, 21 July: 3–7.

Rayment, S. (2004) 'British commanders condemn US military tactics', *Daily Telegraph*, 12 April: 4.

Roberts, L., Lafta, R., Garfield, R., Khudhairi, J. and Burnham, G. (2004) 'Mortality before and after the 2003 invasion of Iraq: cluster sample survey', *Lancet*, 29 October: 1–8.

Said, E. (1978) *Orientalism*, London: Routledge and Kegan Paul.

Said, E. (2003) *Orientalism*, 2003 edition, London: Penguin.

Tuastad, D. (2003) 'Neo-Orientalism and the new barbarism thesis: aspects of symbolic violence in the Middle East conflict(s)', *Third World Quarterly*, 24: 591–9.

Watson T. (2003) 'Introduction: critical infrastructures after 9/11', *Postcolonial Studies*, 6: 109–11.

Wilkins, D. (2003) 'Tough new tactics by US. Tighten grip on Iraq towns', *New York Times*, 7 December, www.nytimes.com.

Williams, R. (2003) 'Terrorism, anti-terrorism and the normative boundaries of the US polity: the partiality of politics after 11 September 2001', *Space and Polity*, 7: 273–92.

Zulaika, J. (2003) 'The self-fulfilling prophecies of counterterrorism', *Radical History Review*, 85: 191–9.

Part IV Rethinking cities in globalization

13 Reading the city in a global digital age

The limits of topographic representation

Saskia Sassen

INTRODUCTION

Understanding a city or a metropolitan region in terms of built topography is increasingly inadequate when global and digital forces become part of the urban condition. What we might call the topographic moment is a critical and a large component of the representation of cities. But it cannot incorporate the fact of globalization and digitization as part of the representation of the urban. Nor can it critically engage today's dominant accounts about globalization and digitization. These accounts evict place and materiality even though particular components of the global and the digital are deeply imbricated with the material and the local and hence with that topographic moment. A key analytic move that bridges between these very diverse dimensions is to capture the possibility that particular components of a city's topography can be spatializations of global and digital dynamics and formations. Such particular topographic components are then one site in a transnational multisited circuit or network. These spatializations destabilize the meaning of the local or the sited, and thereby of the topographic understanding of cities. This holds probably especially for global cities.

A number of scholars have examined the urban condition in today's global and digital era, and raised various methodological and conceptual questions about the city as an object of study (e.g. Castells 1996; Yeung 2000; Paddison 2001; Thrift and Amin 2002; Drainville 2004; Graham 2004; Orum and Chen 2004; Ascher 2005; Veltz 2005). Against this larger background, my particular concern in this essay is to distinguish between the topographic representation of key aspects of the city and an interpretation of these same aspects as spatialized global economic, political, and cultural dynamics.[1] This is one analytic path into questions about cities in a global digital age. It brings a particular type of twist to the discussion on urban topography and cities since globalization and digitization are both associated with dispersal and mobility.

The effort is then to understand what analytic elements need to be developed in order to redress the limits of topography in making legible the possibility that at least some global and digital components get spatialized in cities. Among such components are not only the power projects of major global economic actors but also the political projects of contestatory actors, e.g. electronic activists. A topographic representation of rich and poor areas of a city would simply capture the physical conditions of each – advantage and disadvantage. It would fail to capture the electronic connectivity possibly marking even poor areas as locations on global circuits. Once this spatialization of various global and digital components is made legible, the richness of topographic analysis can add to our understanding of this process. The challenge is to locate and specify the fact of such spatializations and its variability.

This brings up a second set of issues: topographic representations of the built environment of cities tend to emphasize the distinctiveness of the various socio-economic sectors: the differences between poor and rich neighborhoods, between commercial and manufacturing districts, and so on. Although valid, this type of representation of a city becomes particularly partial when, as is happening today, a growing share of advanced economic sectors also employ significant numbers of very low-wage workers and subcontract to firms that do not look as if they belong in the advanced corporate sector. Similarly, the growth of high-income professional households has generated a whole new demand for low-wage household workers, connecting expensive residential areas with poorer ones, and placing these professional households on global care-chains that bring in many of the cleaners, nannies, and nurses from poorer countries. In brief, economic restructuring is producing multiple interconnections among parts of the city that topographically look as if they may have little to do with each other. Given some of the socio-economic, technical, and cultural dynamics of the current era, topographic representations may well be more partial today than in past phases.

The limitations of topography to capture these types of interconnections – between the global and the urban, and between socio-economic areas of a city that appear completely unrelated – calls for analytic tools that allow us to incorporate such interconnections in spatial representations of cities. Some of these interconnections have long existed. What is different today is their multiplication, their intensity, their character. Some elements of topographic representation, such as transport systems and water and sewage pipes, have long captured particular interconnections. What is different today in this regard is the sharpening of non-physical interconnections, such as digital interconnections, perhaps also pointing to a deeper transformation in the larger social, economic, and physical orders.

Topography remains critical, but is increasingly insufficient. One way of addressing these conditions is to uncover the interconnections among urban forms and urban fragments, and among orders – the global and the urban, the digital and the urban – that appear as unconnected. Elsewhere (Sassen 2006: chapters 1 and 7–9) I have developed the notion of emergent assemblages that bring together elements from diverse established orders, thereby unsettling these older orders. These assemblages are getting constituted within multiple insitutional and spatial settings, and across traditional borders, notably those of nation-states. As an analytic category this notion of assemblage can also be used to capture the juxtaposition of conditions and dynamics examined in this essay. Detecting the limits of topography is one more step for understanding what our large cities are about today and in the near future, and what constitutes their complexity.

SPATIALIZED POWER PROJECTS

Cities have long been key sites for the spatialization of power projects – whether political, religious, or economic (Sennett 1990; Lloyd 2005). There are multiple instances that capture this. We can find it in the structures and infrastructures for control and management functions of past colonial empires and of current global firms and markets. We can also find it in the segregation of population groups that can consequently be more easily produced as either cheap labor or surplus people; in the choice of particular built forms used for representing and symbolic cleansing of economic power, as in the preference for 'Greek temples' to house stock markets; and we can find it in what we designate today as high-income residential and commercial gentrification, a process that allows cities to accommodate the expanding elite professional classes, with the inevitable displacement of lower-income households and firms. Finally, we can see it in the large-scale destruction of natural environments aimed at implanting particular forms of urbanization marked by spread rather than density and linked to specific real estate development interests. A familiar example is the uncontrolled strip-development and suburbanization that shaped the Los Angeles region.

Yet the particular dynamics and capacities captured by the terms globalization and digitization signal the possibility of a major transformation in these patterns of spatialization. The most common interpretation posits that digitization entails an absolute disembedding from the material world. Key concepts in the dominant account about the global economy – globalization, information economy, telematics – all suggest that place no longer matters. And they suggest that the type of place represented by major cities may have become obsolete from the perspective of the economy, particularly for leading sectors, which have full access to, and are the most advanced users of, the new information and telecommunications technologies. These accounts privilege the fact of instantaneous global transmission over the concentrations of built infrastructure that make transmission possible; information outputs over the work of producing those outputs, from specialists to secretaries; and the new transnational corporate culture over the multiplicity of cultural environments, including re-territorialized immigrant cultures, within which many of the 'other' jobs of the global information economy take place.[2]

One consequence of such a representation of the global information economy as placeless would be that there is no longer a spatialization of this type of power today as it would supposedly have dispersed geographically and gone partly digital. It is this proposition that I have contested in much of my work, arguing that this dispersal is only part of the story and that we see in fact new types of spatializations of power. How do we reintroduce place in economic analysis? And how do we construct a new narrative about economic globalization, one that includes rather than excludes all the spatial, economic, and cultural elements that are part of the urban global economy as it is constituted in cities, in the increasingly structured inter-city networks, and in digital spaces that are becoming urbanized in distinct and unexpected ways? A topographic reading would introduce place yet, in the end, it would fail to capture the fact that global dynamics and digitized spaces might inhabit localized built environments and thereby unsettle the meaning of the latter.

ANALYTIC BORDERLANDS

As a political economist, addressing these issues has meant working in several systems of representation and constructing spaces of intersection. There are analytic moments when two systems of representation intersect; these are easily experienced as spaces of silence, or of absence. One challenge is to see what happens in those spaces, what operations take place there. In my own work I have had to deal frequently with these spaces of intersection and have come to conceive of them as analytic borderlands – an analytic terrain where discontinuities are constitutive of a space rather than being merely a dividing line. Thus much of my work on economic globalization and cities has focused on these discontinuities and has sought to reconstitute their articulation analytically as borderlands rather than as dividing lines.[3]

Methodologically, the construction of such analytic borderlands for use with the types of cities that concern me here pivots on detecting circuits for the distribution and installation of diverse operations – in the domain of the economy, politics, culture, and so on. I focus on circuits that cut across what are generally seen as two or more discontinuous 'systems,' institutional orders, places, or dynamics. These circuits may be internal to a city's economy or perhaps, at the other extreme, global. In the latter case, a given city can be seen as containing fragments that are one site on one or more circuits that may contain a few or many other such kindred fragments from cities across the world. The operations that get distributed through these circuits can range widely – they can be economic, political, cultural, subjective. Cross-border circuits constituted through kindred urban fragments are assemblages of a new sort – neither fully urban nor fully global, not fully territorial yet anchored in multiple partial territories.

Detecting circuits internal to a city allows us to follow economic activities into territories that lie outside the increasingly narrow borders of mainstream representations of the urban economy and to negotiate the crossing of discontinuous spaces. For instance, it allows us to locate various components of the informal economy (whether in New York, Paris, or Mumbai) on circuits that connect it to what are considered advanced industries, such as finance, design, or fashion. A topographic representation would capture the enormous discontinuity between the places and built environments of the informal economy and those of the financial or design district in a city, but would fail to capture their complex economic interactions and dependencies.

International and transnational circuits allow us to establish the particular networks that connect specific activities in one city with specific activities in cities in other countries. It entails not only disaggregating general categories such as 'the' city into diverse spaces, but also 'the' global economy into the multiple circuits and markets that constitute it. The specialized division of functions in the global economy is partly constituted and implemented through a proliferation of specific inter-city networks. These are specific in a double sense: They often involve particular groups of cities and particular contents. For instance, if I were to track the global market for gold futures, that is to say, financial instruments based on gold, some of the key cities that would appear on my map are Chicago and London. If I were to track the global trade in gold the metal, additional cities would appear, notably Johannesburg, Mumbai, and Dubai. The critical mass of these networks has expanded to include in its aggregate about 40 major and minor global cities. The specialized differences among cities take on renewed value in this organizational infrastructure. For instance, Los Angeles is located on a variety of global circuits (including bi-national circuits with Mexico) which are quite different from those of New York or Chicago. And a city like Caracas can be shown to be located on different circuits from those of Bogota.

Common notions of inter-city competition do not adequately capture these kinds of developments. Global firms and markets need multiple global cities from where they can organize their operations. This network of global cities is much more than a set of cross-border flows connecting cities. It is a complex, highly specialized organizational infrastructure for the management and servicing of the leading economic sectors.

The fact that bits of cities get assembled into specialized inter-city geographies brings to the fore a second important issue. We can think of global cities and urban regions as criss-crossed by these circuits; one image that results is a city as an amalgamation of the local bits of inter-city circuits. Included here, as I discuss later, are also some of the disadvantaged sectors in major cities, which are today forming lateral cross-border connections with similarly placed groups in other cities.

These inter-city geographies function at many different scales. But, whether regional or global, they do not run through a vertically organized framing as do, for instance, the affiliates of a multinational corporation or the country-specific programs of the IMF. For the city, these transnational circuits entail a type of fragmentation that may have always existed in major cities but has now been multiplied many times over and intensified sharply. Topography would fail to capture much of this spatialization of global economic circuits, except, perhaps, for certain aspects of the distribution/transport routes.

SITED MATERIALITIES AND GLOBAL SPAN

It seems to me that the difficulty analysts and commentators have had in specifying or understanding the impact of digitization on cities results from two analytic flaws (Sassen 2002). One of these (especially evident in the US) confines interpretation to a technological reading of the technical capabilities of digital technologies. This is fine for engineers. But when one is trying to understand the impacts of a technology, such a reading becomes problematic.[4] A purely technological reading of the technical capabilities of digital technology inevitably leads one to a place that is a non-place, where we can announce with certainty the neutralizing of many of the configurations marked by physicality and place-boundedness, including the urban.[5]

The second flaw is a continuing reliance on analytic categorizations that were developed under other spatial and historical conditions, that is, conditions preceding the current global digital era. Thus the tendency is to conceive of the digital as simply and exclusively digital and the non-digital (whether represented in terms of the physical/material or the actual, all problematic though common conceptions) as simply and exclusively that. These either/or categorizations filter out the possibility of mediating conditions, thereby precluding a more complex reading of the impact of digitization on physical and place-bound conditions.

One alternative categorization captures imbrications (for a full development of this alternative see Sassen 2006: chapters 7 and 8). Let me illustrate using the case of finance. Finance is certainly a highly digitized activity; yet it cannot simply be thought of as exclusively digital. To have electronic financial markets and digitized financial instruments requires enormous amounts of materiel, not to mention people. This materiel includes conventional infrastructure, buildings, airports, and so on. Much of this materiel is, however, inflected by the digital. Conversely, much of what takes place in digital space is deeply inflected by the cultures, the material practices, the imaginaries of the non-digital world. Much, though not all, of what we think of when it comes to cyberspace would lack any meaning or referents if we were to exclude the world outside cyberspace. In brief, the digital and the non-digital are not mutually exclusive conditions. Digital space is embedded in the larger societal, cultural, subjective, economic, imaginary structurations of lived experience and the systems within which we exist and operate (Sassen 1999; Lovink 2002).[6]

RESCALING THE OLD HIERARCHIES

The complex imbrications between the digital (as well as the global) and the non-digital bring with them a destabilizing of older hierarchies of scale and often dramatic rescalings. As the scale of nation-states loses some of its erstwhile significance, given globalization, privatization, and deregulation of national economies, other scales are gaining novel types of importance. Most especially among these are sub-national scales such as the global city and supranational scales such as global markets or regional trading zones. There is by now a vast scholarship covering a range of these rescaling dynamics and formations (e.g. Pillon and Querrien 1995; Sun 1999; Barry and Slater 2002; Ferguson and Jones 2002; Glasius *et al.* 2002; Schiffer 2002; Taylor *et al.* 2002; Lebert 2003; Brenner 2004; Taylor 2004;

Olesen 2005). Older hierarchies of scale that emerged in the historical context of the ascendance of the nation-state continue to operate; they are typically organized in terms of institutional size – from the national down to the regional, the urban, the local, with the international an aggregation of national scales. Today they are destabilized as multiple, often partial and specialized, rescalings cut across institutional size (e.g. Sun 1999; Urry 2000; Yeung 2000; Brenner 2004) and also across the institutional encasements of territory produced by the formation of national states (Ferguson and Jones 2002; Beck 2006). This does not mean that the old hierarchies disappear, but rather that rescalings emerge alongside the old ones, and that they can often trump the latter.

These transformations entail complex imbrications of the digital and non-digital and between the global and the non-global (Garcia 2002; Krause and Petro 2003; Graham 2004; Taylor 2004; Sack 2005; Sassen 2006: chapters 7 and 8). They can be captured in a variety of instances. For example, much of what we might still experience as the 'local' (an office building or a house or an institution right there in our neighborhood or downtown) actually is something I would rather think of as a 'microenvironment with global span' in so far as it is digitally internetworked. Such a microenvironment is in many senses a localized entity, something that can be experienced as local, immediate, proximate, and hence captured in topographic representations. It is a sited materiality. But it is also part of global digital networks which give it immediate far-flung span. To continue to think of this as simply local is not very useful or adequate. More importantly, the juxtaposition between the condition of being a sited materiality and having global span captures the imbrication of the digital and the non-digital and illustrates the inadequacy of a purely technological reading of the technical capacities of digitization. Such a technological reading would lead us to posit the neutralization of the place-boundedness of precisely that which makes possible the condition of being an entity with global span. And it illustrates the inadequacy of a purely topographic account.

A second example is the bundle of conditions and dynamics that marks the model of the global city. Just to single out one key dynamic: the more globalized and digitized the operations of firms and markets, the more their central management and coordination functions (and the requisite material structures) become strategic. It is precisely because of digitization that simultaneous worldwide dispersal of operations (whether factories, offices, or service outlets) and system integration can be achieved. And it is precisely this combination which raises the importance of central functions. Global cities are strategic sites for the combination of resources necessary for the production of these central functions.[7]

Much of what is liquefied and circulates in digital networks and is marked by hypermobility remains physical in some of its components. Take, for example, the case of real estate. Financial services firms have invented instruments that liquefy real estate, thereby facilitating investment and circulation of these instruments in global markets. Yet part of what constitutes real estate remains very physical. At the same time, however, that which remains physical has been transformed by the fact that it is represented by highly liquid instruments that can circulate in global markets. It may look the same, it may involve the same bricks and mortar, it may be new or old, but it is a transformed entity.

We have difficulty capturing this multivalence through our conventional categories: if it is physical, it is physical; and if it is digital, it is digital. In fact, the partial representation of real estate through liquid financial instruments produces a complex imbrication of the physical and the digital moments of what we continue to call real estate. It is this same dynamic of imbrication, albeit at a more complex level, that contributes to the massive concentrations of material resources in global cities which the leading global sectors require.

Hypermobility and digitization are usually seen as mere functions of the new technologies. This understanding erases the fact that it takes multiple material conditions to achieve this outcome (e.g. Graham and Marvin 2001; Rutherford 2004), and that it takes social networks, not only digital ones (Garcia 2002; Sack 2005). Once we recognize that the hypermobility of the instrument, or the digitization of the actual piece of real estate, had to be produced, we introduce the imbrication of the digital and the non-digital. It takes capital fixity to produce capital mobility; that is to say, it takes state-of-the-art built environments, conventional infrastructures – from highways to airports and railways – and well-housed talent. These are all, at least partly, place-bound conditions, even though the nature of their place-boundedness is going to be different from what it was 100 years ago, when place-boundedness was much closer to pure immobility. Today it is a place-boundednesss that is inflected and inscribed by the hypermobility of some of its components, products, and outcomes. Both capital fixity and mobility are located in a temporal frame where speed is ascendant and consequential. This type of capital fixity cannot be fully captured in a description of its material and locational features, i.e. in a topographic reading.

Conceptualizing digitization and globalization along these lines creates operational and rhetorical openings for recognizing the ongoing importance of the material world even in the case of some of the most 'dematerialized' activities.[8]

THE SPATIALITIES OF THE CENTER

Thus digital technologies have not eliminated the importance of massive concentrations of material resources but have, rather, reconfigured the interaction of capital fixity and hypermobility. The complex management of this interaction has given some cities a new competitive advantage (Sassen 2001). The vast new economic topography that is being implemented through digital space is one moment, one fragment, of an even vaster economic chain that is in good part embedded in non-digital spaces. There is today no fully digitized firm or economic sector. As I suggested earlier, even finance, the most digitized and globalized of all activities, has a topography that weaves back and forth between actual and digital space. To different extents across sectors and types of firms, a firm's tasks now are distributed across these two kinds of spaces. Further, the actual configurations are subject to considerable transformation, as tasks are computerized or standardized, markets are further globalized, and so on.

The combination of new capabilities for mobility along with ongoing concentration of command functions among powerful firms suggests that spatial agglomeration remains as a key feature of advanced economic sectors. But it is not simply a continuation of older agglomeration patterns. Today there is no longer a simple or straightforward relation between agglomeration and such geographic entities as the downtown or the central business district (CBD). In the past, and up to quite recently in fact, agglomeration was synonymous with the downtown or the CBD. The new technologies and organizational forms have altered the spatial correlates of agglomeration. I prefer to use the concept 'centrality' to open up the range of spatial correlates.[9]

Given the differential impacts of the new information technologies on specific types of firms and sectors of the economy, the spatial correlates of the 'center' can assume several geographic forms. Thus the center can be the CBD, as it still is largely for some of the leading sectors, notably finance, or an alternative form of CBD such as Silicon Valley. Yet, even as the CBD in major international business centers remains a strategic site for the leading industries, it is one profoundly reconfigured by technological and economic change (Fainstein 2001; Cicolella and Mignaqui 2002; Schiffer 2002) and by long-term immigration from all parts of the world (e.g. Laguerre 2000). Further, there are often sharp differences in the patterns assumed by this reconfiguring of the central city in different parts of the world (Marcuse and Van Kempen 2000; Schiffer 2002; Gugler 2004).

Second, the center can extend into a metropolitan area in the form of a grid of nodes of intense business activity (e.g. Scott 2001; Parsa and Keivani 2002). One might ask whether a spatial organization characterized by dense strategic nodes spread over a broader region is in fact a new way of organizing the territory of the 'center,' rather than, as in the more conventional view, an instance of suburbanization or geographic dispersal. In so far as these various nodes are articulated through digital networks, they represent a new geographic correlate of the most advanced type of 'center.' This is a partly deterritorialized space of centrality in that it comprises not only territorial nodes but also digital inter-nodal networks.

Third, we are seeing the formation of a trans-territorial 'center' constituted via intense economic transactions in the network of global cities. These transactions take place partly in digital space and partly through conventional transport and travel. The result is a multiplication of often highly specialized circuits connecting sets of cities (Yeung 2000; Taylor *et al.* 2002; Taylor 2004; GaWC 2006; Harvey forthcoming); increasingly we see other types of networks built on those circuits, such as transnational migrant networks and global care-chains (Smith and Guarnizo 2001; Gzesh and Espinosa 2002; Samers 2002; Ehrenreich and Hochschild 2003). These networks of major international business centers constitute new geographies of centrality. The most powerful of these new geographies of centrality at the global level binds diverse global cities across the world: London, New York, Tokyo, Paris, Frankfurt, Zurich, Amsterdam, Madrid,

Los Angeles, Sydney, Hong Kong, Singapore, among others. But this geography now also includes cities such as Bangkok, Seoul, Taipei, São Paulo, Mexico City, Shanghai. In the case of a complex landscape such as Europe's, we see in fact several geographies of centrality, one global, others continental and regional.

Fourth, new forms of centrality are being constituted in electronically generated spaces. For instance, strategic components of the financial industry operate in such spaces. The relation between digital and actual space is complex and varies among different types of economic sectors (Sassen 2006: chapter 7) as well as in non-economic sectors (Avgerou 2002; Pace and Panganiban 2002; Bach and Stark 2005; Latham and Sassen 2005a; Sack 2005).

WHAT DOES LOCAL CONTEXT MEAN IN THIS SETTING?

Firms operating partly in physical places and partly in globe-spanning digital space cannot easily be contextualized in terms of their surroundings. Nor can the networked sub-economies they tend to constitute. The orientation of this type of subeconomy is simultaneously toward itself and toward multiple specialized global circuits. Topographic representations would fail to capture this simultaneously inward and global orientation.

The intensity of transactions internal to such a sub-economy (whether global finance or cutting edge high-tech sectors) is such that it overrides all considerations of the broader locality or urban area within which it exists. These firms and subeconomies develop a stronger orientation towards global markets (e.g. Schiffer 2002; Taylor 2004) than to their immediately surrounding areas. In so far as they are a significant component of today's cities, this global orientation overrides a key proposition in the urban systems literature, to wit, that cities and urban systems integrate and articulate national territory. Such an integration effect may have been the case during the period when mass manufacturing and mass consumption were the dominant growth machines in developed economies and, along with Keynesian policies, thrived on and strengthened national scalings of economic processes. Today, the ascendance of digitized and globalized sectors has diluted that systemic articulation with the larger national economy and the immediate surrounding. In its stead we see the formation of particularized, often highly specialized articulations.

The articulation of these sub-economies with other zones and sectors in their immediate socio-spatial surroundings is of a special sort. To some extent there is interdependence, but it is largely confined to the servicing of the leading sectors, and it is furthermore partly obscured by topographic fragmentation when industrial services firms and workers provide the servicing. The most legible articulation is with the various highly priced services that cater to the workforce, from up-scale restaurants and hotels to luxury shops and cultural institutions, typically part of the visible socio-spatial order of these new sub-economies. Secondly, there are also various low-priced services that cater to the firms and to the households of the workers; these rarely 'look' as if they are part of the advanced corporate economy. The demand by firms and households for these services actually links two worlds that we think of as radically distinct and thus unconnected. Thirdly, these world-market oriented sub-economies have few if any connections to large portions of their urban surroundings, even though they are physically proximate and might even be architecturally similar. It is the second and third instances that engender a question about the insufficiency of topographic representation.

We might start by asking: What is the meaning of locality under these conditions? The new networked sub-economy occupies a strategic, partly deterritorialized geography that cuts across borders and connects a variety of points on the globe. Its local insertion accounts for only a (variable) fraction of its total operations, its boundaries are not those of the city within which it is partly located, nor those of the local area where it is sited. This sub-economy interfaces the intensity of the vast concentration of very material resources it needs when it hits the ground and the fact of its global span and inter-city geography. Its interlocutor is not the surrounding context but the fact of the global.

I am not sure what this tearing away of the context and its replacement with the fact of the global could mean for urban practice and theory. But it is clearly a problem from the perspective of urban topography. The analytic operation called for is not the search for how a (local) urban fragment (in this case, a world-market oriented sub-economy) is connected with the 'surroundings' or the context. It is, rather, detecting the installation of that fragment in a strategic cross-border geography constituted through multiple such local fragments. This type of local fragment now transacts directly with the global, forming cross-border structurations that, though largely horizontal, are global. The global thus constituted is marked by the fact that it inhabits multiple such localities. We see here the incipient formation of particularized assemblages of bits of urban territory and distinct urban cultures (economic, political) that get dislodged from older national and urban institutional framings even as they remain lodged in the corresponding geographic terrains.

Though partial and particular, this entails a profound disruption in the relationship of a city to its topography.

CITIES AS FRONTIER ZONES: THE FORMATION OF NEW POLITICAL ACTORS

A very different type of case can be found in the growth of electronic activism by often poor and rather immobile actors and organizations. Topographic representations that describe fragmentations, particularly the isolation of poor areas, may well obscure the existence of underlying interconnections across poor areas in a city and across cities worldwide. What presents itself as segregated or excluded from the mainstream core of a city can actually be part of increasingly complex interactions with other similarly segregated sectors in cities of other countries. There is here an interesting dynamic whereby top sectors (the new transnational professional class) and bottom sectors (e.g. immigrant communities or activists in environmental or anti-globalization struggles) partly inhabit a cross-border space that connects particular cities. Middle sectors of the economy and of the society tend to be more standardized and far less likely to be part of transnational networks.

Major cities, especially if global, contain multiple low-income communities many of which develop or access various global networks. Through the Internet, local initiatives become part of transnational networks of kindred activisms without losing the focus on specific local struggles (e.g. Cleaver 1998; Mele 1999; Henshall 2000; Tsaliki 2002; Lustiger-Thaler and Dubet 2004; van de Donk *et al.* 2005; Friedman 2005). It enables a new type of cross-border political activism, one centered in multiple localities yet intensely connected digitally. This is in my view one of the key forms of critical politics that the Internet can make possible: a politics of the local with a big difference – these are localities that are connected with each other across a region, a country or the world.[10] That the network is global does not mean that it all has to happen at the global level.

But also inside such cities we see the emergence of specific political and subjective dimensions that are difficult to capture through topographic representations (e.g. Lovink and Riemens 2002; Mills 2002; Sandercock 2003; Poster 2004; Drainville 2004; Bartlett forthcoming). Neither the emergence nor the difficulty is new. But I would argue that there are times when both become sharper – times when traditional arrangements become unsettled. Today is such a time. Global cities become a sort of new frontier zone where an enormous mix of people converges and new forms of politics are possible. Those who lack power, those who are disadvantaged, outsiders, discriminated minorities, can gain presence in global cities, presence vis-à-vis power and presence vis-à-vis each other (Sassen 2000). This signals, for me, the possibility of a new type of politics centered in new types of political actors. It is not simply a matter of having or not having power. There are new hybrid bases from which to act.

The space of the city is a far more concrete space for politics than that of the nation. It becomes a place where non-formal political actors can be part of the political scene in a way that is much more difficult at the national level (e.g. Bartlett forthcoming). Nationally, politics needs to run through existing formal systems: whether the electoral political system or the judiciary (taking state agencies to court). Non-formal political actors are rendered invisible in the space of national politics. The city can accommodate a broad range of political activities – squatting, demonstrations against police brutality, fighting for the rights of immigrants and the homeless, the politics of culture and identity, gay and lesbian politics – and it can transform what may have started as an insult into a politics, e.g. today's queer politics and the queering of gender. Much of this becomes visible on the street. Much of urban politics is concrete, enacted by people rather than dependent on massive media technologies. Street-level politics makes possible the formation of new types of political subjects that do not have to go through the formal political system.

In this context, today's large cities, especially if global, emerge as a strategic site for these new types of political operations. They are a strategic site for global corporate capital. But they are also among the sites where the formation of new claims by informal political actors materializes and assumes concrete forms (Torres, *et al.* 1999; Isin 2000; Lovink and Riemens 2002). The partial loss of power at the national level produces the possibility for new forms of power and politics at the sub-national level.[11] The national as container of social process and power is cracked (Taylor 2000; Mills 2002; Brenner 2004; Beck 2006). This 'cracked casing' then opens up possibilities for a geography of politics that links sub-national spaces and allows non-formal political actors to engage strategic components of global capital.

Digital networks are contributing not only to the production of new kinds of interconnections underlying what appear as fragmented topographies, whether at the global or at the local level, but also to new types of subjectivities among the disadvantaged involved in these networks (e.g. Beck 2006). Political activists can use digital networks for global or non-local transactions *and* they can use them for strengthening local communications and transactions inside a city (e.g. Lovink and Riemens 2002) or rural community (e.g. Garcia 2005). Recovering how new digital technologies can serve to support local initiatives and alliances across a city's neighborhoods is extremely important at a time when the notion of the local is often seen as losing ground to global dynamics and actors, and digital networks are typically thought of as global. What may appear as separate segregated sectors of a city may well have increasingly strong interconnections through particular networks of individuals and organizations with shared interests (*Journal of Urban Technology* 1995; Espinoza 1999; Lovink and Riemens 2002).

Any large city is today traversed by these topographically 'invisible' circuits. This condition is not new but is intensifying, becoming multi-scalar, and incorporating novel contents. Critical for the disadvantaged and the resource-poor in a city is that the mix of place-boundedness and access to digital networks opens up the scope of their struggles and the political subjectivities that are forged in these struggles. That mix is beginning to produce a sort of non-cosmopolitan globality among particular groups in localities across the world. This is partly a subjective condition, but also one with concrete contents. Thus the attachment of an individual's or a group's focus on the particulars of their stuggles need not preclude their being part of a globality that incorporates multiple other individuals and groups around the world, each with their own attachments to the specifics of their struggles and places. This condition is a kind of third option – neither simply local nor ipso facto 'cosmopolitan.' It disrupts the easy conflation of the global with the cosmopolitan, and of the local with the non-global. Thereby it opens up the conceptual possibility that transnational politics and subjectivities can comprise a far larger range of actors and places than is suggested by the more common conflations of the global with the mobile and the cosmopolitan. Immobiltiy and locality can become part of novel types of globalities, and to do so they need not leave behind the thick realities that consume them.

CONCLUSION

Economic globalization and digitization produce a spatiality for the urban that pivots on deterritorialized cross-border networks and territorial locations with massive concentrations of resources. This is not a completely new feature. Over the centuries cities have been at the intersection of processes with supra-urban and even intercontinental scalings. What is different today is the intensity, complexity, and global span of these networks, and the extent to which significant portions of economies are now digitized and hence can travel at great speeds through these networks. Also new is the growing use of digital networks by even poor neighborhood organizations to pursue a variety of both intra- and inter-urban political initiatives. All of this has raised the number of cities that are part of cross-border networks operating at often vast geographic scales, and it has multiplied the types and contents of those networks. One effect is to unsettle the meaning of both the local and the global.

As cities and urban regions are increasingly traversed by non-local, including notably global, circuits, much of what we experience as the local, because locally sited, is actually a transformed condition in that it is imbricated with non-local dynamics or is a localization of global processes. One way of thinking about this is in terms of local spatializations of multiple global projects – economic, political, cultural. This produces a specific set of interactions in a city's relation to its topography.

The particular urban spatiality thus produced is partial in a double sense. It accounts for only part of what happens in cities and what cities are about as it coexists with other older spatialities. And it inhabits only part of what we might think of as the space of the city, whether this be understood in terms as diverse as those of a city's administrative boundaries or in the sense of the multiple public imaginaries that may be present in different sectors of a city's people. If we consider urban space as productive, as enabling new configurations, then these developments point to multiple possibilities.

NOTES

1 These are all complex and multifaceted subjects. It is impossible to do full justice to them or to the literatures they have engendered. I have elaborated on both the subjects and the literatures elsewhere (Sassen 2006: chapters 7 and 8).

2 The eviction of these activities and workers from the dominant representation of the global information economy has the effect of excluding the variety of cultural contexts within which they exist, a cultural diversity that is as much a presence in processes of globalization as is the new international corporate culture.

3 This produces a terrain within which these discontinuities can be reconstituted in terms of economic operations whose properties are not merely a function of the spaces on each side (i.e. a reduction to the condition of dividing line) but also, and importantly, of the discontinuity itself, the argument being that discontinuities are an integral part, a component, of the economic *system*.

4 An additional critical issue is the construct technology. One radical critique can be found in Latour, and his dictum that technology is society 'made durable' (Latour 1991, 1996). My position on how to handle this construct in social science research is developed in Sassen (2002, 2006: chapter 7). More generally see Mansell and Silverstone (1998).

5 Another consequence of this type of reading is to assume that a new technology will *ipso facto* replace all older technologies that are less efficient, or slower, at executing

the tasks the new technology is best at. We know that historically this is not the case. For a variety of critical examinations of the tendency towards technological determinism in much of the social sciences today see Wajcman 2002; Howard and Jones 2004; for particular applications that make legible the limits of these technologies in social domains see, for example, Callon 1998; Avgerou, Ciborra and Land 2004; Cederman and Kraus 2005; for cities in particular see Graham 2004.

6 There is a third variable that needs to be taken account of when addressing the question of digital space and networks, though it is not particularly relevant to the question of the city. It concerns the transformations in digital networks resulting from both technical issues and the use of these networks. (For critical accounts see, for example, MacKenzie and Wajcman 1999; Marres and Rogers 2000; Lovink 2002; Rogers 2004; Mansell and Collins 2005; Dean *et al.* 2006).

7 These economic global city functions are to be distinguished from political global city functions, which might include the politics of contestation by formal and informal political actors enabled by these economic functions. This particular form of political global city functions is, then, in a dialectical relation (both enabled and in opposition) to the economic functions (see Sassen 2000; Bartlett forthcoming).

8 A critical issue, not addressed here, concerns questions of governance and political formats of digital networks (e.g. Bennett 2003; Robinson 2004; Drake 2004; Koopmans 20004; Klein 2004; Mansell and Collins 2005). These networks are not neutral technical events (see also the issues raised in note 6 above).

9 Several of the organizing hypotheses in the global city model concern the conditions for the continuity of centrality in advanced economic systems in the face of major new organizational forms and technologies that maximize the possibility for geographic dispersal. See new Introduction in the updated edition of *The Global City* (Sassen 2001). For a variety of perspectives see, for example, Landrieu *et al.* (1998); Abrahamson (2004); Rutherford (2004).

10 I conceptualize these 'alternative' circuits as countergeographies of globalization because they are deeply imbricated with some of the major dynamics constitutive of the global economy yet are not part of the formal apparatus or of the objectives of this apparatus. The formation of global markets, the intensifying of transnational and translocal business networks, the development of communication technologies that easily escape conventional surveillance practices – all of these produce infrastructures and architectures that can be used for other purposes, whether money laundering or alternative politics.

11 There are, of course, severe limitations on these possibilities, many having to do with the way in which these technologies have come to be deployed. See Graham and Aurigi (1997); Hoffman and Novak (1998); Sassen (1999, 2006: chapter 7); Latham and Sassen (2005b).

REFERENCES

Abrahamson, M. (2004) *Global Cities*, New York: Oxford University Press.

Ascher, F. (2004) Nouveau Principes de l'urbanisme. Paris: Éditions de l'Aube.

Avgerou, C. (2002) *Information Systems and Global Diversity*, Oxford: Oxford University Press.

Avgerou, C., Ciborra, C. and Land, F. (2004) *The Social Study of Information and Communication Technology Innovation, Actors, and Contexts*, Oxford: Oxford University Press.

Bach, J. and Stark, D. (2005) 'Recombinant technology and new geographies of association', in R. Latham and S. Sassen (eds) *Digital Formations: IT and New Architectures in the Global Realm*, Princeton, NJ: Princeton University Press, pp. 37–53.

Barry, A. and Slater, D. (2002) 'Introduction: the technological economy', *Economy and Society*, 31 (2): 175–93.

Bartlett, A. (forthcoming) 'The politics of protest: subjectivity, migration and the new urban order', in S. Sassen (ed.) *Deciphering the Global*, New York and London: Routledge.

Beck, Ulrich (2006) *Cosmopolitan Vision*, Cambridge: Polity Press.

Bennett, W.L. (2003) 'Communicating global activism: strengths and vulnerabilities of networked politics', *Information, Communication & Society*, 6 (1): 143–68.

Brenner, N. (2004) *New State Spaces: Urban Governance and the Rescaling of Statehood*, Oxford: Oxford University Press.

Callon, M. (1998) *The Laws of the Markets*, Oxford: Blackwell Publishing.

Castells, M. (1996) *The Rise of the Network Society*, Boston, MA: Blackwell Publishing.

Cederman, L.-E. and Kraus, P.A. (2005) 'Transnational communications and the European demos', in R. Latham and S. Sassen (eds) *Digital Formations: IT and New Architectures in the Global Realm*, Princeton, NJ: Princeton University Press, pp. 283–311.

Cicollela, P. and Mignaqui, I. (2002) 'The spatial reorganization of Buenos Aires', in S. Sassen (ed.) *Global Networks/Linked Cities*, New York: Routledge, pp. 309–26.

Cleaver, H. (1998) 'The Zapatista effect: the Internet and the rise of an alternative political fabric', *Journal of International Affairs*, 51 (2): 621–40.

Dean, J., Anderson, J. and Lovink, G. (eds) (2006) *Formatting Networked Societies: Information Technology in and as Global Civil Society*, New York: Routledge.

van de Donk, W., Loader, B.D., Nixon, P.G. and Rucht, D. (eds) (2005) *Cyberprotest: New Media, Citizens, and Social Movements*, London: Routledge.

Drainville, A. (2004) *Contesting Globalization: Space and Place in the World Economy*, London: Routledge.

Drake, W.J. (2004) *Defining ICT Global Governance*. SSRC Research Network on IT and Governance, New York: SSRC, http://www.ssrc.org/programs/itic/publications/knowledge_report/memos/billdrake.pdf.

Eade, J. (ed.) (1996) *Living the Global City: Globalization as a Local Process*, London: Routledge.

Ehrenreich, B. and Hochschild, A.R. (eds) (2003) *Global Woman: Nannies, Maids, and Sex-Workers in the New Economy*, New York: Metropolitan Books.

Espinoza, V. (1999) 'Social networks among the poor: inequality and integration in a Latin American city', in B. Wellman (ed.) *Networks in the Global Village*, Boulder, CO: Westview Press.

Fainstein, S. (2001) *The City Builders*, Lawrence, KS: Kansas University Press.

Ferguson, Y.H. and Jones, R.J.B. (eds) (2002) *Political Space: Frontiers of Change and Governance in a Globalizing World*, Albany, NY: SUNY Press.

Friedman, E.J. (2005) 'The reality of virtual reality: the Internet and gender equality advocacy in Latin America', *Latin American Politics and Society*, 47 (1): 1–34.

Garcia, D.L. (2002) 'The architecture of global networking technologies', in S. Sassen (ed.) *Global Networks/Linked Cities*, London: Routledge, pp. 39–70.

Garcia, D.L. (2005) 'Cooperative networks and the rural–urban divide', in R. Latham and S. Sassen (eds) *Digital Formations: IT and New Architecture in the Global Realm*, Princeton, NJ: Princeton University Press, pp. 117–46.

GaWC (Globalization and World Cities study group and network) (2006) http://www.lboro.ac.uk/gawc.

Glasius, M., Kaldor, M. and Anheier, H. (eds) (2002) *Global Civil Society Yearbook 2002*, London: Oxford University Press.

Graham, S. (ed.) (2004) *Cybercities Reader*, London: Routledge.

Graham, S. and Aurigi, A. (1997) 'Virtual cities, social polarization, and the crisis in urban public space', *Journal of Urban Technology*, 4 (1): 19–52.

Graham, S. and Marvin, S. (2001) *Splintering Urbanism: Networked Infrastructures, Technological Mobilites and the Urban Condition*, New York and London: Routledge.

Gugler, J. (2004) *World Cities Beyond the West*, Cambridge: Cambridge University Press.

Gzesh, S. and Espinoza, R. (2002) 'Immigrant communities building cross-border civic networks: the Federation of Michoacan Clubs in Illinois', in H.K. Anheier, M. Glasius and M. Kaldor (eds) *Global Civil Society Yearbook 2002*, Oxford: Oxford University Press, pp. 226–7.

Harvey, R. M. (forthcoming) 'Global cities of gold', in S. Sassen (ed.) *Deciphering the Global*, New York and London: Routledge.

Henshall, S. (2000) 'The COMsumer Manifesto: empowering communities of consumers through the Internet', *First Monday*, 5 (5), http://firstmonday.org/issues5_5/henshall/index.html.

Hoffman, D.L. and Novak, T.P. (1998) 'Bridging the racial divide on the Internet', *Science*, 280 (17): 390–1.

Howard, P.N. and Jones, S. (eds) (2004) *Society Online: The Internet in Context*, London: Sage.

Isin, E.F. (ed.) (2000) *Democracy, Citizenship and the Global City*, London: Routledge.

Journal of Urban Technology (1995) special issue: 'Information technologies and inner-city communities', 3 (19).

Klein, H. (2004) 'The significance of ICANN', SSRC Information Technology & International Cooperation Program, New York: SSRC, http://www.ssrc.org/programs/itic/publications/knowledge_report/memos/kleinmemo4.pdf.

Koopmans, R. (2004) 'Movements and media: selection processes and evolutionary dynamics in the public sphere', *Theory and Society*, 33 (3): 367–91.

Krause, L. and Petro, P. (eds) (2003) *Global Cities: Cinema, Architecture, and Urbanism in a Digital Age*, New Brunswick, NJ: Rutgers University Press.

Laguerre, M.S. (2000) *The Global Ethnopolis: Chinatown, Japantown and Manilatown in American Society*, London: Macmillan.

Landrieu, J., May, N., Spector, T. and Veltz, P. (eds) (1998) *La Ville éclatée*, La Tour d'Aigues: Editiones de l'Aube.

Latham, R. and Sassen, S. (eds) (2005a) *Digital Formations: IT and New Architectures in the Global Realm*, Princeton, NJ: Princeton University Press.

Latham, R. and Sassen, S. (2005b) 'Introduction: Digital formations: constructing an object of study', in R. Latham and S. Sassen (eds) *Digital Formations: IT and New Architectures in the Global Realm*, Princeton, NJ: Princeton University Press, pp. 1–34.

Latour, B. (1991) 'Technology is society made durable', in J. Laws (ed.) *A Sociology of Monsters*, London: Routledge.

Latour, B. (1996) *Aramis or the Love of Technology*, Cambridge, MA: Harvard University Press.

Lebert, J. (2003) 'Writing human rights activism: Amnesty International and the challenges of information and communication technologies', in M. McCaughey and M. Ayers (eds) *Cyberactivism: Online Activism in Theory and Practice*, London: Routledge, pp. 209–32.

Lloyd, R. (2005) *NeoBohemia: Art and Bohemia in the Postindustrial City*, London and New York: Routledge.

Lovink, G. (2002) *Dark Fiber: Tracking Critical Internet Culture*, Cambridge, MA: MIT Press.

Lovink, G. and Riemens, P. (2002) 'Digital city Amsterdam: local uses of global networks', in S. Sassen (ed.) *Global Networks/Linked Cities*, New York: Routledge, pp. 327–46.

Lustiger-Thaler, H. and Dubet, F. (eds) (2004) 'Social movements in a global world', special issue of *Current Sociology*, 52 (4).

MacKenzie, D. and Wajcman, J. (1999) *The Social Shaping of Technology*, Milton Keynes: Open University Press.

Mansell, R. and Collins, B.S. (eds) (2005) *Trust and Crime in Information Societies*, Northampton, MA: Edward Elgar.

Mansell, R. and Silverstone, R.(1998) *Communication by Design: The Politics of Information and Communication Technologies*, Oxford: Oxford University Press.

Marcuse, P. and van Kempen, R. (2000) *Globalizing Cities. A New Spatial Order*, Oxford: Blackwell Publishing.

Marres, N. and Rogers, R. (2000) 'Depluralising the web, repluralising public debate: the case of GM food on the web', in R. Rogers (ed.) *Preferred Placement: Knowledge Politics on the Web*, Maastricht: Jan van Eyck Editions, pp. 113–26.

Mele, C. (1999) 'Cyberspace and disadvantaged communities: the Internet as a tool for collective action', in M. A. Smith and P. Kollock (eds) *Communities in Cyberspace*, London: Routledge, pp. 264–89.

Mills, K. (2002) 'Cybernations: identity, self-determination, democracy, and the "Internet Effect" in the emerging information order', *Global Society*, 16: 69–87.

Olesen, T. (2005) 'Transnational publics: new space of social movement activism and the problem of long-sightedness', *Current Sociology*, 53 (3): 419–40.

Orum, A. and Chen, X. (2004) *World of Cities*, Malden, MA: Blackwell.

Pace, W.R. and Panganiban, R. (2002) 'The power of global activist networks: the campaign for an international criminal court', in P.I. Hajnal (ed.) *Civil Society in the Information Age*, Aldershot: Ashgate, pp. 109–26.

Paddison, R. (2001) 'Introduction', in R. Paddison (ed.) *Handbook of Urban Studies*, London: Sage, pp. 11–3.

Parsa, A. and Keivani, R. (2002) 'The Hormuz corridor: building a cross-border region between Iran and the United Arab Emirates', in S. Sassen (ed.) *Global Networks/Linked Cities*, New York and London: Routledge, pp. 183–208.

Peraldi, M. and Perrin, E. (eds) (1996) *Reseaux productifs et territoires urbains*, Toulouse: Presses Universitaires du Mirail.

Pillon, T. and Querrien, A. (eds) (1995) 'La Ville-monde aujourd'hui: Entre virtualité et ancrage', special issue of *Futur Anterieur*, 30–2.

Poster, M. (2004) 'Consumption and digital commodities in the everyday', *Cultural Studies*, 18 (3): 409–23.

Robinson, S. (2004) 'Towards a neoapartheid system of governance with IT tools', SSRC IT & Governance Workshop, New York: SSRC, http://www.ssrc.org/programs/itic/publications/knowledge_report/memos/robinsonmemo4.pdf.

Rogers, R. (2004) *Information Politics on the Web*, Cambridge, MA: MIT Press.

Rutherford, J. (2004) *A Tale of Two Global Cities: Comparing the Territorialities of Telecommunications Developments in Paris and London*, Aldershot: Ashgate.

Sack, W. (2005) 'Discourse, architecture, and very large-scale conversation', in R. Latham and S. Sassen (eds) *Digital Formations: IT and New Architectures in the Global Realm*, Princeton, NJ: Princeton University Press, pp. 242–82.

Samers, M. (2002) 'Immigration and the global city hypothesis: towards an alternative research agenda', *International Journal of Urban and Regional Research*, 26 (2): 389–402.

Sandercock, L. (2003) *Cosmopolis II: Mongrel Cities in the 21st Century*, New York and London: Continuum.

Sassen, S. (1999) 'Digital networks and power', in M. Featherstone and S. Lash (eds) *Spaces of Culture: City, Nation, World*, London: Sage, pp. 49–63.

Sassen, S. (2000) 'New frontiers facing urban sociology', *British Journal of Sociology*, 51 (1): 143–59.

Sassen, S. (2001) *The Global City: New York, London, Tokyo*, 2nd edn, Princeton, NJ: Princeton University Press.

Sassen, S. (2002) 'Towards a sociology of information technology', *Current Sociology*, 50 (3): 365–88.

Sassen, S. (2006) *Territory, Authority, Rights: From Medieval to Global Assemblages*, Princeton, NJ: Princeton University Press.

Schiffer, S.R. (2002) 'São Paulo: articulating a cross-border regional economy', in S. Sassen (ed.) *Global Networks/Linked Cities*, New York: Routledge, pp. 209–36.

Scott, A.J. (2001) *Global City-Regions*, Oxford: Oxford University Press.

Sennett, R. (1990) *The Conscience of the Eye*, New York: Knopf.

Smith, M.P. and Guarnizo, L. (2001) *Transnationalism from Below*, Piscataway, NJ: Transaction Publishers.

Sum, N.-L. (1999) 'Rethinking globalisation: re-articulating the spatial scale and temporal horizons of trans-border spaces', in K. Olds, P. Dicken, P.F. Kelly, L. Kong and H.W.C.-C. Yeung (eds) *Globalization and the Asian Pacific: Contested Territories*, London: Routledge, pp. 129–45.

Taylor, P.J. (2000) 'World cities and territorial states under conditions of contemporary globalization', *Political Geography*, 19 (5): 5–32.

Taylor, P.J. (2004) *World City Network: A Global Urban Analysis,* London: Routledge.

Taylor, P.J., Walker, D.R.F. and Beaverstock, J.V. (2002) 'Firms and their global service networks', in S. Sassen (ed.) *Global Networks/Linked Cities*, New York: Routledge, pp. 93–116.

Thrift, N. and A. Amin (2002) *Cities: Reimagining the Urban*, Cambridge: Polity Press.

Torres, R.D., Inda, J.X. and Miron, L.F. (1999) *Race, Identity, and Citizenship*, Oxford: Blackwell.

Tsaliki, L. (2002) 'Online forums and the enlargement of the public space: research findings from a European project', *The Public*, 9: 95–112.

Urry, J. (2000) *Sociology Beyond Societies: Mobilities for the Twenty-first Century*, New York: Routledge.

Veltz, P. (2005) *Mondialisation, Villes et Territoires: l'economie d'archipol*, Paris: Presses Universitaires de France.

Wajcman, J. (ed.) (2002) 'Information technologies and the social sciences', special issue of *Current Sociology*, 50 (3).

Yeung, Y.-M. (2000) *Globalization and Networked Societies*, Honolulu, HI: University of Hawai Press.

14 Poststructuralism, power and the global city

Richard G. Smith

INTRODUCTION

In 2003 in the journal *Urban Geography* an article by the sociologist David R. Meyer was published. Meyer's article amounts above all to an observation that theory on the global network of cities remains undeveloped. Meyer rightly observes that there is a 'theoretical lacuna' that needs to be addressed 'if scholars are to make greater progress in understanding the global network of cities' (2003: 308). Meyer contends that we need to 'deepen' the theory of the global network of cities and so implies that we need to refine existing political-economic and world-systems approaches. However, my argument is somewhat different. This chapter further develops a line of argument (see Smith 2003a,b, 2005, 2006) that is moving away from, not deepening, political economy and world-systems perspectives on the global network of cities. This chapter further develops the argument that one way to make progress in our theoretical and empirical research on urban networks is to engage with the ideas of Deleuze (and other poststructuralists), and the new philosophies of connection – inspired by Deleuze's writings – of actor-network theory (ANT) and non-representational theory.

Meyer's article fails to acknowledge that recently some alternative theoretical progress has begun to be made on theorising urban networks. A 'new urbanism'[1] has started to emerge which is trying to reimagine the connectivity of cities through ideas from poststructuralism, ANT and non-representational theory (Amin and Thrift 2002; Doel and Hubbard 2002; Latham 2002; Smith 2003a,b). This recent theoretical effort fully takes on board the idea that the concept of networks may be enhanced through poststructuralist relational thinking and it is to this project that this chapter contributes by outlining both a poststructuralist-inspired approach to cities and a spatial conceptualisation of the power of highly connected cities through their actant networks.

This chapter argues that it is the very fabric – the filamental material*ity* – of cities that produces their power. The fabric of highly connected (global) cities is one of networks of actants (humans and non-humans) that produce affective power through their interactive capacity to make and modify relations. To make this argument the chapter is in two halves. In the first half a new approach to urban studies is broadly outlined, as a poststructuralist and ANT-inspired approach to cities is contrasted with the intrinsic and extrinsic approaches that dominate urban studies. In the second half of the chapter an account of urban power as a co-product of actant-networks (associations of many different actants) that is decentred (non-individual) and disseminated through networks (the very fabric of cities and globalisation) is outlined. Thus, it is shown how power in urban networks can be conceptualised through the extension of Foucault's poststructuralist ideas by ANT (though writers such as Latour and Law) *where everything is taken away* to clear the path for an alternative analysis of power.

A POSTSTRUCTURALIST-INSPIRED APPROACH TO CITIES

> Everything now returns to the surface.
>
> Deleuze (1990: 7)

To my knowledge the first poststructuralist-inspired book on the contemporary city was *Zone 1/2* (Crary *et al.* 1986)[2], published twenty years ago. The opening paragraph to the book is rhizomatic – 'and . . . and . . . and . . .' (Deleuze and Guattari 1987: 25) – and sets the scene for the development of a poststructuralist and ANT approach to cities and their networks[3];

> To draw a carp, Chinese masters warn, it is not enough to know the animal's morphology, study its anatomy or understand the physiological functions vital to its existence. They tell us that it is also necessary to consider the reed against which the carp brushes each morning while seeking its nourishment, the oblong stone behind which it conceals itself, or the rippling of water when it springs toward the surface. These elements should in no way be treated as the fish's environment, the milieu in which it evolves or the natural background against which it can be drawn. They belong to the carp itself, insofar as it is not defined as a distinct form capable of a set of movements or as a particular organism performing a series of functions. Instead, the carp must be apprehended as a certain power to affect and be affected by the world. In other words, rather than a formed and organized individual, the brush should sketch a life, since a life is constituted simply by traces left behind and imprints silently borne.
>
> (Feher and Kwinter 1986: 10)

The aim of the editors of *Zone 1/2* was to be faithful to the precepts of the Chinese masters and let the city emerge 'as a specific *power to affect* both people and materials – a power that modifies the relations between them' (11). For them the city is not a fixed background against which things occur, or an environment in which things occur, but has a power to affect that is 'neither a side-effect nor an attribute of a city-substance', but 'is itself the very fabric of the city's consistency'

(11). Feher and Kwinter are mindful of the poststructuralist argument that the city (like anything) is not to be found in a specific place, it is never simply present, but rather obeys a logic of supplementarity to always be both present and absent. However, they also anticipate an ANT-like argument that would see the city as an event diffused through actant-networks. That is to say the argument that the city has a filamental materiality (see Smith 2003b) populated by actants. We know from ANT that an actant is anything with the capacity to act and so can be human or non-human such as an object, a technology or a knowledge (e.g. a fax machine, a written document, money, a mobile electronic gadget, a computer system, a skill, an organisation – the list could be almost endless). What is more, we know that actants join together to form heterogeneous networks where each influences and strengthens the other in a process of mutual influence (actants are enrolled as allied to give strength and power). To quote Latour (1988: 159), 'No actant is so weak that it cannot enlist another. Then the two join together and become one for a third actant, which they can therefore move more easily. An eddy is formed, and it grows by becoming many others.' It is through that process of making actant networks – the fabric of urban life – that cities come to have a power to affect. Let me now explain in detail how poststructuralist-inspired thinking has dramatic consequences for the intrinsic and extrinsic approaches that have shaped urban studies in the twentieth century.

AN ALTERNATIVE TO INTRINSIC AND EXTRINSIC APPROACHES TO THE CITY

Broadly speaking, to gloss over complications and various nuances, there are two dominant approaches to the city (see Feher and Kwinter 1986). First of all there are *intrinsic* approaches that reduce the city to its innards in a way that seems analogous to a surgeon dissecting a body (complexity is reduced by shutting out the wider world). The most obvious example is the Chicago School (which was still influential into the 1960s), whose 'members' codified and introduced morphological and physiological approaches to the city (for a flavour of this work see Theodorson 1982). Second are the numerous *extrinsic* approaches to the city in geography, sociology and political economy that do not treat the city like a body but rather seek to explain the city from without as formed and shaped by exterior forces such as general socio-economic laws (e.g. the forces of capitalism, patriarchy or globalisation). Perhaps the most influential extrinsic approach in urban studies is the neo-Marxist approach of writers such as Harvey (1973) and Castells (1972). Their conceptualisation of the city as a product of capital and class was highly influential with, for example, Friedmann (1986) drawing on their ideas (and those of Wallerstein's world-systems analysis) to directly connect the study of cities to the world economy through his seminal 'world city hypothesis'.

Intrinsic approaches to the city are increasingly unpopular as even 'common sense' provides a powerful critique. In the popular press there is much talk of globalisation and worldwide networks – interconnection through wider and wider sets of relations – which questions how feasible it is to somehow delimit and demarcate a city (or anything) as a discrete unit, separate from the relations that run through it (there are no 'worlds within worlds'). To quote Guattari (1986: 460) 'The city no longer exists as an entity. It is only a node at the core of a multidimensional network – within the spatial web of urbanization'. In short, a global perspective reveals how a city cannot be studied in isolation but has to be understood as belonging to a network of cities that stretch across the world. In contrast, extrinsic approaches where the city is seen as being overlaid by a capitalist dynamic still dominate thinking about the city, but I think this is only because a 'common sense' critique has not been made in the urban literature. In the social sciences extrinsic approaches have been effectively critiqued by postmodernism, poststructuralism and also ANT writers such as Latour, who points out that external forces are

> attributed by their critics to actors who did not ask for them. Take some small business-owner hesitatingly going after a few market shares, some conqueror trembling with fever, some poor scientist tinkering in his lab, a lowly engineer piecing together a few more or less favourable relationships of force, some stuttering and fearful politician; turn the critics loose on them, and what do you get? Capitalism, imperialism, science, technology, domination – all equally absolute, systematic, totalitarian. In the first scenario, the actors were trembling; in the second, they are not. The actors in the first scenario could be defeated; in the second, they no longer can. In the first scenario, the actors were still quite close to the modest work of fragile and modifiable mediations; now they are purified, and they are all equally formidable.
>
> (1993: 125–126)

Thinking through Latour it seems to me that notions such as the post-industrial city, the postmodern or post-Fordist city, the late capitalist city, the global city (and all the others) 'only perpetuate a hidden dependence on systems ballasted by an infrastructure' (Guattari 1986: 460). That is to say they only accentuate the myth of something happening outside the city that presses in upon it, and consequently *eclipse the power of the city to affect the world*. In other words, rather than suffer the sort of people who in a local meeting 'bring up international capitalism every time you try to have a discussion about rubbish-collection' (Massey 1993: 66), the sort of people that think it makes sense to distinguish between global and local, foreground and background, or 'bottom-up' and 'top-down' views of the world, a new approach to cities is needed which realises that 'big explanations have to be replaced by little networks' (Latour 1996: 134). And this is why we should be inspired by Deleuze and Guattari (1983, 1987) to develop an *immanent approach* that thinks of cities as 'bodies without organs' (BwO), as continuums of flows and intensities held together only by those forces that compose them, because the BwO is not an extrinsic explanation of desire, but 'is what remains when you take everything away' (Deleuze and Guattari 1987: 151). In other words, desire is produced without reference to 'any exterior agency, whether it be a lack that hollows it out or a pleasure that fills it' (ibid. 154).

THE CITY AS A BWO

Thinking of the city as a BwO means abandoning all intrinsic and extrinsic approaches for an immanent approach where everything is at the surface, connected and in movement. The BwO is a stunning critique of extrinsic approaches such as structuralism and therefore of organicism and the idea of the organism. And that is interesting when we come to think about cities because the story of the history of urban planning[4] is one of the 'urban body' being turned into the 'urban organism'. That is to say that the city has not only been progressively dissected and differentiated into functional organ-ised zones, but has also come to be represented and rationalised as a circulatory system of all kinds of flows. Gille (1986) has examined the transition from the idea of the city as a body to that of the city as an organism. In the late eighteenth century the hygienist movement was calling for remedies to the apparent sickness – the *malaria urbana* – of the urban body. They argued that the city was ill because it was suffering from confusion and stasis, and consequently to be made well the indistinct had to be differentiated and the stagnant (fixed) had to be put into circulation. Thus, for the hygienists the key motifs for urban planning – to be inscribed in the urban landscape – were differentiation and circulation. However, what is crucial to realise is that these two motifs are not separate. The hygienists realised that 'untangling the city's imbroglios and turning them into "functional" units would virtually amount to doing away with the city, cutting it up into isolated monads', consequently the 'installation of a network of "circulation" is a fundamental stage in modern urban surgery' (1986: 231). In other words, dissecting the city into small and disjoined units would kill the city because such units are 'incapable of self-sufficiency outside of the urban network that has established them', and it is the network of circulation that 'becomes the purest and most essential expression of urban unity' (ibid.). Thus, through administrating appropriate 'medicine' the city was 'cured', but the consequence was that the hygienist's city became the norm and model for every city. And that is a key point because whilst 'We thought we were dealing with potions, remedies, panaceas . . . we now have to admit that it was all a matter of law, principle and structure' (ibid. 235). In fact, the success of the hygienist movement for purifying the city is somewhat uncertain, with Gille arguing that hygiene was appealed to as a way to be heard when in fact many of their ideas had more 'to do with architecture or social reform than with health problems' (ibid.).

One of the points of Gille's work is to explicate how hygienism – with the fundamental principle of prioritising circulation over stagnation – became a discourse that was successfully concretised in urban procedures and led to the urban body being transformed slowly, but surely, into an urban organism. Gille discusses an

opening address entitled 'Circulation or Stagnation?' by the English hygienist
F.O. Ward to an international congress of hygiene in 1852. Gille's reading reveals
how Ward extends – uses as an analogy – Harvey's discovery of circulation in the
body of the individual to naturalise the idea of circulation in the social body. Gille
is quite right in explaining Ward's logic:

> For the city has finally discovered its intrinsic nature: it was because it was a
> body that it was sick, that it was not really a city; and it is because it is finally
> an organism that its eternal truth is revealed. An organism is a supposedly
> living being whose meaning nevertheless can be entirely reduced to the struc-
> ture that accounts for it, whose organs are actualisations of abstract functions,
> whose essential finality is the preservation (and in this case, the extension) of
> its structure, a being whose concrete existence has no meaning as such.
>
> (1986: 239)

This conversion of the city from a body to an organism is what would concern
Deleuze and Guattari (1987) because 'The BwO is opposed not to the organs
but to that organization of the organs called the organism' (1987: 158). And that
is the crucial point because it is not the body that the BwO opposes,[5] but rather
– remember your structuralism – the organic organisation of the organs called
the organism. The organism (structuralism) is the enemy because 'it is a stratum
on the BwO, in other words, a phenomenon of accumulation, coagulation, and
sedimentation that, in order to extract useful labour from the BwO, imposes upon
it forms, functions, bonds, dominant and hierarchized organizations, organized
transcendences' (ibid. 159). Not turning the city into an organism is what this
new approach to cities is all about. Rather than insisting that the city must be organised, must be an organism, must be conditioned by an exterior force, Deleuze
and Guattari provide the inspiration for an approach that understands the city as
'a complex surface of activities and interactions' (Shields 1996: 242). And that
shift away from intrinsic and extrinsic approaches to the city to focus instead on
the city as a surface of practices and relations is the basis for any understanding
of the city as actant with a power to affect. Furthermore, conceptualising cities
as BwOs, as immediately invested by positive and constitutive desire, reminds
us of Foucault's work on how power is implicit in everyday interaction. It is no
coincidence that Deleuze and Guattari's account of desire (the BwO) is in many
ways analogous to Foucault's version of power (and desire) in action as discourse,
and consequently ANT's conceptualisation of power as outlined by writers such
as Latour and Law. As Baudrillard (1987: 17) rightly observes, '*it's simply that in
Foucault power takes the place of desire*' (italics original); in both Deleuze–Guattari and Foucault one has 'the deployment and the positive dissemination of flows
and intensities'. To develop this point, and that of a new approach to global cities
as actant, let's draw on Foucault and Latour to discuss an ethnography of power
by the ANT sociologist Law where abstract discussions of power take on flesh.

POSTSTRUCTURALISM AND ANT: FROM MICHEL FOUCAULT TO BRUNO LATOUR AND JOHN LAW

In recent years there has been a shift towards thinking about the connections within and between cities, a shift to thinking about cities as – and in – networks, but what has tended to happen is that the networks within which cities are entangled have remained hidden as emphasis has concentrated on nodes (the cities) rather than links. In fact, it is only with poststructuralism that we realise that the nodes are the links and vice versa (see Smith 2006). What is more, ANT resonates with poststructuralism because it brings into consideration not only the networks in which cities are entangled, but also the actant power of those networks.

Perhaps the best way to begin this discussion of the city's power to affect is to start with something that is like the city because it is also an object with the power to affect relations (but at least seems smaller). Massey's outline of an ANT explanation of the affect of a microphone in a seminar in Heidelberg is particularly audible:

> Just take a microphone, for example, and look at it as the product of a whole massive intersection of social relations, of relations between the microphone and other things, and all the networks that it took to make that thing, from the people who dug up the metals in the first place through to the people who did the advertising, through to the ways in which this particular brand of microphone came to be the one that got bought here, which was to do with all kinds of rhetorics of advertising and sales promotion and a million things, and to see it as the kind of materialised result of that network, a massive network, an impossibly complicated network, but also to see that it isn't just a static thing as a node in that network, it has its own effectivity, which is to say it isn't just a product. . . . The existence of that microphone has effects, it's going to affect what happens to the distribution of this seminar later and therefore affect other social relations, it's going to get into other classrooms, it's going to make more people hear what we all say – so certain debates will get promulgated a little bit further as a result of the physical existence of that result called the microphone and a whole other set of relations. The whole world can be seen as these occasional temporary manifestations, materialisations of sets of relations in which not only human things are active conveyors of relations.
>
> (Massey 1999: 57)

The microphone is an actant, both produced by and active in a wide network of relations. An argument that ANT theorists have been making about relational networks is that non-humans such as machines and gadgets have affective power. In other words, both humans and *non-humans have affects without necessarily having any intention to do so*. This observation immediately broadens any conventional or social conceptualisation of networks. If we now think about actant-networks then the whole world becomes the materialisation of 'sets of relations in which not only human things are active conveyors of relations' (ibid.). Like the microphone the city is an actant with a power to affect because, whilst 'Affect is often thought of as just a posh word for emotion . . . it is meant to point to something which is non-individual, an impersonal force resulting from the encounter, an ordering of the relations between bodies which results in an increase or decrease in the potential to act' (Thrift 2003: 104). The city (like place) is a crucial actor in producing affects because it modifies relations and changes the affective connections that are made, remade and unmade. Thus, the important question to ask is how we can be consistent with Deleuze's and Guattari's (1987) rhizomatic BwO to provide both a new apprehension of power and a practical analysis of power? The answer is to turn to Foucault to realise how power is decentred, and to Latour and other ANT writers such as Law to see how power is disseminated through materials because 'it's so *obvious* that the world and its relations are made of materials' (Law and Hetherington 2000: 35).

Law (1994, 1996; Law and Hetherington 2000) in the spirit of Foucault provides an ANT account of how 'Andrew' (the Managing Director of Daresbury SERC Laboratory in the UK) has power. A distal (see Cooper and Law 1995) or conventional view would see Andrew as the source of power, at the head of and in charge of a large organisation. However, a proximal or closer look reveals a different account of power in the laboratory. Law wonders what would happen if Andrew's office materials, the non-humans, were removed (remember the earlier quotation from Deleuze and Guattari (1987: 151)). If he had no computer the consequences would be great. It would mean no spreadsheets, no means of calculation, no budgets or projections, and so no knowledge of the laboratory's finances and economic viability. Furthermore, he would have no email, and if phone and fax were removed his ability to communicate beyond his office would be difficult to say the least. The point then is that when both humans and non-humans are stripped away (no technology, no secretary, no pocket calculator, no stationery, no postal service etc.), and Andrew stands alone in an empty building, we are not left with a powerful man. And that means that we must recognize that Andrew's power as a director is necessarily extended, distributed and produced through the arrangements of the organisation. 'This, then, is the lesson. We are all spread out. We are nothing more than a network of social and technical relations. We are *made* by our organisational relations. Power resides elsewhere. It is always deferred. It is always a *product*. It is always an *effect*' (Law 1996: 4). In other words, Law's story, which in many ways spells out a detailed explanation of Latour's quotation (1993: 125–6), is that agency and power are decentred and disseminated – produced in between all kinds and number of actants – and consequently are within networks that can be of any size or length. Law's ethnography contends that despite power seeming to be attributed in advance it does not belong to an individual (such as a manager, company director or politician), or any group of people (with a unitary intent) perhaps 'running' an organisation who are the apex of a hierarchy, but is dispersed, disseminated, displaced and produced through a network composed of all manner of human and non-human actants that enable the network to act (and, with globalisation, to act at considerable distance). In other words, it is the successful enrolment and involvement of actants in networks that makes a capacity and ability to act over distance, to project power.

Meanwhile, Andrew has all his technologies back, his secretary too, and everything else that was taken away. But it is now clear that he manages only '*within the relations made in his organisation*' (Law 1996: 6, italics original). In fact, it becomes clear that Andrew has very little choice in his decisions if he is to serve the organisation. 'He is the *creature* of the organisation. He is the *expression* of the organisation. He *performs* the organisation. Its relations. Its projects. Its desires. Its goals' (ibid.). Andrew is but a moment in the relations of the organisation, and it is those relations that work to remove his discretion, his ability to make free choices that do not serve the organisation. But something else becomes clear: Andrew has little choice but he does not simply follow 'orders'. Andrew has a lot of responsibilities. He is a scientist, a manager, an administrator and an accountant at the very least. He is multiple because he is made up of these, and no doubt many other, organisational logics. Andrew is one of the many actants that make up the non-coherent organisation of the science laboratory, a laboratory that is replete with different logics and voices, full of negotiable responsibilities, and a world away from the notion of science as an unchallengeable force of domination (remember Latour 1993: 125–6). Or so you might think from reading Law's imaginative ethnography. But the closer you look the stronger Andrew is. Andrew is strong, the organisation is strong, because weaving, twisting and folding produces sturdiness, resistance and strength. As Latour (1998: 2) puts it, 'Strength does not come from concentration, purity and unity, but from dissemination, heterogeneity and the careful plaiting of weak ties'.

To begin to think of power in this new way you must reject the critique that is sometimes levied against Foucault, Deleuze–Guattari, and now Latour and ANT in general. They have been criticised for concentrating on details and the local scale (the microphysics of power, the micropolitics of desire, the microrelations of actant networks) to the detriment of the big context and the global scale where the 'real' or 'deep' causes of large inequalities can be found (the ever-enduring structures of capitalism, patriarchy, imperialism, neoliberal and neoconservative globalisation etc.). The approach outlined here is different because the argument being made is that there is no such thing as scale (see Smith 2003a) when your social theory and ontology is made of connections, associations, relations and networks that are more or less long, more or less durable, and more or less asymmetrical. The argument being made is even against those arguing for the social construction of scale precisely because the argument is that *it is highly disempowering to shift to the big picture (think again of that quotation from Latour (1993: 125–6)) rather than follow a network, precisely because the political point is to question the relations that produce asymmetries.*

In short, as poststructuralists we do not distinguish between processes and things (just like the BwO) and that means that the city (like anything) is both present and absent, here and there, near and far, local and global, precisely because it is made of relations. But crucially I am making the argument that something needs to be added from ANT to that basic poststructuralist argument. Thinking through Law's discussion of 'Andrew' we are beginning to realise how the very fabric that makes up an office's, a building's, a city's consistency is one of relations and networks that are replete with human and non-human actants. As such a city is a *hybrid collectif* (Callon and Law 1995) with a power to affect.

CONCLUSION: RETHINKING THE POWER OF CITIES

> ... the metropolis is not a center and has no center: made up of networks, it is itself caught up in a network of cities through which the flux of the world economy circulates. This transnational network is relatively independent of international boundaries.
>
> Querrien (1986: 219)

> To think the city, one must leave it and see it as part of the world, *extra-muros*.
>
> Hénaff (1997: 61)

The point we have now reached is to realise that a poststructuralist and ANT-inspired relational and networked perspective can transform research on power,[6] and consequently any understanding of the power of cities. Let us think about the idea of 'global cities' (Sassen 1991). Those who use that shibboleth never seem to unpack its core thesis that certain cities (New York, London, Tokyo) are centres that 'command and control' the world economy (see Sassen 1991). Nowhere in the literature does any author try to explain in any detail the difference that networks make to the power of these cities to affect.[7] Some of the key authors in world and global cities research, for example, simply talk about how the headquarters of transnational corporations (Friedmann 1986) or advanced producer service firms (Sassen 1991, 1994), are located in certain world or global cities and therefore those cities are *ipso facto* powerful. There is no argument as such, just an intuitive leap of faith (see Smith and Doel, 2006). In contrast, ANT would support the idea of world and global cities as 'switchers' or, more accurately, as actants with the power to affect and be affected through a networking, rather than a command, logic (an argument I pointed to in Smith 2003b). It takes just a little imagination to ask what it would mean for our study of the power of world or global cities[8] if we did some ethnographic work that like the BwO abandons any extrinsic explanation in order to follow the constant movement of socio-technical (human and non-human) networks and so pay attention to how global reach and influence is accomplished, rather than to continue to think of networks and influence as somehow static and pre-given (as in Beaverstock *et al.* 2000; Castells 1996; Taylor 2001; etc.). Extrinsic approaches must be resisted because as Latour points out:

> actors know what they do and we have to learn from them not only what they do, but how and why they do it. It is *us*, the social scientists, who lack knowledge of what they do, and not *they* who are missing the explanation of

why they are unwittingly manipulated by forces exterior to themselves and known to the social scientist's powerful gaze and methods.

(1999: 19)

If 'Andrew' were say the CEO of a transnational corporation, an international banker, London's or New York's mayor or the managing partner of a global law firm, then immediately our understanding of the power of world and global cities would be spatialised, broader and much richer than any account that continued to pretend that power comes from concentration, agglomeration, a centre, a global (Sassen 1991) or world (Friedmann 1986) city as they are currently defined.

NOTES

1 The idea of a 'new urbanism' should not be confused with the new urban revanchism described and identified (or created) by Smith (1999) or the 'new urbanism' architectural design movement with key figures such as Andres Duany and Elizabeth Plater-Zyberk.
2 This book is rare, collected, and consequently hard to find and expensive to purchase.
3 ANT started to emerge at about the same time as *Zone 1/2* was published. The Paris group of science and technology studies started using the idea of actor-networks in the 1980s (see Callon *et al.* 1986). It is not surprising that the quotation from Feher and Kwinter resonates with ANT because these writers have been heavily influenced by poststructuralism and writers such as Deleuze in particular (see for example Kwinter 2002).
4 I think that the power of urban planning to produce cities has been highly exaggerated over the years. In contrast, the question of how desire constitutes the urban is hardly mentioned at all. The works of Deleuze and Guattari can perhaps guide us through the initial stages of the important task of thinking about how cities are emotional and affectual.
5 That is the fundamental mistake that Blake (2004) makes in her discussion of Deleuze. She reads Deleuze as opposing the body to the BwO and that error leads her towards political economy and away from that poststructuralist and ANT thinking that has been transforming our thinking about the spatialities of power (see Law 1991; Allen 2003).
6 See Allen (2003) to further understand this point.
7 A conceptual inversion is required. In fact, why is there a literature on global and world cities if capitalism and industrialisation are prioritised over urbanization – are deployed as the explanation of urbanism in the first and last instance? If all we have to do to account for the city is to understand the forces, processes and structures of capitalism then a literature on global and world cities seems rather unnecessary. We must resist the drive to portray cities as special effects of structures to ask: what is important about cities? What is important about city networks, rather than global economic, cultural, political and social networks per se?
8 The phrase and idea of the 'global city' is now commonly attributed to Sassen after her influential work of that title (1991). However, writing back in 1970 Lefebvre (translation published in 2003) develops the idea of the global city after noting that the idea originates from Maoism and perhaps Mao Tse-tung himself.

REFERENCES

Allen, J. (2003) *Lost Geographies of Power*, Oxford: Blackwell.

Amin, A and Thrift, N (2002) *Cities: Reimagining the Urban*, Cambridge: Polity Press.

Baudrillard, J. (1987) *Forget Foucault*, New York: Semiotext(e).

Beaverstock, J.V., Smith, R.G. and Taylor P.J. (2000) 'World city network: a new metageography', *Annals of the Association of American Geographers*, 90 (1): 123–34.

Blake, C. (2004) *Apparatus of Capture: The Use of Deleuzian Thought and Actor-Network Theory to Conceptualise Urban Power Relations*. GaWC Research Bulletin 111.

Callon, M. and Law, J. (1995) 'Agency and the hybrid collectif', *South Atlantic Quarterly*, 94 (2): 481–507.

Callon, M., Law, J. and Rip, A. (1986) *Mapping the Dynamics of Science and Technology*, London: Macmillan.

Castells, M. (1972) *La Question urbaine*, Paris: Maspero.

Castells, M. (1996) *The Rise of the Network Society*, Oxford: Blackwell.

Cooper, R. and Law, J. (1995) 'Organization: distal and proximal views', in S. Bacharach, P. Gaghardi and B. Mundell (eds) *Research in the Sociology of Organisations*, Greenwich, CT: JAI Press, pp. 237–74.

Crary, J., Feher, M., Foster, H. and Kwinter, S. (eds) (1986) *ZONE 1/2: The Contemporary City*. New York: Zone Books.

Deleuze, G. (1990) *The Logic of Sense*, New York: Columbia University Press.

Deleuze, G. and Guattari, F. (1983) *Anti-Oedipus*, Minneapolis: University of Minnesota Press.

Deleuze, G. and Guattari, F. (1987) *A Thousand Plateaus*, Minneapolis: University of Minnesota Press.

Doel, M. and Hubbard, P (2002) 'Taking world cities literally: marketing the city in a global space of flows', *City*, 6 (3): 351–68.

Feher, M. and Kwinter, S. (1986) 'Foreword', in J. Crary, M. Feher, H. Foster and S. Kwinter (eds) *ZONE 1/2: The Contemporary City*, New York: Zone Books, pp. 10–13.

Friedmann, J. (1986) 'The world city hypothesis', *Development and Change*, 17 (1): 69–83.

Gille, D. (1986) 'Maceration and purification', in J. Crary, M. Feher, H. Foster and S. Kwinter (eds) *ZONE 1/2: The Contemporary City*, New York: Zone Books, pp. 226–81.

Guattari, F (1986) 'Questionnaire 17' in J. Crary, M. Feher, H. Foster and S. Kwinter (eds) *ZONE 1/2: The Contemporary City*, New York: Zone Books, p. 460.

Harvey, D. (1973) *Social Justice and the City*, London: Edward Arnold.

Hénaff, M. (1997) 'Of stones, angels and humans: Michel Serres and the global city', *SubStance*, 83: 59–80.

Kwinter, S. (2002) *Architectures of Time*, Cambridge: MIT Press.

Latham, A. (2002) 'Retheorizing the scale of globalisation: topologies, actor-networks, and cosmopolitanism', in A. Herod and M.W. Wright (eds) *Geographies of Power: Placing Scale*, Oxford: Blackwell, pp. 115–44.

Latour, B. (1993) *We Have Never Been Modern*, New York: Harvester Wheatsheaf.

Latour, B. (1996) *Aramis, or the Love of Technology*, Cambridge, MA: Harvard University Press.

Latour, B. (1998) 'On actor-network theory: a few clarifications', online available at http://amsterdam.nettime.org/Lists-Archives/nettime-1-9801/msg00019.html (accessed on 9 January 2003).

Latour, B. (1999) 'On recalling ANT', in J. Law and J. Hassard (eds) *Actor-Network Theory and after*, Oxford: Blackwell, pp. 15–25.

Law, J. (1991) 'Power, discretion and strategy', in J. Law (ed.) *The Sociology of Monsters*, London: Routledge, pp. 165–91.

Law, J. (1994) *Organizing Modernity*, Oxford: Blackwell.

Law, J. (1996) 'The manager and his powers', paper presented to the Mediaset Convention, Venice, 12 November (accessed on 17 September 2001 and available at: http://www.comp.lancs.ac.uk/sociology/stslaw1.html).

Law, J. and Hetherington, K. (2000) 'Materialities, spatialities, globalities', in J. Bryson, P.W. Daniels, N. Henry and J. Pollard (eds) *Knowledge, Space, Economy*, London: Routledge, pp. 34–49.

Lefebvre, H. (2003) *The Urban Revolution*, Minneapolis: University of Minnesota Press.

Massey, D. (1993) 'Power-geometry and a progressive sense of place', in J. Bird, B. Curtis, T. Putnam, G. Robertson and L. Tickner (eds) *Mapping the Futures: Local Cultures, Global Change*, London: Routledge, pp. 59–69.

Massey, D. (1999) *Power-Geometries and the Politics of Space–Time*, Heidelberg: Department of Geography.

Meyer, D.R. (2003) 'The challenges of research on the global network of cities', *Urban Geography*, 24 (4): 301–13.

Querrien, A. (1986) 'The metropolis and the capital', in J. Crary, M. Feher, H. Foster, S. Kwinter (eds) *ZONE 1/2: The Contemporary City*, New York: Zone Books, pp. 219–21.

Sassen, S. (1991) *The Global City: New York, London, Tokyo*, Princeton, NJ: Princeton University Press.

Sassen, S. (1994) *Cities in a World Economy*, London: Pine Forge Press.

Shields, R. (1996) 'A guide to urban representation and what to do about it: alternative traditions of urban theory', in A.D. King (ed.) *Re-Presenting the City: Ethnicity, Capital and Culture in the 21st Century Metropolis*, London: Macmillan, pp. 227–52.

Smith, N. (1999) 'Which new urbanism? New York City and the revanchist 1990s', in R.A. Beauregard and S. Body-Gendrot (eds) *The Urban Moment: Cosmopolitan Essays on Late 20th Century City*, London: Sage, pp. 185–208.

Smith, R.G. (2003a) 'World city actor-networks', *Progress in Human Geography*, 27 (1): 25–44.

Smith, R.G. (2003b) 'World city topologies', *Progress in Human Geography*, 27 (5): 561–82.

Smith, R.G. (2005) 'Networking the city', *Geography: An International Journal*, 90 (2): 172–6.

Smith, R.G. (2006) 'Place as network', in I. Douglas, R. Hugget and C. Perkins (eds) *Companion Encyclopaedia of Geography*, 2nd edn, London: Routledge.

Smith, R.G. and Doel, M.A. (2006) 'Global cities and the geography of global command: exposing a fundamental construct validity problem', unpublished manuscript available from the authors.

Taylor, P.J. (2001) 'Specification of the world city network', *Geographical Analysis*, 33 (2): 181–94.

Theodorson, G.A. (ed.) (1982) *Urban Patterns: Studies in Human Ecology*, University Park, PA: Penn State University.

Thrift, N. (2003) 'Space: the fundamental stuff of human geography', in S.L. Holloway, S.P. Rice and G. Valentine (eds) *Key Concepts in Geography*, London: Sage, pp. 95–107.

15 The mismatch between concepts and evidence in the study of a global urban network

Ben Derudder

INTRODUCTION

Throughout the last two decades, researchers have analyzed the emergence of a global urban network centered on a number of key cities in the global economy. Taken together, these studies are loosely united in their observation that cities such as New York and London derive their importance from a privileged position in transnational networks of capital, information, and people. There is, in other words, a widespread consensus that under conditions of contemporary globalization an important city 'is no longer identifiable for its stable embeddedness in a given territorial milieu. It is instead a changing connective configuration with variable actors which can be thought of as "nodes" of local and global networks' (Dematteis 2000: 63).

However, despite this broad agreement, there are equally obvious differences in the way in which this transnational urban network has been conceptualized. For instance, it is clear that the influential 'global city' concept is presented as a specific analytical construct rather than as an attempt to refine existing concepts. In the revised edition of *The Global City*, Sassen (2001a: xxi) maintains that '[w]hen I first chose to use [the term] global city I did so knowingly – it was an attempt to make a difference.' This attempt to discriminate is further specified vis-à-vis two other important theorizations, i.e. John Friedmann's 'world cities' and Allen Scott's 'global city-regions'. Sassen (2001a: xxi) stresses, for instance, that although it may be the case that 'most of today's major global cities are also world cities' there may just as 'well be some global cities today that are not world cities in the full, rich sense of that term.' And, while global cities and global city-regions may '[a]s categories for analysis . . . share key propositions about economic globalization,' it is clear that they 'overlap only partly in the features they each capture' (ibid. 351). As a consequence, they have 'distinctive theoretical and empirical dimensions' (Sassen 2001b: 78).

Although naming cities as nodes in a global urban network is thus clearly not a trivial matter of semantics, there is a tendency in the literature to downplay the various analytical and empirical differences between key concepts. Olds and

Yeung (2004: 515) and Brenner (1998: 29), for instance, commence their articles with an endnote in which they state that the terms 'global city' and 'world city' will be used interchangeably. The refusal to distinguish between concepts may sometimes be acceptable for heuristic or pedagogic reasons, but I will argue that there are equally instances in which conceptual differences cannot be dismissed. To drive my point home, this chapter is divided into two parts. First, I discuss the main tenets of three of the most important conceptualizations of cities in the context of a global urban network, i.e. world cities, global cities, and global city-regions. The main objective of this section is to single out the core ideas behind each concept in order to contrast them as clearly as possible. In the second section, I show why it is problematic to condense the various terms into interchangeable notions that share a basic connotation of 'important cities'. This is done through a general overview of the blurry discourse in recent empirical studies and a more detailed review of some inappropriate analytical claims in a recent study by Alderson and Beckfield (2004).

MAIN THEORETICAL APPROACHES

In this section, I review the main ways in which cities have been conceptualized in the context of a transnational urban network, i.e. as world cities, as global cities, and as global city-regions. Based on this overview, I contrast some of the alleged characteristics of cities within each of these conceptualizations.

World cities

The world city concept can be traced back to two interrelated papers by Friedmann and Wolff (1982) and Friedmann (1986).[1] Both texts framed the rise of a global urban network in the context of a major geographical transformation of the capitalist world-economy. This restructuring, most commonly referred to as the 'New International Division of Labor', was basically premised on the internationalization of production and the ensuing complexity in the organizational structure of multinational enterprises (MNEs). This increased economic-geographical complexity requires a limited number of control points in order to function, and world cities are deemed to be such points. The territorial basis of a world city is hereby more than merely a CBD, since '[r]eference is to an economic definition. A city in these terms is a spatially integrated economic and social system at a given location or metropolitan region. For administrative purposes the region may be divided into smaller units which underlie, as a political or administrative space, the economic space of the region' (Friedmann 1986: 70).

Friedmann (1986) tries to give theoretical body to his 'framework for research' by subsuming it under the heading of Wallerstein's world-systems analysis, hence the title of Knox and Taylor's (1995) *World Cities in a World-System* and the hyphen in 'world-economy' (see also Saey 1996; Taylor 2000). As is well known, Wallerstein (1979) envisages capitalism as a system that involves a hierarchical and a spatial inequality of distribution based on the concentration of relatively monopolized and therefore high-profit production in a limited number of 'core' zones. The division of labor that characterizes this spatial inequality is materialized through a tripolar system consisting of core, semiperipheral, and peripheral zones. The prime purpose of world city research, now, is that it seeks to build an analytical framework that searches to deflect attention from the role of territorial states in the reproduction of this spatial inequality (Brenner 1998: 4). Territorial states have, of course, been prime actors in the unfolding of this uneven development, but drawing on the work of Mann (1986) and Dodgshon (1998) it can be put forward that the world-economy is radially rather than territorially managed. This means that the economic and political power of core territories is in fact spatially structured along well defined routeways that link centres of control via available authoritative and allocative resources. Hence, what is commonly labelled as 'core' in world-systems analysis does not necessarily consist of a series of 'strong' territorial states, but of a hierarchy of major and lesser centres (i.e. world cities) that thereupon diffuse their status and function over a wider area and at different scales (Dodgshon 1998: 56).

In other words: despite 'being largely studied through its mosaic of states . . . the modern world-system is defined by its networks' (Taylor 2000: 20), and world cities are the nodes in such networks of power and dominance. Apart from being the economic power houses of the world-system, world cities are also locales from which other forms of command and control are exercised, e.g. geopolitical and/or ideological–symbolical control over specific (semi)peripheral regions in the world-system. Miami's control position over Central America is a case in point here (Grosfoguel 1995). Friedmann (1986: 69) reminds us, however, that 'the economic variable is likely to be decisive for all attempts at explanation,' whereby major importance attaches to corporate headquarters and international financial institutions and agencies. Although the presence of a business services sector and/or a well developed infrastructure seems to be required, these are conceptually less important, since they are necessary but not sufficient conditions in the formation of a network of world cities.

Global cities

The global city concept can be traced back to the publication of Saskia Sassen's *The Global City* in 1991. Sassen proposes to look afresh to the functional centrality of cities in the global economy, and does so by focusing on the attraction of producer service firms to major cities that offer knowledge-rich and technology-enabled environments. In the 1980s and 1990s, many such service firms followed their global clients to become important MNEs in their own right, albeit that service firms tend to be more susceptible to the agglomeration economies offered by city locations. These emerging producer service complexes are at the root of global city formation, which implies a shift of attention to the advanced servicing of worldwide production. Hence, from a focus on formal command power in the world-system, the emphasis shifts to the *practice* of global control: the work of producing and reproducing the organization and management of a global production system and a global market-place for finance . . . . Power is essential in the organization of the world economy, but so is production: including the production of those inputs that constitute the capability for global control and the infrastructure of jobs involved in this production.

(Sassen 1995: 63–4, her emphasis)

Through their transnational, city-centered spatial strategies, producer service firms have created worldwide office networks covering major cities in most or all world regions, and it is exactly the myriad connections between these service complexes that provide, according to Sassen (2001a: xxi), a means for the 'formation of transnational urban systems.' This urban network, Sassen (1994: 4) argues, results in a new geography of centrality that may very well cut across existing north/south divides. Hence, rather than reproducing existing core/periphery patterns in the world-economy, this network may break through these divides.

The focus on urban agglomeration economies has a major implication for the territorial demarcation of global cities. Rather than being structured in mutual dependence to a hinterland, the functional centrality of global cities becomes 'increasingly disconnected from their broader hinterlands or even their national economies' (Sassen 2001a: xxi). To territorially demarcate global cities, Sassen (2001b: 80) opts 'for an analytical strategy that emphasizes core dynamics rather than the unit of the city as a container – the latter being one that requires territorial boundary specification.' This does not necessarily imply that the functional centrality in global cities is a simple continuation of older centrality patterns as in New York City, since the territorial basis can consist of 'a metropolitan area in the form of a grid of nodes of intense business activity, as we see in Frankfurt and Zurich' (Sassen 2001a: 123). It is nonetheless clear that the proper unit of analysis may very well be smaller than the 'metropolitan region'. Tokyo as a global city, for instance, is the 'Tokyo Metropolis' rather than the larger 'Tokyo Metropolitan Region' or the 'National Capital Region' (ibid. 371).

Global city-regions

Despite earlier contributions by Petrella (1995) and Veltz (1996), the global city-region concept is most commonly traced back to the work of Allen Scott (2001a,b), who has conceptualized global city-regions as the new key territorial units in a post-Fordist global economy. Following this lead, Scott maintains that the role of nation-states as chief territorial–organizational nexuses is increasingly being overtaken by an extensive archipelago of global city-regions. Perhaps ironically, a major driving force in this organizational transition can be found within the territorial scale that global city-regions are thought to replace. That is, the growth of global city-regions is crucially conditioned by a re-scaling of the national frameworks in which they are embedded. This re-scaling has been induced by new economic-geographic strategies of transnationally oriented capitalist firms, who are increasingly trying to circumvent and restructure the nationally organized Fordist–Keynesian regimes of accumulation with their nationally organized forms of social monetary and labor regulation (Brenner 1999: 68).

In simple geographic terms, global city-regions consist of dense megalopolises that are bound up in intricate ways in intensifying and far-flung transnational relationships. They represent an outgrowth of large metropolitan areas or contiguous sets of metropolitan areas, together with surrounding hinterlands of variable extent which may themselves be sites of scattered urban settlements. Although global city-regions are invariably characterized by large populations, size is not their defining criterion. Rather, it is the extent to which the development of individual settlements is a function of the gravitational power of major regions of the world that defines whether they are contributing to the build-up of a global city-region. The archetypical example of a global city-region is the Pearl River Delta Area, which functionally connects Hong Kong, Shenzhen, Guangzhou, Zhuhai, Macau, and other small towns in the region to the global economy.

Scott *et al.* (2001: 17) stress that an identification of global city-regions in terms of population size is therefore not very adequate, but find it nonetheless feasible to refer to the world map of large metropolitan areas, because '[p]roduction and performance are . . . raised by urban concentration in two ways. First, concentration secures overall efficiency of the economic system. Second, it intensifies creativity, learning, and innovation both by the increased flexibility of producers that makes it possible and by the enormous flows of ideas and knowledge that occur alongside the transactional links within localized production networks.' This hesitant crisscrossing between urban morphology and regionalized production networks make it unclear how a global city-region may be demarcated, but is clear that an initial threshold of morphological concentration must be surpassed to initiate a growth pole-like process.

Strictly speaking, there is no conceptual necessity to speak of a *metropolitan* region, since the requirement of a focal point is nowhere discussed, let alone demonstrated. Hence, although most global city-regions are constructed around a major city (in the functional and the morphological sense), there may as well be global city-regions *without* a major city. Henton (2001: 398–9), for instance, addresses Silicon Valley as a 'global city-region', and paints thereby a picture that is reminiscent of a region rather than a city:

> There were over twenty-seven local jurisdictions in the region and little cooperation. Silicon Valley has become so big and complex it had trouble dealing with regional challenge. No city by itself had the resources or authority to meet the big challenges.

This makes 'global city-regions' look very much like the 'regions' in Scott's (1998) earlier work, of course, and it is therefore no surprise that some of the chapters in Scott's (2001b) book refer to regions rather than to cities or city-regions (e.g. Ohmae 2001; Porter 2001). The most important point for the present discussion is that a global city-region may stretch well beyond the confines of a metropolitan region.

The transnational network in which global city-regions are embedded has an 'archipelago' rather than a 'core/periphery' structure. Indeed, although 'not all large metropolitan areas are equally caught up in processes of globalization' (Scott 2001b: 1), it can be noted that '[t]he processes of urban and regional development we are describing here are not limited to the wealthiest countries.' As a consequence, Scott (2001a: 822) sees

> no reason – with due acknowledgment of the enormous difficulties posed by the vicious circles in which they are often caught – why at least some [metropolitan areas in the developing world] cannot benefit from the processes of urbanization and economic growth described above. These processes suggest that some of the more urbanized regions in these countries will eventually accede as dynamic nodes to the expanding mosaic of global city-regions, just as city-regions like Seoul, Taipei, Hong Kong, Singapore, Mexico City, São Paulo, and others, have done, and are doing, before them.

Overview

Table 15.1 summarizes the gist of the main theoretical approaches. Although each concept has been refined and/or revised in other contributions, it seems fair to

Table 15.1 Main theoretical approaches in the study of a transnational urban network

	World cities	*Global cities*	*Global city-regions*
Key author	Friedmann	Sassen	Scott
Function	Power	Advanced servicing	Production
Key agents	Multinational corporations	Producer service firms	Firms embedded in post-Fordist production networks
Structure of the network	Reproduces (tripolar) spatial inequality in the capitalist world-system	New geography of centrality and marginality cutting across existing core/periphery patterns	Archipelago structure replacing existing core/periphery patterns
Territorial basis	Metropolitan region	Traditional CBD or a grid of intense business activity*	(Metropolitan) region**

Notes
* The spatial demarcation depends on the specific *form* of the territorialization of the core dynamics behind global city-formation. This implies that both the continuation of traditional CBDs (New York) and a new pattern centered on a grid of intense business activity (Zürich) are possible. However, the proper unit of analysis is clearly smaller than the 'metropolitan region' as a whole (see body of text for further elaboration).
** Although most global city-regions have one or more major cities at their core, there is no conceptual need for functional centrality (see body of text for further elaboration).

state that the table gives a balanced overview of the conceptual core of each term: (i) Friedmann's world cities are centers of dominance and power, (ii) Sassen's global cities are production centers for the inputs that constitute the capability for global control, and (iii) Scott's global city-regions are production nexuses in a global economy dominated by a post-Fordist accumulation regime. These different starting points thereupon give way to diverging perspectives on the main features of a city-as-node in transnational networks: the city's prime function, the key agents in the urban network, the alleged structure of the network as a whole, and the territorial basis of the city-as-node.

One can argue back and forth on the profoundness of the differences summarized in Table 15.1, but it seems clear that there is an unambiguous need to distinguish between the three concepts. The territorial demarcation of the units of analysis, for instance, may differ significantly. While a global city is most often a single focal point that operates separately from its hinterland, a global city-region may consist of multiple cities and their hinterlands that may themselves be subject to urbanization processes. Put succinctly: the Pearl River Delta may very well be a global city-region, but it can most certainly not be a global city (and vice versa for Hong Kong). Another important difference lies in the anticipated structure of the overall network. While a network of world cities is expected to reproduce core/periphery patterns across the world-economy, a network of global cities and a network of global city-regions are expected to cut across such divides. In other words: it is not unlikely that 'semiperipheral' cities such as Mexico City, São Paulo, and Seoul are well connected global service centers (i.e. global cities) without being major power centers in the world-economy (i.e. world cities). Hence, rankings of world cities and global cities may be expected to diverge rather than converge.

CONCEPTUAL BACKDROP OF EMPIRICAL STUDIES

In the previous section, I discussed how some of the commonly employed terminologies in the study of a global urban network cover specific analytical constructs. As a consequence, these terminologies require a particular conceptual treatise and a specific empirical grasp. In the present section, I aim to show that recent empirical studies have in general not lived up to this requirement. It is obvious that the potential problems are manifold, and I will therefore restrict this discussion to two discrepancies between theory and measurement. First, I present a simple overview of the conflation of terms in recent empirical studies. Second, I focus on the way in which this loose use of terms is merely a surface-level feature of a more profound conceptual conflation.

Confusion in terminology

Recent empirical researches have tried to assert their analytical relevance by referring to various concepts. In this paragraph, I discuss to what degree they have employed the appropriate terminology. I begin by summarizing the assumptions behind the main empirical approaches, and then contrast the appropriate and the employed terminologies in these studies.

Main empirical approaches

Empirical researchers have relied on a wide variety of data sources to assess the networked importance of cities, but generally speaking the production of these databases has been premised upon two foundations, which may respectively be labeled (i) the corporate organization approach and (ii) the infrastructure approach.[2]

The corporate organization approach starts from the observation that firms pursuing global strategies are the prime agents in the formation of urban networks. Two leading examples are the research pursued by the Globalization and World Cities group and network (GaWC, http://www.lboro.ac.uk/gawc) and a recent analysis by Alderson and Beckfield (2004). The GaWC researchers have developed a methodology for studying transnational urban networks based on the assumption that advanced producer service firms 'interlock' cities through their intra-firm communications of information, knowledge, plans, directions, advice, etc. to create a network of global service centers (Taylor 2001). Building on this specification, information was gathered on the location strategies of 100 global service firms across 315 cities (Taylor *et al.* 2002). Applying the formal social network methodology set out in Taylor (2001), this information was converted in a 315×315 matrix, which was then analyzed with standard network analytical tools (Derudder and Taylor 2005). Using a similar methodological approach, Alderson and Beckfield (2004) have analyzed links between 3,692 cities based on the organizational geographies of 446 of the largest multinational firms and their subsidiaries for the year 2000. Despite some methodological differences, both studies base their city-centered spatial analysis on an assessment of the location strategies of firms with transnational fields of activity. In other words, it is suggested that a meaningful measurement of transnational inter-city relations can be derived from intra-firm connections between different parts of a firm's holdings: Alderson and Beckfield (2004: 813–14) consider this to be a 'key relation' in 'an MNE-generated city system', while Taylor (2004: 9) argues that it is 'firms through their office networks that have created the overall structure of the world city network.' The main difference between the approaches lies in the type of firms used throughout their analysis: GaWC researchers focus on the location strategies of producer services firms, Alderson and Beckfield (2004) use information on the geography of multinational corporations irrespective of the exact nature of their activities (e.g. their table of the distribution of firms 'across industries', 821).

The gist of the infrastructure approach lies in the observation that advanced telecommunication and transportation infrastructures are unquestionably tied to key cities in the global economy. The most important cities also harbor the most important airports, while the extensive fiber backbone networks that support the Internet have likewise been deployed within and between major cities, hence creating a vast planetary infrastructure network on which the global economy has come to depend almost as much as physical transport networks (Rutherford *et al.* 2004). These enabling (tele)communication and transportation networks are the foundation on which the connectivity of key cities is built, and it is therefore no surprise that the geography of these networks has been used to invoke a spatial imagery of a transnational urban network. Smith and Timberlake (2002: 139), for instance, have sought to describe the spatial patterning of a transnational urban network 'as indicated by their interrelations in the air passenger networks' (see also Derudder and Witlox 2005), and Townsend (2001: 1700) has illustrated 'how global cities have fared in the rapid and massive deployment of Internet networks.' Both analyses claim that it is possible to devise an urban network based on the geography of infrastructure networks, whereby the main difference between both approaches lies in the type of infrastructure, i.e. physical transport versus telecommunications.

Appropriate and employed terminology

Table 15.2 summarizes the employed concepts and the appropriate terminologies within these empirical approaches. References to the employed concepts are, of course, primarily based on the directly used discourse, but some references are

Table 15.2 Employed terminologies in recent empirical studies

Empirical approach	Corporate organization		Infrastructure	
	Producer service firms	*Multinational enterprises*	*Telecommunications*	*Physical transportation*
Employed terminology	Beaverstock *et al.* (1999): world cities and global cities	Rodríguez-Pose and Zademach (2003): global cities	Graham (2002): global cities and global city-regions	Keeling (1995): world cities
	Taylor *et al.* (2002): world cities and global cities	Alderson and Beckfield (2004): world cities and global cities	Malecki (2002): world cities, global cities and 'large cities'	Smith and Timberlake (2001, 2002, 2005): world cities and global cities
	Derudder et al. (2003): world cities and global cities		Townsend (2001): global cities	Derudder and Witlox (2005): world cities
	Derudder and Taylor (2005): world cities and global cities			
	↓	↓	↓	↓
Appropriate terminology	Global cities	World Cities	(Global City-Regions)	

based on more implicit allusions. A case in point is Derudder *et al.* (2003), who present a global urban analysis of 234 cities based on the data gathered in Taylor *et al.* (2002). With the exception of a single mention of 'global cities' (Derudder *et al.* 2003: 885), the authors employ the term 'world city', but it is quite clear that 'global cities' are deemed to be at least as relevant as a concept. For instance, Derudder *et al.* (2003: 877) ascertain that they follow 'Sassen in her treatise of cities as global service centers – locales where advanced producer services are concentrated for servicing their global corporate clients.'

The studies carried out within the corporate organization approach have a comprehensible conceptual background, and it is therefore fairly straightforward to pin down what terms should be referred to. Studies drawing on the location strategies of producer service firms pertain to Sassen's global city formation, whereas studies employing data on the organizational structure of large MNEs refer to Friedmann's world city formation. In contrast to data on corporate organization, the relevance of infrastructure data appeals more to common sense than to a precise concept. For instance, a recent volume edited by Sassen (2002) features two consecutive chapters on infrastructure networks, whereby Graham's (2002: 89) fiber network analysis refers to 'global cities' and 'global city-regions', whereas Smith and Timberlake's (2002: 139) airline network analysis refers to 'world cities'. It remains unclear, however, how infrastructure data can distinguish between these concepts. The employed terminologies in these infrastructure network-based studies are therefore not so much a starting point for a precise specification, but more a vague attempt to suggest a credible conceptual backdrop.

If there *is* a meaningful analytical connection between theoretical concepts and infrastructure-based measurements, then the most relevant concept seems to be Scott's global city-regions. The main reason for this is that the connectivity of cities such as Frankfurt and Amsterdam cannot straightforwardly be traced back to a single, central location. Rather, the airports of these cities are in practice serving 'an emergent form in the global urban hierarchy' (Smith and Timberlake 2001: 1671), which is a city-region that consists of one or more major cities together with their hinterlands. It is therefore misleading to conceive Frankfurt and Amsterdam as the 'real' units of analysis in their airline-based studies. Rather, it seems more fruitful to speak of the 'Rhine–Main conurbation' (ibid.) and the 'Randstad conurbation (which also includes Rotterdam and The Hague)' (1672). The main point is that this suggests that a broader region is directly implicated in the connectivity within airline networks, since it is a function of both the city *and* its surrounding urban field. Hence the possible link to global city-regions, albeit that there is a clear necessity for further elaboration on the analytical connections between this concept and infrastructure data.

The confrontation of the employed and the appropriate discourse in Table 15.2 reveals that empirical studies only seldom refer to the appropriate terminologies. Admittedly, a blurry discourse is not necessarily a sign of a sweeping conceptual conflation. For instance, despite favoring the term 'world city', GaWC researchers unambiguously refer to Sassen's work on place and production in a global economy.[3] Thus, there can be little doubt that GaWC studies present an analysis of global cities, and passing judgment on the basis of an erroneous terminology would therefore be outright excessive. In the next section, I shall therefore move beyond this 'surface-level' critique, and show how this vague use of terminologies is merely a symptom of a more profound conceptual confusion. To this end, I shall discuss some claims in a recent empirical study by Alderson and Beckfield (2004).

Conceptual confusion

In their article, Alderson and Beckfield (2004: 814) address two very concrete questions. First, the authors seek to 'determine which cities are in fact central to the MNE-generated city-system.' Second, having established such a ranking of cities in terms of network centrality, they 'examine precisely what sort of "system" these cities form.' The latter question involves searching out the predominant structure of the network. Possible answers include a simple linear hierarchy, a core/periphery structure, a structure exemplified by cohesive subgroups bounded by trading blocs, etc. To answer both questions, Alderson and Beckfield rely upon an inter-city matrix that paints a detailed picture of an MNE-generated urban network.

The first research question is answered through an analysis of three measures of 'point centrality'. In their paper, Alderson and Beckfield devote most attention to the so-called 'outdegree centrality' of a city, which is calculated as the sum of all ties sent from corporate headquarters in that city to non-headquarter cities. This indicator is, according to Alderson and Beckfield (ibid. 822, 827–8), an 'unambiguous indicator of world city-ness.' After having computed this measure for all cities, Alderson and Beckfield (ibid. 828–9) assess whether the emerging ranking differs substantially from 'other' world city rankings. One of these rankings is Beaverstock *et al.*'s (1999) *Roster of World Cities*, a study based on an inventory of advanced producer services in cities. Alderson and Beckfield (ibid. 829) report that there are 'notable discrepancies' between the two lists: only 46 percent of the cities featuring in the Beaverstock *et al.* (1999) ranking have a significant position in the ranking based on the 'outdegree centrality.' Furthermore, this overlap is almost exclusively confined to the most connected cities in the network. The latter observation is no surprise, according to Alderson and Beckfield (ibid. 829), since leading researchers such as Friedmann and Sassen have argued 'that cities such as London, New York, and Tokyo sit at the top of the world city system.' However, despite the comforting parallels at the apex, the authors equally mention some 'surprises at the top' (ibid.). First, Tokyo surpasses London and New York in this ranking, although it 'is typically viewed as being eclipsed by London and New York in power' (ibid.). Second, 'Paris emerges as a city of the first rank' (ibid.), which is noted as in broad agreement with Smith and Timberlake's (2001) airline network analysis. Third, two prominent names do not appear in their list: Singapore and Miami. The absence of these two cities, it is claimed, may be due to their importance in 'regional city-systems' (Alderson and Beckfield 2004: 829–30), but there is also a more overarching concern about the extent to which these findings 'may be biased by the type of data we employ' (ibid. 829).

Alderson and Beckfield's (ibid. 825–26) second research question is addressed by 'measuring' the overall structure of the network through a range of sophisticated network-analytical tools. The overarching conclusion is that this MNE-generated urban network exhibits a more or less distinct core/periphery structure whereby '[t]he average rank of cities located in semiperipheral countries is lower than that of core cities, whereas that of cities located in peripheral countries is lower still' (ibid. 844), which leads to the conclusion that this is not consistent with 'Sassen's vision of a world city system in the grips of substantial global restructuring. Rather than cutting across the hierarchy of states in the interstate system, the contemporary urban hierarchy appears to map onto it fairly well.'

Despite the many merits of Alderson and Beckfield's network analysis, one can discern a number of problems because of the failure to distinguish between different concepts. This lack of rigor is in turn legitimized by an inadequate discourse in the Beaverstock *et al.* (1999) and Smith and Timberlake (2001) studies. The low overlap between these rankings is not so much 'biased' by the type of data, but can and should be traced back to the observation that these data sources capture different concepts that may or may not converge empirically: Beaverstock *et al.* (1999) present an inventory of *global cities*, Alderson and Beckfield analyze a network of *world cities*, and Smith and Timberlake (2001) essentially measure a network of *global city-regions*. Thus, the observation that cities such as São Paulo, Seoul, and Mexico City rank fairly high in Beaverstock *et al.* (1999) and fairly low in Alderson and Beckfield is not necessarily a surprise; it merely reveals that these 'semiperipheral' cities are well connected service centres without being the (economic) powerhouses of the world-economy. Moreover, the clear-cut core/periphery structure of this MNE-generated urban network cannot contradict Sassen's claims of a transnational urban network 'in the grips of substantial global restructuring,' since her suggestion pertains to a network of global service centers. To assess Sassen's proposition, one needs to apply Alderson and Beckfield's network-analytical framework to a GaWC-like dataset. For one thing, the observation that São Paulo, Seoul, and Mexico City rank fairly high in Beaverstock *et al.* (1999) and Derudder *et al.* (2003), but fairly low in Alderson and Beckfield, is in fact broadly supportive for Sassen's (1994: 4) claim that a network of global cities cuts across core/periphery divides. Furthermore, the fact that Paris is a well-connected global city-region (as measured through airline networks) can hardly be used to 'support' the observation that Paris ranks higher as a world city than it ranks as a global city. And finally, the 'remarkable' importance of Tokyo in an MNE-generated city-system is in fact by no means surprising; it only appears that way when this dominance is discussed in terms of another concept (i.e. global cities). The fact of the matter is that Tokyo's dominance in corporate HQs is Sassen's starting point to call for another approach to functional centrality in the global economy. Indeed, Sassen (2001a: 108) admits outright that Tokyo is the leading city in terms of corporate HQs, and states that this can be traced back to the fact that 'participation in the global economy for [Japanese multinational] firms still means going through a lot of government channels given a fairly regulated economy. Hence, location in Tokyo is crucial.' One cannot claim that Tokyo's remarkable dominance in a network of world cities contrasts with Sassen's assessment in the context of a network of global cities, especially if this observed dominance is in fact Sassen's motive to call for a different perspective.

CONCLUSION

Peter Taylor (2004: 33) has recently argued that the 'world city literature' as a whole has been characterized by 'theoretical sophistication and empirical poverty,' whereby one effect of this 'evidential crisis has been the failure for there to emerge any agreement on just which cities are world or global cities and which fail to qualify' (ibid. 39). This clearly comes to the fore in a comparison of 16 different rankings of 'world cities, global cities, and international financial centres from different sources' (ibid. 39–41). Taylor (ibid. 39) notes that there are only four cities that all 16 studies agree upon (London, New York, Paris, and Tokyo), while there are 78 other cities that at least one source names in its ranking. This profound disagreement, Taylor suggests, reflects the failure of this literature to provide precise empirical specifications of the various concepts. In this chapter, however, I have tried to show that this 'failure' cannot be asserted on the basis of such disagreement. Taylor's list contains specifications of different concepts, and there is therefore no reason why their geographies should converge.

Although I have restricted the discussion to the inapt discourse in recent empirical studies and some dubious claims in one of these studies, my argument has a broader appeal. Alderson and Beckfield's (2004) and Taylor's (2004: 39–41) decision not to distinguish between different concepts is representative of a state of affairs where it is taken for granted that the various 'approaches are perhaps not quite as distinct as they may seem' (Hall 2001: 61). However, I have argued that the observation that cities such as London, New York, Tokyo, and Paris invariably feature at the apex of the various rosters of world cities, global cities, and global city-regions does not imply that these concepts are interchangeable.

I am aware that one might turn the rationale of this chapter on its head. Why is it that, although the concepts are distinct, most of today's leading global cities are also major world cities? Why is it that a ranking of global cities all in all maps fairly well onto existing core/periphery patterns? These are important and pertinent questions, but such questions cannot be solved empirically; they need to be confronted theoretically. Parnreiter *et al.* (2005) and Brown *et al.* (2005) have recently undertaken such studies. Simply stated, both papers propose to analyze the (as yet unidentified) ways in which the spatial equality created through global commodity chains is facilitated by advanced servicing from a 'nearby' global city – Parnreiter *et al.* (2005: 1) aptly speak of a 'missing link between global commodity chains and global cities.' If meaningful regional links between (i) firms that are part of an MNE-structure and (ii) a 'nearby' producer service complex can be revealed, then we can make a start on addressing the parallels between world city and global city formation. The main point here is that such an uncovering cannot be based on some superficial parallels in empirical rankings.

NOTES

1   There are earlier uses of this term, but Brenner (1998: 5) notes that these uses reflected the 'territorialization of the urbanization process on the national scale: the cosmo-politan character of world cities was interpreted as an expression of their host states' geopolitical power.'

2   There are also a limited number of empirical studies that cannot be subsumed under this bifurcation, such as Beaverstock *et al.* (2000). The general purpose of Beaverstock *et al.* (2000: 43), however, is not to present a definitive mapping of a transnational urban network but to begin addressing the 'poverty of data by showing that appropri-ate data can be identified and analysed to study relations between world cities.'

3   The straightforward designation of GaWC studies as research into 'global cities' should, however, be somewhat nuanced. It can, for instance, be noted that the em-pirical rationale of most GaWC research starts from a critique of Sassen's global city concept for its bias towards a limited number of cities. Therefore their use of the terms 'world cities,' 'globalizing cities,' and 'cities in globalization' may be considered as a deliberate move to enable a larger set of cities. Thus although Sassen's process is used in GaWC studies, it can be said that they do try to bypass her concept of 'global cities.' Furthermore, it is relevant that Friedmann (1986) mentions finance and 'high level business services', so there is some overlap between world cities and global cities in the original sources.

REFERENCES

Alderson, A.S. and Beckfield, J. (2004) 'Power and position in the world city system', *American Journal of Sociology*, 109: 811–51.

Beaverstock, J.V., Smith, R.G. and Taylor, P.J. (1999) 'A roster of world cities', *Cities*, 16(6): 445–58.

Beaverstock, J.V., Smith, R.G., Taylor, P.J., Walker, D.R.F. and Lorimer, H. (2000) 'Glo-balization and world cities: some measurement methodologies', *Applied Geography*, 20(1): 43–63.

Brenner, N. (1998) 'Global cities, glocal states: global city formation and state territorial restructuring in contemporary Europe', *Review of International Political Economy*, 5(1): 1–37.

Brenner, N. (1999) 'Beyond state-centrism? Space, territoriality and geographical scale in globalization studies', *Theory and Society*, 28: 39–78.

Brown, E., Derudder, B., Parnreiter, C., Pelupessy, W. and Taylor, P.J. (2005) *Spatialities of Globalization: Towards an Integration of Research on World City Networks and Global Commodity Chains*. GaWC research bulletin 151.

Dematteis, G. (2000) 'Spatial images of European urbanization', in A. Bagnasco and P. Le Galès (eds) *Cities in Contemporary Europe*, Cambridge: Cambridge University Press, pp. 48–73.

Derudder, B. and Taylor, P.J. (2005) 'The cliquishness of world cities', *Global Networks*, 5(1): 71–91.

Derudder, B. and Witlox, F. (2005) 'An appraisal of the use of airline data in assessing the world city network: a research note on data', *Urban Studies*, 42(13): 2371–88.

Derudder, B., Taylor, P.J., Witlox, F. and Catalano, G. (2003) 'Hierarchical tendencies and regional patterns in the world city network: a global urban analysis of 234 cities', *Regional Studies*, 37(9): 875–86.

Dodgshon, R.A. (1998) *Society in Time and Space: A Geographical Perspective on Change*, Cambridge: Cambridge University Press.

Friedmann, J. (1986) 'The world city hypothesis', *Development and Change*, 17: 69–83.

Friedmann, J. and Wolff, G. (1982) 'World city formation: an agenda for research and action', *International Journal of Urban and Regional Research*, 3: 309–44.

Graham, S. (2002) 'Communication grids: cities and infrastructure', in S. Sassen (ed.) *Global Networks, Linked Cities*, London: Routledge, pp. 71–91.

Grosfoguel, R. (1995) 'Global logics in the Caribbean city system: the case of Miami', in P.L. Knox and P.J. Taylor (eds) *World Cities in a World-System*, Cambridge: Cambridge University Press, pp. 156–170.

Hall, P. (2001) 'Global city-regions in the twenty-first century', in A. Scott (ed.) *Global City-Regions: Trends, Theory, Policy*, Oxford: Oxford University Press, pp. 59–77.

Henton, D. (2001) 'Lessons from Silicon Valley: governance in a global city-region', in A. Scott (ed.) *Global City-Regions: Trends, Theory, Policy*, Oxford: Oxford University Press, pp. 391–400.

Knox, P.L. and Taylor, P.J. (eds) (1995) *World Cities in a World-System*, Cambridge: Cambridge University Press.

Mann, M. (1986) *The Sources of Social Power. Vol. I: A History of Power from the Beginning to AD 1760*, Cambridge: Cambridge University Press.

Ohmae, K. (2001) 'How to invite prosperity from the global economy into a region', in A. Scott (ed.) *Global City-Regions: Trends, Theory, Policy*, Oxford: Oxford University Press, pp. 33–43.

Olds, K. and Yeung, H.W.-C. (2004) 'Pathways to global city formation: a view from the developmental city-state of Singapore', *Review of International Political Economy*, 11(3): 489–521.

Parnreiter, C., Fischer, K. and Imhof, K. (2005) *The Missing Link between Global Commodity Chains and Global Cities: The Financial Service Sector in Mexico City and Santiago de Chile*. GaWC research bulletin 156.

Petrella, R (1995) 'A global agora versus gated city-regions', *New Perspectives Quarterly*, Winter: 21–2.

Porter, M.E. (2001) 'Regions and the new economics of competition', in A. Scott (ed.) *Global City-Regions: Trends, Theory, Policy*, Oxford: Oxford University Press, pp. 139–157.

Rutherford, J. Gillespie, A. and Richardson, R. (2004) 'The territoriality of pan-European telecommunications backbone networks', *Journal of Urban Technology*, 11(3): 1–34.

Saey, P. (1996) 'Het wereldstedennetwerk: de nieuwe Hanze?', *Vlaams Marxistisch Tijdschrift*, 30(1): 120–3.

Sassen, S. (1991) *The Global City: New York, London, Tokyo*, Princeton, NJ: Princeton University Press.

Sassen, S. (1994) *Cities in a World Economy*, Thousand Oaks, CA: Pine Forge Press.

Sassen, S. (1995) 'On concentration and centrality in the global city', in P.L. Knox and P.J. Taylor (eds.) *World Cities in a World-System*, Cambridge: Cambridge University Press, pp. 63–78.

Sassen, S. (2001a) *The Global City: New York, London, Tokyo*, 2nd edition, Princeton, NJ: Princeton University Press.

Sassen, S. (2001b) 'Global cities and global city-regions: a comparison', in A. Scott (ed.) *Global City-Regions: Trends, Theory, Policy*, Oxford: Oxford University Press, pp. 78–95.

Sassen, S. (ed.) (2002) *Global Networks, Linked Cities*, London: Routledge.

Scott, A.J. (1998) *Regions and the World Economy: The Coming Shape of Global Production, Competition, and Political Order*, Oxford: Oxford University Press.

Scott, A.J. (2001a) 'Globalization and the rise of city-regions', *European Planning Studies*, 9(7): 813–26.

Scott, A.J. (ed.) (2001b) *Global City-Regions: Trends, Theory, Policy*, Oxford: Oxford University Press.

Scott, A.J., Agnew, J., Soja, E.W. and Storper, M. (2001) 'Global city-regions', in A. Scott (ed.) *Global City-Regions: Trends, Theory, Policy*, Oxford: Oxford University Press, pp. 11–32.

Smith, D.A. and Timberlake, M. (2001) 'World city networks and hierarchies, 1977–1997: an empirical analysis of global air travel links', *American Behavioral Scientist*, 44(10): 1656–78.

Smith, D.A. and Timberlake, M. (2002) 'Hierarchies of dominance among world cities: a network approach', in S. Sassen (ed.) *Global Networks, Linked Cities*, London: Routledge, pp. 117–41.

Taylor, P.J. (2000) 'World cities and territorial states under conditions of contemporary globalization', *Political Geography*, 19: 5–32.

Taylor, P.J. (2001) 'Specification of the world city network', *Geographical Analysis*, 33(2): 181–94.

Taylor, P.J. (2004) *World City Network: A Global Urban Analysis*, London: Routledge.

Taylor, P.J., Catalano, G. and Walker, D.R.F. (2002) 'Measurement of the world city network', *Urban Studies*, 39(13): 2367–76.

Townsend, A.M. (2001) 'Network cities and the global structure of the Internet', *American Behavioral Scientist*, 44: 1697–1716.

Veltz, P. (1996) *Mondialisation, villes et territoires: l'economie d'archipel*, Paris: Presses Universitaires de France.

Wallerstein, I. (1979) *The Capitalist World-Economy*, Cambridge: Cambridge University Press.

16 Cities within spaces of flows

Theses for a materialist understanding of the external relations of cities

Peter J. Taylor

PREFACE

In the conference from which this volume is derived, I used a format of 48 theses with the intention of producing a 'punchy' presentation. But there were two more basic reasons for this choice of mode of argument. First, I was at an early stage in my thinking on the subject matter and laying out a skeletal argument was both necessary and as far as my thinking had reached. Second, and related to the latter, the subject matter I had chosen encompassed a scope of ideas that was unusually wide-ranging: the method of short, sharp statements as theses was a way of compressing the tentative argument into a manageable whole for the audience. All these conditions still exist so I have continued with this format in this chapter. However, as well as revising, extending and reorganizing the theses as a result of further thinking, I have added this preamble to indicate how I have come to the subject matter at hand.

The spark for the train of thought that is the theses below came from observing the same basic statement from two outstanding urban theorists writing half a century apart: in 1921 Max Weber (1958: 65) argued that the city is not simply 'a large locality' and in 1969 Jane Jacobs (1970: 129) argued that 'a city is not a large town' – I used the latter quote as my title for the conference presentation. Of course, these theorists subsequently diverge in their respective paths to city definition: Weber focuses on functions, Jacobs understands cities as a process (Jacobs 1970: 50). Move on another quarter century and we find Jacobs echoed in Castells' (1996: 386) influential work on network society: 'the global city is . . . a process'. Although this last is based upon another different theory drawing on ideas of global cities atop a world urban hierarchy, Jacobs and Castells share a similar relational view of cities. The materialist understanding I try to develop below attempts to bring together key aspects of Jacobs' and Castells' ideas on cities and space.

But I do not divorce their ideas from traditional geographical concerns for cities and space. Going back as far as the geographer-anarchist Elisée Reclus (1895), I have long been intrigued by his postulating a flat (i.e. isotropic) plain to derive a regular geometric pattern of cities long before Christaller (1966) published his treatise on central place theory in 1933. On a superficial reading of Reclus I had thought he presaged the central place model but he actually proposes a quite different basis for the distribution of cities 'regulated . . . by the step of the traveller' (251). Here is a model of the spacing of cities not as central places but as destinations. I draw directly on this possibility of different spatial orderings of urban places to develop theses on how cities differ from towns. Thus central place theory appears in my theses but I am able to curtail its extrapolation into 'city systems' that are 'national urban hierarchies'. This requires a means of separating economic space production from political space production. For this I go to another part of Jacobs' oeuvre that I relate to Arrighi's (1994) emphasis on separating profit-orientated business enterprises from power-orientated government organization in the rise of the capitalist world-economy. Commonly 'political economy' approaches emphasize the need to integrate politics and economics and this is achieved through the myth of the territorial coincidence of 'national economy' with nation-state. I eschew this approach. Critical to my thinking here has been Arrighi's (1994: 109–27) intriguing argument that, in the initial manifestation of a modern world–system in the sixteenth century, the formative powers were not geographically coincident: the business enterprises of Genoa combined symbiotically with the protective power of Castile in a domination of the incipient capitalist world-economy that is difficult to comprehend through modern territorialist lenses. It was by thinking through the implications of this early modern city/state collaboration that I was able to link it to Jacobs' (1992) ideas on alternative ways of making a living: one essentially economic, the other political. Thus the theses that follow respect a separation between the economic and the political: the argument is against their study through integration *à la* conventional political economy, but rather through their interrelations as distinctive social activities.

As will be gathered from the above, the theses draw on ideas covering many decades. This may seem an unfashionable approach to take; there is a tendency today to use a set of ideas for a short while before moving on to the next set that seem to be more attractive in a sort of 'up-to-date' way. But Berman (1988) has warned us that, in any realm of social activity, the fashion-conscious are always in danger of becoming objects of modernity rather than subjects in modernity. For social scientists, in particular, being objects rather than subjects must be totally debilitating. For instance, 'modern intellectual fashion police' should have objectified my work by forcing the abandonment of world-systems analysis sometime in the 1980s; in current undergraduate textbooks that my students use for understanding 'development', world-systems analysis is conflated with the 'dependency approach', which superseded the 'modernization approach' in the 1970s but has itself been superseded by several new inventions of development since. In the theses below I keep to 'old-fashioned' world-systems ideas since my reasons for adopting this approach (relating to continuing world poverty) remain as strong as ever. The hurry to flip between discourses may be intellectually stimulating but I fear it loses key insights in dismissal of 'old ideas'. My argument is not so much about 'building paradigms' or avoiding perennial 'reinventing the wheel', it is simply a plea to understand that profound thinkers are not just of historical interest – 'when were their ideas superseded?' is not necessarily the best question to ask. Rather, where the same ideas appear in several contexts it might just mean that we are in the presence of some fundamental notions, perhaps even truly profound ideas, that need to be cherished and used rather than discarded at the behest of the latest fashion.

I should make it clear that my theses are not a summary of past ideas; rather I try to use those ideas and juxtaposition them in new ways and with new ideas to create fresh views of inter-city relations. I have organized the theses into four sets of ideas that constitute my argument: first, I present the foundation premises underpinning my position; second, I lay out my theses on city as process and differentiate 'city-ness' from 'town-ness'; third and fourth, I develop what are largely corollary theses that are transhistorical and specifically modern respectively.

FUNDAMENTAL PREMISES

Cities

1 Cities have been in existence for some thousands of years across many regions of the world; an understanding of these social entities, therefore, must be transhistorical and geographical in nature.
2 Cities always come in packs, never alone; it follows that urban theories and theories of 'the city' that deal with these social entities as singular are inherently inadequate.
3 Cities must be studied relationally as inter-city relations; case studies, comparative studies and even 'throughput' studies (for instance, Harvey's (1982) circulation/fixed capital, Lefebvre's (1996) particle/wave analogue, Cox and Mair's (1988) local coalitions capturing flows and Massey's (1993) progressive place) are inadequate conceptual bases for understanding cities.

Cities and states

4 Cities are social entities primarily created by economic processes; it is the study of work carried out in and through cities that will provide the necessary starting point for a materialist theory of cities.
5 States are social entities primarily created by political processes; behind the work of states there always lies the threat of force.
6 Relations between cities and states constitute an intertwined geohistory of cooperation and competition concerning wealth and protection; ultimately cities cannot be understood without exploring their relations with states, for which Jacobs (1992) provides a basic starting point.

Strategies of social reproduction and their 'moral syndromes'

7 Following Jacobs (1992), there are only two strategies of social reproduction: raiding/taking/protecting that leads to states, and trading/making/producing that leads to cities. Over millennia the activities towards these strategies have

each created a distinctive ethic, which Jacobs calls their moral syndrome. Each moral syndrome is based upon precepts of acceptable behaviour (see Table 16.1). Too many people diverging from the precepts will destroy the basis for that particular social reproduction (e.g. too many pirates will ultimately destroy trading).

8 In the guardian moral syndrome the primary ethical precept of the agents is loyalty, the behavioural premise is a zero-sum game and the key institutions are force and balance of power. For instance, a breakdown in loyalty causes catastrophe for an army, and its territorial losses will be mirrored by its enemy's territorial gains.

9 In the commercial moral syndrome the primary ethical precept is honesty, the behavioural premise is a win–win scenario (what Hicks (1969: 44) calls the 'All-round Advantage' of voluntary trade) and the key institutions are contracts and markets. For instance, a bank can only operate on the basis of honesty for the mutual benefit of lenders and borrowers, and a breakdown in this trust causes catastrophe for bank and customers.

Spaces of social reproduction

10 Each moral syndrome underpins behaviour that makes and uses particular forms of space that are Castells' (1996) spaces of places and spaces of flows.

11 The dominant space of guardian activities is a space of places, territories to be won and protected (examples range from hunter-gatherer territories to the

Table 16.1 Jacobs' moral syndromes as clusters of precepts

Commercial Syndrome	*Guardian Syndrome*
1. *Basic cluster*	1. *Basic cluster*
Shun force	Shun trading
Come to voluntary agreements	Exert prowess
Be honest	Be obedient and disciplined
2. *Operating cluster*	Adhere to tradition
Collaborate easily with strangers	Respect hierarchy
Compete	Be loyal
Respect contracts	2. *Action cluster*
Use initiative and enterprise	Take vengeance
3. *Progress cluster*	Deceive for the sake of the task
Be open to inventions and novelty	3. *Lifestyle/representation cluster*
Be efficient	Make rich use of leisure
Promote comfort and convenience	Be ostentatious
Dissent for the sake of the task	Dispense largesse
4. *Capital cluster*	Be exclusive
Invest for productive reasons	Show fortitude
Be industrious	4. *Life cluster*
Be thrifty	Be fatalistic
5. *Life cluster*	Treasure honour
Be optimistic	

Source: derived from Jacobs (1992).

many varieties of states). The familiar mosaic map of modern nation-states is a particularly powerful contemporary space of places.

12   The dominant space of commercial activities is a space of flows, movement of commodities for production and trade (examples range from ancient trading networks to many varieties of financial markets). Relations between 'global cities' as 'international financial centres' are a particularly powerful contemporary space of flows.

THESES ON CITY AS PROCESS

Cities, towns and 'urban'

13   A city is not a large town; neither settlement type can be defined by the simple attribute of demographic size because both are best conceptualized as processes. This is the key thesis of the argument.

14   The concept of 'urban' as contrasted to rural is a chaotic conception because it conflates town and city. Note that urbanization, a demographic territorial attribute, is typically used as a derogatory anti-city concept indicating a threat to a space of place that is rural. If a territorial term is required, it might be best to replace urbanization by de-ruralization.

15   'Urban' implies a concentration of population – a node in a space of flows – but the spaces created by towns and cities are inherently different; the contrast is between local hierarchies and non-local networks respectively.

Towns as central places

16   Towns are constituted by the processes that generate 'urban hinterlands'; the town–hinterland relationship may vary geohistorically but in essence towns are based upon local relations as described by phrases such as 'country towns' and 'market towns'.

17   Christaller's (1966) central place theory defines such relations and shows formally how towns are arranged evenly and hierarchically.

18   Guardian thinking on cities has used central place theory to define 'urban systems' as constituting national urban hierarchies that can be 'regulated' (Bourne 1975). But this is to incorporate leading national cities in this town-making process, with hinterlands becoming 'regional' (and ultimately 'national' for the top city) rather than strictly local.

Beyond town-ness

19   A key limitation of central place theory is that it treats cities as large towns in violation of our thesis (16). I am not arguing that cities do not have local hinterlands but that they are not to be defined as simply the towns with the largest hinterlands. It is necessary to consider cities beyond this one particular

relational theory.

20   A major advantage of defining both towns and cities as processes (thesis 16) is that their 'town-ness' and 'city-ness' can operate simultaneously through the same settlement. Thus it is that cities service a hinterland and in doing so they exhibit their 'town-ness'.

21   In addition, of course, cities also exhibit a primary 'city-ness', a completely different process. This is the reason why central place theory's prediction of spatial regularity tends to work much better at lower levels of the hierarchy (towns) than upper echelons where dominating city-ness processes tend to override central place imperatives of town-ness.

Cities and Jacobs' import replacement theory

22   Jacobs (1970; 1984) defines a city as a settlement that has experienced one or more rapid spurts of new work, adding to old work, and thus producing a more complex and diverse economy than a town.

23   But cities do not arise alone; the most important new work is import replacement whereby a city begins to produce what it previously imported from other cities in a network of cities.

24   This is not a zero-sum game because the generation of new exports to replace lost exports, and conversion of import replacements to become new exports, both lead to a dynamic city network based upon evolving mutualities in an overall expansion of economic life.

City-ness and town-ness contrasted

25   City-ness is net-work, the development of a network of cities; town-ness is hinter-work, the development of a hierarchy of towns and their 'service areas'.

26   Being part of a hierarchy is very different from being a member of a network. Town-ness is to know your place of dependence on settlements higher in the hierarchy. The import replacement of city-ness is a declaration of economic autonomy from such relations through the mutualities of a city network.

27   Since these are two completely different processes it follows that there can be no simple settlement evolution theory – farm, village, town, city – with the latter as the most 'developed' settlement. This has important geohistorical implications because it means that towns and cities have different geohistories creating different spaces.

TRANSHISTORICAL THESES

Two processes, two 'urban revolutions'

28   In the orthodox materialist interpretation of the origins of cities there is only one urban revolution (Childe 1950) – 'the' urban revolution based

upon the agricultural surpluses consequent upon the expansion of farming into productive riverine environments (notably Mesopotamia); but, if there are two distinctive 'urban' processes, if follows that there should two such 'revolutions, one for towns and one for cities.

29   The traditional agricultural explanation of the 'urban revolution' can now be interpreted as a creation of town-ness; increased yields per area unit provided local agricultural surpluses enabling some settlements to become towns by carving out hinterlands, and with a settlement hierarchy process no doubt encouraged by guardian agents through political control of food storage. Of course, the riverine environment was also highly conducive to trade expansion and it is through this process that cities will have developed.

30   But trading long predates not only the traditional 'urban revolution' but also agriculture itself. Hence, Jacobs' (1970) controversial thesis of 'cities first, then agriculture' is relevant here: the 'city revolution' may well be before the Neolithic, hunters and gatherers becoming long-distance traders prior to the emergence of local farmers. Thus cities could have predated towns.

Cities and trade

31   The existence of trade, however, does not inevitable mean a network with nodes as potential cities – frontier trade, relay trade and peddling trade do not require cities. Diasporic trade, on the other hand, is a sure sign of city-ness; this results from communities of permanent 'foreign' traders acting as factors, agents, facilitators and knowledge sources for their kith and kin from other cities within the network.

32   Trade is stimulated by a real differentiation; initially largely natural (resource) differences would be the stimulus, but social (production) differences have come to be more important; unlike central place theory such processes do not lead to spatial regularity – natural differences will presumably create erratic city locations, social differences for complementary trading may well create regionally clustered patterns of cities.

33   Creators of new net-work, foreign agents, convert initial entrepôt (warehouse functions) into vibrant cosmopolitan cities by interlocking cities into a network.

Net-work

34   Net-work occurs in two basic forms: in Blaut's (1976) terms producing a commodity for market by change-of-form (manufacture) or by change-of-place (exchange).

35   These create two spaces of flows: in Jacobs' (1970) terms the 'little movements in the hubs' (intra-city (or intra-regional) relations that are economic clusters) and the 'great wheels of economic life' (inter-city relations that are city networks).

36   These both contribute to the expansion of economic life as externalities,

increased returns through non-market mechanisms (Hicks 1969). Internal 'cluster externalities' and external 'network externalities' are critical to vibrant net-work and consequent economic expansion.

Interlocking networks in time

37 City networks are dynamic but not so all cities, all of the time; cities in networks rise and fall relative to each other.

38 One of the most important properties of city networks is that at any one time, while some cities are relatively economically stagnant (producing no new work), others continue to generate wealth – producing new net-work that provides the conditions for the stagnant to become dynamic again through a new round of import substitution.

39 City networks exhibit historical layers of new work investment patterns across cities (Massey 1984); it is this dynamic geography – never reducing to no new work – that maintains the expansion of economic life.

City-states and city leagues

40 Part of city wealth creation is used to buy protection from states, but this guardian activity has also been provided by the cities themselves through both city-states and city leagues; territorial states are not necessary for wealth creation but in practice protection has usually been outsourced to such states.

41 The Mesopotamian 'urban revolution' provides the first example of 'city-states', independent cities with their own governments. Note, however, that this political independence – developing its own guardians – is not a requirement for city success; what is needed is a degree of economic autonomy to participate fully in the interlocking network.

42 City leagues were a common medieval European political structure that aided in creation of city group monopolies but, again, these city guardian activities are not necessary for city success; there were many ad hoc arrangements between cities and their merchants across the world that provided the required city autonomy to prosper in interlocking networks. The *lex mercatoria* as a legal system for merchants by merchants remains the classic example of not outsourcing to guardians.

MODERN THESES

Cities in nation-states

43 Modern nation-states are unique among states for their attempts to build bounded 'national economies' (initially by emulating city 'mercantilist' policies (Heckscher 1955)); the result has been economically arbitrary spatial

amalgams of city economies (Jacobs 1984).

44 These attempts to tame economic spaces of flows through curtailing them in political spaces of places have succeeded in distorting the former to the degree that 'national urban systems' are commonly identified (Bourne and Simmons 1978); in reality, bounded economic protectionism has only operated successfully over limited time periods and in large states where there are enough cities to constitute a viable national interlocking sub-network (e.g. the USA, Germany, Japan).

45 The general effect of this 'nationalization' of (cosmopolitan) cities has been to favour a few leading cities, often just the capital city, so that the patch of the world-economy controlled by the state exhibits severe hierarchical city relations (superficially conforming to central place predictions).

World city network

46 Today there is an interlocking network of cities wherein the diasporic feature (expats) has been reorganized because of communication–computing technologies that provide instantaneous worldwide communication; this facilitates new capacities for management and servicing of multi-locational production worldwide.

47 This interlocking network is most clearly expressed through the new net-work associated with advanced producer service firms in professional, creative and financial work (Sassen 2001); major firms have worldwide office networks to provide a 'global seamless' service to clients (Taylor 2004).

48 This new layer of investment (offices in the ubiquitous world city tower blocks) provides infrastructure for the new work to create the world city network, a critical component of economic expansion in contemporary globalization.

Cities in globalization

49 Nation-states have not disappeared with globalization; rather they are key components of a much more complex social organization. In terms of inter-city relations this may be reflected in accentuating 'national hierarchies' as one city captures the 'gateway' function between the 'national' economy and the world-economy: the classic case is Osaka's relative decline because of its 'Tokyo problem' (Hill and Fujita 1995) – other 'second city losers' may be Melbourne, Rio de Janeiro, Lyons, St Petersburg and all British cities that are not London.

50 However, contemporary intertwining of cities and states is much more complicated than this 'gateway' outcome suggests because the enabling distance-reducing or -eliminating technologies allow complex mixtures of centralization and decentralization (Sassen 2000).

51 'Gateways' can be easily bypassed using new communication technologies and therefore it is too soon to predict simple hierarchical enhancement

since all cities are partially 'released' from national bounds through global networking.

Cities and the world-economy

52   The world-economy's core/periphery pattern is usually described and depicted as combinations of nation-states. However, since this is an economic process, it is perhaps better not to describe it through guardian institutions, but through city economies in their networks. Wallerstein's (1979) description of core processes can be interpreted as city-making processes (both produce spatially clustered 'high-tech' outcomes), and peripheral processes – the development of underdevelopment – can be similarly interpreted as Jacob's (1984) malign projection of core-city economic processes.

53   In the contemporary world-economy, therefore, the core is defined by the processes of new work that are constituting the world city network, and the periphery is the rest of the world beyond the world city network.

54   The semiperiphery is defined in Wallersteinian terms as locales where core and periphery processes are approximately balanced; these are cities in the erstwhile 'third world' that are now part of the world city network but are also 'mega-cities' (pernicious population 'town' growth that is a periphery process). This is what makes cities such as São Paulo, Mexico City, Mumbai, Johannesburg and Bangkok among the most interesting settlements in the first decades of the twenty-first century.

REFERENCES

Arrighi, G. (1994) *The Long Twentieth Century*, London: Verso.

Blaut, J.M. (1976) 'Where was capitalism born?', *Antipode*, 8(2): 1–11.

Bourne, L.S. (1975) *Urban Systems: Strategies for Regulation*, Oxford: Clarendon.

Bourne, L.S. and Simmons, J.W. (eds) (1978) *Systems of Cities*, New York: Oxford University Press.

Castells, M. (1996) *The Rise of Network Society*, Oxford: Blackwell.

Childe, V.G. (1950) 'The urban revolution', *Town Planning Review*, 21(1): 3–17.

Christaller, W. (1966(1933)) *Central Places in Southern Germany*, translated by C.W. Baskin, Englewood Cliffs, NJ: Prentice Hall.

Cox, K.R. and Mair, A. (1988) 'Locality and community in the politics of local economic development', *Annals of the Association of American Geographers*, 78: 307–25.

Harvey, D. (1982) *The Limits to Capital*, Oxford: Blackwell.

Heckscher, E.F. (1955) *Mercantilism*, 2nd edition, London: George Allen and Unwin.

Hicks, J. (1969) *A Theory of Economic History*, Oxford: Clarendon.

Hill, R.C. and Fujita, K. (1995) 'Osaka's Tokyo problem', *International Journal of Urban and Regional Research*, 19: 181–91.

Jacobs, J. (1970) *The Economy of Cities*, New York: Vintage.

Jacobs, J. (1984) *Cities and the Wealth of Nations*, New York: Vintage.

Jacobs, J. (1992) *Systems of Survival*, New York: Vintage.

Lefebvre, H. (1996) *Writings on Cities*, translated by E. Kofman and E. Lebas, Oxford: Blackwell.

Massey, D. (1984) *Spatial Divisions of Labour*, London: Macmillan.

Massey, D. (1993) 'Power geometry and a progressive sense of place', in J. Bird, B. Curtis, T. Putnam, G. Robertson and L. Tickner (eds), *Mapping the Futures*, London: Routledge.

Reclus, E. (1895) 'The evolution of cities', *Contemporary Review*, 67(2): 246–64.

Sassen, S. (2000) *Cities in a World Economy*, Thousand Oaks, CA: Pine Forge.

Sassen, S. (2001) *The Global City*, Princeton, NJ: Princeton University Press.

Taylor, P. J. (2004) *World City Network: A Global Urban Analysis*, London: Routledge.

Wallerstein, I. (1979) *The Capitalist World-Economy*, Cambridge: Cambridge University Press.

Weber, M. (1958) *The City*, translated by D. Martindale and G Neuwirth, New York: Free Press.

17 How cities scientifically (do not) exist

Methodological appraisal of research on globalizing processes of intercity networking

Pieter Saey

INTRODUCTION

In what sense does networking turn a city into a world or global city? This is not an easy question to answer, and the attempted responses testify to a fundamental divergence of epistemological, metatheoretical and methodological views. This diversity of approach is no surprise. Almost from the very beginning, research on world cities has been characterized by a great variety of perspectives, ranging from a quantitative-positivist approach and neo-Marxist regulation theory to flirtation with postmodernism. In his review of research on this topic nine years after the announcement of the world city hypothesis (Friedmann 1986), Friedmann (1995: 43) arrived at the conclusion that the hypothesis had 'been fleshed out into a solid research paradigm'. A 'counter-narrative' to 'the meta-narrative of capital', synthesizing 'what would otherwise be disparate and diverging researches', was born.

However 'advisedly' Friedmann might have used the term 'paradigm', his conclusion was an optimistic overstatement at best and a very misleading assessment at worst. Simon (1995: 133) was nearer the truth in writing that 'much of the literature on world cities has originated from very different, often eclectic or politically naïve, perspectives without any reference to the world city hypothesis or its conceptual underpinnings'. King (1995: 219–20) also hit the mark in distinguishing two paradigms, one focusing on questions of culture, identity and meaning, and the other centred on political economy, 'irrespective of the fact that both theorizations would be denied as totalizing meta-narratives by disciples of the post-modern'. After a further ten years we observe a similar range of opposite methodologies of approach with respect to resolving the question posed at the outset.

These opposite methodologies are the subject of this chapter. The source material consists of a restricted number of publications, which are Alderson and Beckfield (2004), Camagni (2001), Friedmann (1986), Smith (2003a,b) and Taylor (2000, 2004). These publications are selected because we believe they are representative of a specific methodological standpoint. No appreciation regarding content is implied. They will be used as stepping stones in an implicit argument leading to the conclusion that a methodological appraisal depends on the ethical implications of the research in question. We start by explaining our view of science and then successively deal with method, epistemology and metatheory.

Although the authors come from different scientific backgrounds and world city research is by definition interdisciplinary, we have chosen to read the publications as an exercise in human geography. This makes it possible to clarify the issues of the methodological discussion by connecting them to some of the fundamental debate that has been held within that discipline. Human geography investigates how human beings have spread over the earth's surface and settled in the various parts of the world by making use of the physical environment and organizing themselves spatially in a great variety of ways. One way in which humans organize spatially is through the establishment of urban communities and the development of horizontal interconnections. Such intercity relationships with an eventual global reach are achieved by a network that awards world or global status. This interaction should not be envisaged as being between integrated societies living within bounded spaces; rather it is the product of actors enrolling fellow performers within a network. Nevertheless, the generalizing word 'spatial' is used to avoid the risk of reification that is implied in the contrast of spaces of flows with spaces of places.

THE BEARING OF SOCIAL CONSTRUCTIVISM

Let us start by stating the obvious. People do not necessarily interact or communicate more intensively with people nearby than with those further away. In other words, proximity in geographical space, that is location on the earth's surface, does not imply proximity in the space of interaction or communication. Likewise people experience time in many different ways according to their circumstances. They act on the basis of memory and impression, in which events from different periods are mixed and sometimes set in reverse order. Midnight occurs at different times depending upon the place. The life cycle of a political system differs in length from that of an economic system. In other words, brevity and orderly succession in Newtonian time do not imply the same in behavioural time. These are merely common sense observations but, formulated as rejections of the notion of linear time and abstract space, they are used by Smith (2003a,b) as an argument to advance a completely new way of thinking about world cities.

Smith welcomes 'the shift from the idea of a hierarchy of world cities developed by writers such as John Friedmann [looking at what makes world cities thick], to the idea of a world city network developed by writers such as Peter Taylor', who recognized that 'the ability of cities to be a success depends to a large extent upon their participation within networks' (Smith 2003a: 25, 28). However, despite the relational perspective of this type of world city network research, 'the importance of a conceptual and empirical shift still needs to be promoted' (29). Further progress could be made by substituting post-structuralist actor-network theory and non-representational theory for the political-economic approach that characterizes most world city network research. This is a shift away from quantitative empirical research in order to be able to 'think about cities as networks of relations that are alive and brimming with movements, practices, performances and contingencies' (38). In this way room can be made for concepts such as transnational cities, e.g. the conceptualization of NY-LON, New York–London, as a flux of intensities. World cities are to be visualized 'as having a fibrous, thread-like, wiry, stringy, ropy, capillary character that is never captured by the notions of levels, layers, territories, spheres, categories, structure, systems' (Latour quoted in Smith 2003b: 572). Smith drifts knowingly and willingly away 'from categorizing cities at all. The categories of "world" or "global" city are becoming more and more meaningless' (ibid. 578). How can common sense observations lead to such far-reaching propositions?

The answer is straightforward: we are talking about an undiluted, postmodernist elaboration of social constructivism. A constructivist geography has developed in the Anglo-Saxon world particularly, mainly as a response to the overemphasis of social structure in the explanation of human behaviour that allegedly characterized the political-economic approach of radical geography. Its characteristics are holography, ethnography, institutionalism and social constructivism (Kesteloot and Saey 2003).[1] Social constructivism means that reality is a human product; even physical reality is in some basic sense a human product. This in itself is not a particularly new vision; the Marxist concept that man knows reality only as his objectified practice is a kind of social constructivism *avant la lettre*. However, the postmodernist version of social constructivism seems to overstate the case.

Smith wants to work with a continuum and explicitly refuses 'to ossify or freeze the flow of the world into unities' (Smith 2003b: 574). He rejects the possibility of drawing boundaries, because cities are 'porous trans-local sites that are criss-crossed by the multiple lines of networks that are more or less long and more or less durable' (572). This looks very much like a resurrection of the discussion upon which the final word was said over four decades ago, that is whether or not a geographical region is an object, and whether geography is a science of relationships.

Blaut (1962) has argued that in these discussions the existence of some irreducible objects or discrete, concrete entities interacting by means of a relationship is taken for granted. Even those who explicitly reject the existence of these entities are trapped in a metaphysical discourse. The problem disappears when a process mode, comprising systems of interacting, interpenetrating part-processes, replaces the object concept. A system is

> discrete and bounded only to the extent that boundary processes are unimportant in relation to internal processes. . . . Insofar as any group of processes interacts, that is, behaves like a system, it can be called an object or whole. This applies to things as large as the solar system or as small as the atom, as tangible as a tree or as tenuous as an airline route. The system approach completely redefines the concept of object. We apply the term object to any system of processes just as long as it pays to do so . . . . The same holds for

relationship. An empirical relationship is a process, an interaction between two or more objects, which are themselves processes. Whether we choose to call the intervening process a relationship or a separate object in its own right is largely a matter of convenience.

(Blaut 1962: 2)

In the philosophy of science, social constructivism has questioned how scientists reach agreement about what counts as truth and about the validity of scientific method. This often leads to the conclusion that the production of scientific knowledge is a consensus-building process among scientists rather than a process of discovering how nature works (Labinger and Collins 2001). We do not agree with this idea. Leaving aside the question of whether, in view of the perennial discussion and difference of opinion among scientists, it would not be better to talk of fields of conflict rather than of communities of consensus (Rouse 1996), we think it is still justifiable to conceive of science as being an effort to discover how nature works even though the production of scientific knowledge is consensus-building. In our opinion, science solves problems on the basis of factual evidence derived from controlled testing in order to have an idea of what is true and what is false. Science tries to explain facts.

Science explains a fact if it is able to prove that the fact was possible at time T1 and place P1, but impossible at time T2 and place P1, and at time T1 and place P2. A fact is neither an object nor a linguistic entity. A fact is that in virtue of which a true statement is true. In other words, it is a condition in the world that makes a statement true, not the mind of the observer. The statement expresses a truth condition, whereby there is something in the world required, namely a fact. Therefore, facts are conditions in the world, things required to satisfy the truth conditions, requirements expressed by statements. For example, the statement 'the cat is on the mat' expresses the truth condition as a requirement. If the statement is true, there will be a condition in the world that meets the requirement, and that condition is the fact that the cat is on the mat (Searle 1995: 211).

THE PERENNIAL DIVIDE: AESTHETIC OR EMPATHIC UNDERSTANDING VERSUS EXPLANATION

Smith (2003a: 31) contends that actor-network theory 'and non-representational theory can give us a richer poststructuralist inspired conceptualization of networks and provide the rationale for the adoption of a whole range of qualitative methods ... for understanding the actual organization of world city networks'. In actor-network theory, actors and networks become one and the same. Non-representational theory is 'a radical attempt to wrench the social sciences and humanities out of their current emphasis on representation and interpretation by moving away from a view of the world based on contemplative models of thought and action towards theories of practice which amplify the potential of the flow of events' (Thrift 2000). Actor-network theory is one of these theories of practice. The networks are hybrid, not composed of human beings but of actants that are sets of human, non-human and technical elements interacting. In non-representational theory the knowledge contained in aesthetic judgements and bodily movements is recognized. The cognitive conception of knowledge is exposed as a discursive regime of truth that makes it difficult to envisage other ways of thinking. It is criticized because it denies expressive understanding (Allen 2000). It is therefore inadequate as a basis for the development of theories concerning human behaviour in a city.

The application of actor-network theory and non-representational theory in empirical research looks, however, very much like a poststructuralist version of the prestructuralist *genre-de-vie* geography. The point here is not whether such geographers of long ago did or did not anticipate present-day concepts, but whether their conceptual apparatus would impede them accommodating ideas that can be formulated only by means of actor-network and non-representational theory, as maintained by modern authors. There are no indications of such impediments. *Genre-de-vie* geographers were well aware of the 'deadly' character of purely formal descriptions or classifications that pretend to capture all reality, and they knew the danger of imposing categories of thinking upon the world (Sorre 1952: 7). They knew that 'the geographical facts, whether physical or human, are facts in continuing transformation and are to be investigated as such'. This is 'the principle of activity' of Brunhes (1925: 6). Indeed, 'capitalism as *genre de vie*' would have been an appropriate subtitle to Thrift's most recent book *Knowing Capitalism* (2005), and NY-LON would have been called a kind of urban transhumance.[2]

The above reasoning does not imply only minor differences between *genre-de-vie* geography and postmodernist constructive thought. Smith (2003b: 565) agrees with the Deleuzean view that 'all folds are equally important; there are no masters and no servants. Folds cannot be distinguished in terms of the essential and the inessential, the necessary and the contingent, or the structural and the ornamental', whereas Brunhes (1925: 29–30) shares with idiographic geography in general the conviction that by detecting the links of connection between dissimilar elements geography reduces the part of reality that is of a purely contingent nature (see Brunhes's principle of connectivity (1925: 18): 'the facts of geographical reality are closely interrelated and are to be investigated in their multiple connections'). One suspects here a divergence of ontological views. However, despite this divergence, postmodernist constructivism and idiographic geography, of which *genre-de-vie* geography became a part, adopt a similar oppositional attitude to nomothetic geography. The device of idiographic geography says 'regional synthesis is our reason of existence'; that of nomothetic geography proclaims 'by our theories you shall know us' (Harvey 1969, concluding sentence). The implied chasm can be interpreted as a basic contrast between two scientific attitudes, which may be respectively considered as phenomenology and analysis.[3]

In principle phenomenology is lived and encountered reality, using the descriptions of perceptions and experiences instead of observations or surveys. Phenomenology studies the world as it seems. It does not consider reality as it is, because there is no other than the one that appears (van den Bersselaar 1997: 63–101).

Idiographic geography may be called 'phenomenology'. It does not subscribe to its view of reality, but it shares the aim to discover a fundamental essence. It strives to separate accidental from essential properties (Saey 1970: 23). The sense of this, for example, may be in the connections between dissimilar elements that constitute a traditional geographical region, this being the idiographic application of Brunhes's principle of connectivity. Phenomenology tries to reduce an experience to its objective essence by means of variation and comparison. 'Analysis', on the other hand, is associated with Ockham's razor: 'plurality should not be posited without necessity', or 'it is useless to do with more than what can be done with less' (Kaye and Martin 2001: 6), which is the principle of parsimony better known in the version of 'do not unnecessarily multiply entities'; that is, keep your explanations as simple as possible.

Alderson and Beckfield (2004) applied network analysis to detect the hierarchical, core–periphery, or regionalized type of world city system generated by the activities of multinational enterprises. They obtained a theoretical-explanatory value by using these descriptive categories to test the contradicting hypotheses of Hymer, Friedmann and Sassen. This is a typical example of nomothetic geographical analysis, here undertaken by sociologists.

Nomothetic geography was heavily influenced by the Popper–Hempel doctrine of explanation by subsumption under covering laws, the structural equivalence of explanation and prediction, and the associated hypothetico-deductive method in which empirical laws can be tested because they hypothetically forbid certain configurations and cannot be inferred from descriptions of configurations.

Another kind of analysis has been developed by world-systems analysis in its attempts to transcend the idiographic-nomothetic dichotomy (Saey and Saey 1995a,b). World-systems analysis is one of the grand theories frequently referred to in the literature on world cities, but it would be a mistake to call it the foundation of this research, which largely borrows the idea of a tripartition core, semiperiphery and periphery without any analytical cross-fertilization. A notable exception is Taylor (2004), in which we see the method of triangulation in operation to identify mechanisms that generate events.[4]

Triangulation is the counterpart to the idiographic practice that specifies uniqueness and nomothetic replication within the framework of theory corroboration. It means coherence or convergence of knowledge-producing procedures. It was first proposed in 1840 by William Whewell, who called it 'consilience'. That is, explanations embedded in different theories that assume the same unobservable entities to explain observations (van Brakel 1998: 26–7). Here the term is used following Hudson (2003), to refer to the mutual enrichment of, and cross-fertilization between, different grand theories, allowing the use of evidence of varying origin, nature and quality. The knowledge created within one grand theory is related constructively to that created within the framework of other such theories. 'Postparadigmatic' may be an apt term for this practice. It boils down to an application of Ockham's razor inverted: 'do not forget necessary entities'. In other words, do not leave out any of the conditions needed for a satisfactory explanation of what you are studying.

Out of dissatisfaction with the kind of analysis above, postmodernist authors are seduced by the enticement of phenomenology. They are infected by the radical form of social constructivism, based on the writings of mainly French thinkers, in which the boundary between ontology (reality) and epistemology (knowledge of reality) has disappeared or been made irrelevant (Saey 2002), which is commented upon later. As potential phenomenologists, they show a tendency to consider the evocation of the multifaceted experience of the multifaceted city as being the core business of social science. Science should not restrict itself to explaining why my dinner tastes burnt; it should also evoke the experience of having to eat a burned dinner. We quote the first sentence of the ending of Smith (2003b: 574): 'This paper has argued that cities are heterogeneous assemblages of differential relations which vibrate making, remaking, and unmaking the multiple spaces and times of urban networks'. In the electronic version of the paper, the i and t of vibrate are written in superscript, the b is written in subscript.

Let us be clear about this matter. In our opinion social science is tasked to produce representational information on human beings who act and behave according to a mix of knowledge. As scientists are human, their practice is not exclusively cognitive. Many epistemic virtues, which are methodological rules to separate the wheat from the chaff, are aesthetic judgments (van Brakel 1998: 37). Is Ockham's razor a cognitive or an aesthetic criterion? However, this does not imply that scientific knowledge should strive for some sort of expressive understanding of empirical phenomena by resorting to the evocative force of metaphors. After all, a statue does not become part of science because the sculptress has studied manly musculature in order to be able to create a work of art.

REALITY AND KNOWLEDGE OF REALITY

The outside world as a human product

Actor-network and other poststructuralist theories call into question the concepts of linear time and abstract or geographical space as crucial ingredients to the idea of a world waiting to be discovered. They argue the discursive nature of all reality. A world awaiting discovery is replaced by worlds that are steadily emerging from the actions of hybrid networks. This approach to research suggests that distances and spaces are made such that they do not exist by themselves as part of the order of things independent of human existence: distances and spaces are literally made. This applies to the action space of the actors enrolling others into the network of inter-city relationships and, to take an example from another realm of geographical research, the sixteenth-century action space of the Portuguese. It also applies to the geographical space of surveyors and cartographers because geographical space could only appear as part of reality through the work of these experts working in a particular set of societal conditions (Law and Hetherington 2000).

The idea of an outside world waiting to be discovered is indeed untenable, and it is simply true that reality is a human product, but that does not imply that all reality is discursive or the result of practice. As Searle (1995) argues, the impossibility of knower-independent knowledge of an external real world, which rules out the possibility of a *matching* representation of that world, is compatible with the view of truth as correspondence to the facts. It is true that geographical space can only appear as a piece of reality within a particular setting of societal conditions. But there seems to be a dichotomy here: on the one hand, the fact that observation always involves manipulation and, on the other hand, the creation of a piece of social or hybrid reality. It should be clear that the network of inter-city relationships or the Portuguese action space can be undone in a way that geographical space cannot be undone, just as within an actor network it can be decided that a banknote is not money in a way that it cannot be decided that the note is not made of paper.

Rationality and uncertainty

Since at least the late eighteenth century modernity has been accompanied by intellectual countermovements reflecting its negative aspects. Romanticism is one example, postmodernism another. These movements give a similar definition of the problem, but their answers are diverse. Modernity, or the Enlightenment, is accused of untenable reductionism because it constricts rationality to being intellectual, technical, instrumental and means–end. Usually a broader rationality is then put forward to save what this reductionist concept denies, oppresses or excludes (Braeckman 2002). Postmodernism goes even further. It opposes the hegemony of any single grand theory. Its deconstruction of the Enlightenment discourse is aimed at giving voice to the otherwise excluded alternative ways of thinking. By rejecting any metanarrative, it advocates radical and principled uncertainty. Uncertainty is an ontological quality for postmodernism that compels science to adopt a phenomenological attitude. Alternatively, for postnomothetic analytical geography, uncertainty is an epistemological quality and, as a consequence, science is compelled to the postparadigmatic application of triangulation. In nomothetic research uncertainty is a property of the actors whose behaviour is the object of study.

Taylor (2000) presents the world city network as a contribution to a new network metageography, replacing the old territorialist mosaic of states. A metageography is the set of spatial structures through which people order their knowledge of the world: the often unconscious frameworks that organize studies of history, sociology, anthropology, economics, political science or even natural history. Every global consideration of human affairs deploys a metageography, whether acknowledged or not (Lewis and Wigen 1997: ix). The territorialist metageography was state-centric and inevitably prioritized the geographical scale of the nation. Second and third prioritizations were analysis over synthesis in the sense of the antithesis of transdisciplinary research, not of phenomenology, and the priority of attributes over relationships (Taylor 2000: 4).

The world city network is based on relationships; however, the prioritization of these rather than attributes is not necessarily relational thinking. There is a difference between a study that measures relationships between objects and a study that uses relational definitions in which an object is defined in terms of the context as a set of conditions relating to the object (Hendriks 1986). In this sense Smith (2003b: 572–3) may have been right in his criticism of those authors who reason in terms of relationships between two cities and do not envisage something like NY-LON, but his postmodernist stance compels him to misguidedly use the notion of geographical scale. Smith (2003a) argues that non-linear experiences such as simultaneity and juxtaposition require a conceptualization in terms of a skein of networks. This would make geographical scale meaningless. However, just as time should not be confused with the measurement of time (Smith 2003b: 568), the analytical use of the notion of geographical scale should not be confused with its possible reification. Scale is just another word for short or long in terms of portions of the earth's surface. It refers to the fact that processes of integration and fragmentation within a certain area involve actors whose radius of action is larger than that area. The concept is not meaningless; it is the non-relational definition of it that is wrong.

In his reflections on decision-making processes under conditions of dynamic uncertainty engendered by globalization and intercity competition, the economist Camagni (2001) remains within the contours of nomothetic geography. He conceives of the city as an operator that reduces uncertainty through the transcoding of information and the *ex ante* coordination of private decisions. According to Camagni, uncertainty can and should be theorized in terms of procedural rationality requiring a symbolic approach to decision-making based on routines, *ex ante* coordination and shared representation. This theorizing of uncertainty complements the theorizing of substantive rationality, requiring a functional approach to decision-making geared to optimization. Both functional and symbolic approaches are the components of a cognitive logic, which, integrated with the territorial and network approaches of a spatial logic, provide a new understanding of the role of global city-regions, where they are clusters, interconnections, milieus and symbols (Camagni 2001: 96–107). Uncertainty is not a problem for Camagni; it is an epistemological quality of the object of study, and has to be investigated just like other aspects of reality.

METATHEORY

If we consider the role of cities at a more abstract level, two issues about our basic question – 'in what sense are cities turned into world cities or global cities by networking?' – emerge from the relevant literature. These are centrality and power.

The issue of centrality

Friedmann's world city hypothesis started from the concept of cities as centres of production, control and finance (Friedmann 1986). Taylor's world city network (Taylor 2004) uses another conception introduced by Sassen (2001) in which global cities are host to production complexes of advanced producer services, in other words central places. This raises a problem that has bedevilled nomothetic geography. Central place was a coherent concept from a spatial point of view, but a chaotic concept from the genetic point of view.

The well-known traffic, administrative and market principles combine to form the Christaller hierarchical network of central places (Christaller 1933). Christaller's starting point for the market principle was the transcendental economic concept of least effort, but it can easily be reformulated in terms of self-organization by combining threshold and market strategy to compete from a location as close as possible to, or as far as possible from, your rival. In both formulations central place is a chaotic concept simply because centrality has heterogeneous sources of activity seen as the political and cultural system of government institutions and mass media; the reproduction of labour power through education and health care; transportation of products and people and communication services; and circulation of capital through retailing and consumer and producer financing.

The chaotic character can be, and has been, minimized by considering central functions solely from the standpoint that they serve households. That leaves the researcher with many unresolved questions such as: What about producer services and their spatial organization? How are central functions related to growth poles in any sense, including the sense of poles of development and their spatial organization? In which way are the possible contributions of central functions to growth pole or development mechanisms dependent on local, regional, central state and interstate intervention in the fields of law, general conditions of production involving infrastructure and public utilities, reproduction and regulation of the labour power, and protection of the level of capital accumulation?

In his paper on 'Christaller for a global age', Hall (2002) reaches the conclusion that 'the Christaller hierarchy now needs to be supplemented by at least two and perhaps three additional levels, producing a hierarchy of perhaps six or seven levels'. The additions would be the alpha, beta and gamma global and sub-global cities identified by Beaverstock *et al.* (1999). This seems to be a misinterpretation, however, as the nodes of the world city network are central places in a different manner from those of central place theory. They are not locations of central functions servicing the hinterland households; they are rather the locations of advanced producer services servicing a hinterworld. The reach of the world cities is global and the key to the spatial organization involved is the degree of connectivity, the level at which a city participates in advanced producer service relationships between actors with a global radius of action.

The world city network presents connectivity as the areal functional principle (Philbrick 1957) of the world economy. It is not a hierarchical network, it is an interlocking network. Nevertheless, it remains a network of central places. Questions similar to those in nomothetic central place theory consequently arise. One of these questions has received a tentative answer. Taylor (2004) equates Sassen's concept of global cities with Jacobs's perception of dynamic cities. As a result, producer services occupy the key position in a development or growth pole mechanism. A dynamic city develops as the product of a virtuous circle of import-replacing and export multipliers. The locomotive element is new work when firms in the city start to produce, through innovative imitation, things that formerly were imported (Jacobs 1970). This extraverted growth pole concept makes the development of a city dependent on the development of a network of cities and is understandably more useful to Taylor's theory construction, in which producer services perform the Jacobsean new work, than the more introverted classical growth pole theories. But neither the spatial organization of these mechanisms nor the role of state and interstate intervention is thereby clarified, although Jacobs (1985) might provide an insight here (Taylor 2004: 42–52). A new twofold question arises: To what degree does the key position of producer services entail a coordinating or even organizing role of these services in capital accumulation? Does the concept of centrality remain coherent or does it become chaotic when new global central functions of a political, social, ideological or cultural nature are added to the advanced producer services?

The issue of power

The reference to the coordinating or organizing role of producer services in capital accumulation raises the question of power. Whereas the traditional idea of command is integral to Friedmann's concept of world 'headquarter' cities, the assignment of such a role to producer services seems to imply an idea of network power. This idea is hardly compatible with the ontological status of power as an almost tangible substance/attribute/resource in Friedmann's traditional conception. It makes us think of power as the outcome of social interaction in which resources are mobilized in a time- and space-specific way, as a mediated transformation of a capacity rather than the capacity itself. Allen (2003) convincingly argues that the conception of power as a relational effect also applies to commanding power and gives it a Foucaultian twist.[5]

Taylor (2004: 87–88) tries to capture this difference in power by using another pair of concepts, 'power over', that constrains in the sense of 'bending the will of others', and 'power to', that enables in the sense of 'together we are strong', but in the end he declines to make a choice between the two. He has borrowed the terms from Allen, but it is not obvious whether he would agree with the ontological status Allen ascribes to them.

Taylor's reluctance to make a choice is the result of his rather rudimentary treatment of these two concepts. The problem is twofold. On the one hand, the location of production complexes of advanced producer services in a city implies a kind of attractive force exercised by the metropolis. This is interpretable as a constraint on the capacity of other cities to realize the same development. On the other hand, cities play the part, in the description by Allen (2003: 113) of 'a throughput to enable resources to flow and to settle where directed', which is interpretable as a constraint, but also as an act of collaboration. Allen, making critical use of Sassen's work, provides the way out by an analysis of the concrete exercise of power in its various guises of coercion, domination, authority, manipulation, seduction, negotiation and persuasion. A similar criticism applies to Alderson and Beckfield, who conflate capacity and power. These authors link power to centrality, which conveys the idea of command, but one of their measurements of centrality ('betweenness') appears to be an indication of gatekeeper power: 'actors with high betweenness have greater power in the sense that they serve as brokers and can control the flow of information through the network' (Alderson and Beckfield 2004: 833). Smith (2003b: 576) has a more definite gatekeeper conception of power: 'The capacity of a global or world city to "command and control" is governed by its participation as a "switcher" in networks because power is only exercised through any actant's ability to enrol and mobilize others to perform in "their" network.'

Postmodernism confronts us with the conundrum that 'the world can be made and remade, but it cannot be justified on any solid grounds' (Webster 2000: 229). That is why postmodernists have asked the question: When, where, by whom and for which purpose has the world city hypothesis been formulated? That is why they have contributed to the insight that for a few decades the alignment of power and knowledge in modern disciplinary societies has been gradually replaced by the ideology of the market in postmodern consumerist societies. That is why they have introduced the Foucaultian conception of power. These are important questions, insights and conceptions. However, their meaning and fruitfulness are undermined by the tendency to lapse into the doctrine of symmetry or equivalence. This doctrine states that all views can lay an equal claim on credibility whatever their truth value or epistemological and moral content.

The doctrine of symmetry has been enlarged to include the thesis that there is no difference between human and non-human actors. Human beings, institutions and artefacts are all actors. There is no sharp distinction between what organizes and what is organized. All that can be done is to study hybrid networks of actors within which essences and facts are constructed. Take away the relationships between things and there are no things any more (van Brakel 1998: 91–4). There are two problems with this doctrine. The first problem is the claim, whether implicit or explicit, to ultimate value-free research; the second problem is the radical relativism with regard to the possibilities of action and the moral responsibility in the construction of social reality by human beings. One might even wonder whether the doctrine is able to formulate the issue of moral responsibility in a consistent way. The consequence is a choice between plain voluntarism and fatalism.

Neo-Marxist radical geography has developed as a response to the positivism of nomothetic geography and its potential for social control. Radical geographers have asked the same questions as the postmodernists, but their conundrum is a different one: 'men make their own history, but not of their own free will; not under the circumstances they themselves have chosen' (Marx 1973: 146). Human beings do not command, control, rule or run cities in some absolute or possibilist way. They transform, and their transformations can and will have unexpected and undesired effects. Initiatives can and will turn against the initiators. As a consequence the elaboration of an antisubject discourse, even in the form of the extravagant Althusserian structuralism or an overstretching poststructuralism, is a legitimate enterprise. Nevertheless it remains a fact that human beings organize themselves in such a way that they are accountable for their choices and actions. The existence of hybrid networks entails the non-existence of societies composed solely of human beings. But the fact that there are no societies does not imply that the social does not exist. Neither is the non-existence of the social implied by the inseparability of the economical, the political, the cultural and the technical. The social exists because human beings are social creatures in a different manner from the other animals that form collectivities: they have an idea of responsibility for their own actions.

CONCLUSION

The world city network is a modality of the development of capitalist society. This is a society in which socially produced products are privately appropriated. Some appreciate this kind of society, others detest it. In our opinion the task of scientists who are intellectuals concerned about the ethical implications of their research is to investigate social reality in such a way that it becomes clear under what conditions possibilities of transformation can be created. When nomothetic geography considers a city to be an uncertainty-reducing operator, a symbol of territorial control and a producer of symbols, codes and languages, does it thereby improve our insight into these conditions? World-systems analysis has revealed the inadequacies of state-centred thinking. The concept of a world city network aims to reveal the inadequacies of territorial thinking.

Question: Does the network power generated by the development of that world city network create the preconditions for deterritorialized democracy; is that really a possibility? Or, on the contrary, does this network power increase the influence of the ideology of the market, through which consumerism is elevated to the status of key element in identity formation?

The above is a question about the relationship between politics and economics. How are we to answer it when even the analytical value of this type of differentiation is contested?

NOTES

1 Holography — Every part contains information about the whole and information about the whole is enfolded in each part (key words: multi-scalar analysis/globalization; culturally embedded capitalism; politics, economy, culture, technology blend into each other).

Ethnography — The symbolic order (patriarchy, heterosexuality, theocracy, secularism etc.) expresses itself in the spatial order, at all scales, from the kitchen to the world (key words: identity formation; postcolonial geography; feminist geography; gender; everyday activities; symbolic constructions).

Institutionalism — Social practices are institutionalized and institutions are constituted through social practices (key words: access to factors of production; efficiency/equity; agency/structure; micro/meso/macro level).

Social constructivism Reality is a human product (key words: power of language;
                    postcolonialism; feminism; power of politics).

2 *Genre de vie* was a pivotal concept in some branches of mainstream human geography during the twentieth century until about the 1960s. It refers to the complexes of
practices of a material and social-organizational nature that characterize each human
group and serve as the foundations of its existence. There appears to be a great simi-
larity between the way in which Thrift observes capitalism and the way in which Vidal
de la Blache, the father of *genre-de-vie* geography, observes, for example, bands of
hunters: 'The instruments that man puts into operation at the service of its conception
of existence derive from intentions and efforts co-ordinated in view of a genre de vie.
They constitute a whole, they form links in a chain and display a kind of filiation.
One application provokes another. To perfect his throwing weapons, boomerangs,
assegais and javelins, bow and arrow, the hunter introduces modifications: he rebends
or lengthens his bow according to the span he has to obtain; he protects his arm that
may be injured by the recoil of the bowstring by a bracer; he decorates the arrow
with feathers that regulate its flight; he blunts the head when he fears to damage the
plumage of the bird he wants to catch. He arms with a shield that resists the attack.
The shield, light and handy against throwing weapons, is made longer and heavier as
the pike and the lance are increasingly used to stand firm against the assault of the
enemy or the wild animal. When the African Negro of the tropical zone practices iron
metallurgy, he realizes in the forms of knives, their refinements and chasing, their
beards, a variety that aims at an equal number of diverse applications. . . . Is it just the
stimulus of practical utility that prevails in these combinations? . . . Man has pursued
intentions, has realized art through the materials provided by nature, sometimes in
spite of their resistance or their insufficiency. Obeying his impulses and tastes, he has
humanized the surrounding nature to his use. . . . These rudimentary civilisations . . .
already are an achievement, a result of progress in which initiative, will and artistic
sense are deployed' (Vidal de la Blache 1922: 201–2, translation by present author).

3 The phenomenological attitude invites researchers to call hermeneutic interpretation,
comprehending experiences and reflexive analysis of values, goals and interests the
core business of social science whereas understanding the significance of action is its
purpose. The analytical attitude compels social science researchers to look upon all
this as being a prerequisite to explain behaviour.

4 Identification of mechanisms that generate events is the practice of critical realist research and also characterizes much neo-Marxist research. Critical realism distinguish-
es ontological domains that justify the conception of a stratified reality. The domain
of the empirical (experiences) is contained in the domain of the actual (experiences
and events), and both are contained in the domain of the real (experiences, events, and
mechanisms).

5 Power is something intangible that is distributed over society and not a possession
commanded at will by the powerful. It does not relate in an external way to economic
or other practices, but is integral to them; it is immanent. Repressive power structures
are nothing other than the collective appearance of microscopic power working within
the whole.

REFERENCES

Alderson, A. and Beckfield, J. (2004) 'Power and position in the world city system', *American Journal of Sociology*, 109(4): 811–851.

Allen, J. (2000) 'Power/economic knowledge: symbolic and spatial formations', in J.R.
Bryson, P.W. Daniels, N. Henry and J. Pollard (eds) *Knowledge, Space, Economy*, London: Routledge, pp. 15–33.

Allen, J. (2003) *Lost Geographies of Power*, Oxford: Blackwell.

Beaverstock, J.V., Smith, R.G. and Taylor, P.J. (1999) 'A roster of world cities', *Cities*, 16(6): 445–58.

van den Bersselaar, V. (1997) *Wetenschapsfilosofie in veelvoud, fundamenten voor professioneel handelen*, Bussum: Dick Coutinho.

Blaut, J.M. (1962) 'Object and relationship', *Professional Geographer*, 14(6): 1–7.

Braeckman, A. (2002) 'Lyotard and Habermas. Twee visies op uitsluiting en hoe eraan te verhelpen', in P. Saey, F. Mestrum, R. Tinnevelt and A. Braeckman (eds) *Ruimte voor het verhaal van de andere. Over het in- en uitsluitend vermogen van het denken en het discours*. Monografieën over interculturaliteit 8, Mechelen: CIMIC, pp. 95–109.

van Brakel, J. (1998) *De wetenschappen. Filosofische kanttekeningen*, Leuven: Universitaire Pers.

Brunhes, J. (1925) *La géographie humaine I. Les faits essentiels groupés et classés, principes et examples*, 3rd edition, Paris: Félix Alcan.

Camagni, R. (2001) 'The economic role and spatial contradictions of global city-regions: the functional, cognitive and evolutionary context', in A. Scott (ed.) *Global City-Regions: Trends, Theory, Policy*, Oxford: Oxford University Press, pp. 96–118.

Christaller, W. (1933) *Die zentralen Orte in Süddeutschland*, Jena: Gustav Fischer.

Friedmann, J. (1986) 'The world city hypothesis', *Development and Change*, 17(1): 69–84.

Friedmann, J. (1995) 'Where we stand: a decade of world city research', in P.L. Knox and P.J Taylor (eds) *World Cities in a World-System*, Cambridge: Cambridge University Press, pp. 21–47.

Hall, P. (2002) 'Christaller for a global age: redrawing the urban hierarchy', in A. Mayr, M. Meurer and J. Vogt (eds) *Stadt und Region: Dynamik von Lebenswelten, Tagungsbericht und wissenschaftliche Abhandlungen*, Leipzig: Deutsche Gesellschaft für Geographie, pp. 110–28.

Harvey, D. (1969) *Explanation in Geography*, London: Edward Arnold.

Hendriks, P. (1986) *De relationele definitie van begrippen; een relationeel realistische visie op het operationaliseren en representeren van begrippen*. Nederlandse Geografische Studies 24, Nijmegen: K.N.A.G./Universiteit Nijmegen.

Hudson, R. (2003) 'Fuzzy concepts and sloppy thinking: reflections on recent developments in critical regional studies', *Regional Studies*, 37(6/7): 741–6.

Jacobs, J. (1970) *The Economy of Cities*, New York: Random House.

Jacobs, J. (1985) *Cities and the Wealth of Nations: Principles of Economic Life*, New York: Random House.

Kaye, S.M. and Martin, R.M. (2001) *On Ockham*, Belmont: Wadsworth.

Kesteloot, C. and Saey, P. (2003) 'The nature of changes in human geography since the 1980s: variation or progress?', *Belgeo*, 2: 131–43.

King, A.D. (1995) 'Re-presenting world cities: cultural theory/social practice', in P.L. Knox and P.J. Taylor (eds) *World Cities in a World-System*, Cambridge: Cambridge University Press, pp. 215–31.

Labinger, J.A. and Collins, H. (eds), (2001) *The One Culture? A Conversation about Science*, Chicago: University of Chicago Press.

Law, J. and Hetherington, K. (2000) 'Materialities, spatialities, globalities', in J.R. Bryson, P.W. Daniels, N. Henry and J. Pollard (eds) *Knowledge, Space, Economy*, London: Routledge, pp. 34–49.

Lewis, M.W. and Wigen, K.E. (1997) *The Myth of the Continents: A Critique of Metageography*, Berkeley, CA: University of California Press.

Marx, K. (1973) *Surveys from Exile: Political Writings*, vol. 2, Harmondsworth: Penguin Books.

Philbrick, A.K. (1957) 'Principles or areal functional organization in regional human geography', *Economic Geography*, 33(4): 299–336.

Rouse, J. (1996) *Engaging Science: How to Understand its Practices Philosophically*, Ithaca, NY: Cornell University Press.

Saey, M. and Saey, P. (1995a) 'Een transdisciplinair kader voor geografie-beoefening', *Tijdschrift van de Belgische Vereniging voor Aardrijkskundige Studies*, 1: 129–49.

Saey, M. and Saey, P. (1995b) 'Voorbij idiografie en nomothetiek', *Tijdschrift van de Belgische Vereniging voor Aardrijkskundige Studies*, 1: 151–62.

Saey, P. (1970), *De nieuwe oriëntatie van de aardrijkskunde: oorsprong, ontwikkeling en beoordeling*, Gent: Seminaries voor Menselijke, Ekonomische, Historische Geografie der Rijksuniversiteit Gent.

Saey, P. (2002) 'Book review: *Knowledge, Space, Economy*', *Tijdschrift voor Economische en Sociale Geografie*, 93(5): 577–9.

Sassen, S. (2001) *The Global City*, Princeton, NJ: Princeton University Press.

Searle, J.R. (1995), *The Construction of Social Reality*, London: The Penguin Press.

Simon, D. (1995) 'The world city hypothesis: reflections from the periphery', in P.L. Knox and P.J. Taylor (eds) *World Cities in a World-System*, Cambridge: Cambridge University Press, pp. 132–55.

Smith, R.G. (2003a) 'World city actor-networks', *Progress in Human Geography*, 27(1): 25–44.

Smith, R.G. (2003b) 'World city topologies', *Progress in Human Geography*, 27(5): 561–82.

Sorre, M. (1952) *Les Fondements de la géographie humaine. Vol. 3: L'Habitat/Conclusion générale*, Paris: Armand Colin.

Taylor, P.J. (2000) *A Metageographical Argument on Modernities and Social Science*. GaWC bulletin 29.

Taylor, P.J. (2004) *World City Network: A Global Urban Analysis*, London: Routledge.

Thrift, N.J. (2000) 'Non-representational theory', in R.J. Johnston, D. Gregory, G. Pratt and M. Watts (eds) *The Dictionary of Human Geography*, 4th edition, Oxford: Blackwell, p. 556.

Thrift, N. (2005) *Knowing Capitalism*, London: Sage Publications.

Vidal de la Blache, P. (1922) *Principes de géographie humaine*, Paris: Armand Colin.

Webster, F. (2000) 'Virtual culture: knowledge, identity and choice', in J.R. Bryson, P.W. Daniels, N. Henry and J. Pollard (eds) *Knowledge, Space, Economy*, London: Routledge, pp. 226–41.

Index

328   *Index*